T0235994

CAMBRIDGE LIBRARY COLLECTION

Books of enduring scholarly value

Travel and Exploration

The history of travel writing dates back to the Bible, Caesar, the Vikings and the Crusaders, and its many themes include war, trade, science and recreation. Explorers from Columbus to Cook charted lands not previously visited by Western travellers, and were followed by merchants, missionaries, and colonists, who wrote accounts of their experiences. The development of steam power in the nineteenth century provided opportunities for increasing numbers of 'ordinary' people to travel further, more economically, and more safely, and resulted in great enthusiasm for travel writing among the reading public. Works included in this series range from first-hand descriptions of previously unrecorded places, to literary accounts of the strange habits of foreigners, to examples of the burgeoning numbers of guidebooks produced to satisfy the needs of a new kind of traveller - the tourist.

A Voyage Round the World. Performed by Order of His Most Christian Majesty, in the Years 1766–1769

This is an English translation from 1772 of the famous *Voyage Autour du Monde* (1771) by Louis de Bougainville (1729–1811), French admiral and explorer. The contemporary fascination with global circumnavigation created demand for this translation by John Reinhold Forster (1729–98), which includes many annotations and comments from the translator himself on de Bougainville's observations. Describing all of de Bougainville's adventures on the voyage (which took place between 1766 and 1769) as well as his descriptions of local flora and fauna, the gripping tale includes such interesting passages as the unmasking of the botanist's valet as a woman (the first known to have circumnavigated the globe); de Bougainville's famous descriptions of Tahitian society; and graphic descriptions of the discomforts and perils of sea voyaging in the eighteenth century. It includes a copy of the original eighteenth-century plot of the route, and several plates representing original sketches from the trip.

Cambridge University Press has long been a pioneer in the reissuing of out-of-print titles from its own backlist, producing digital reprints of books that are still sought after by scholars and students but could not be reprinted economically using traditional technology. The Cambridge Library Collection extends this activity to a wider range of books which are still of importance to researchers and professionals, either for the source material they contain, or as landmarks in the history of their academic discipline.

Drawing from the world-renowned collections in the Cambridge University Library, and guided by the advice of experts in each subject area, Cambridge University Press is using state-of-the-art scanning machines in its own Printing House to capture the content of each book selected for inclusion. The files are processed to give a consistently clear, crisp image, and the books finished to the high quality standard for which the Press is recognised around the world. The latest print-on-demand technology ensures that the books will remain available indefinitely, and that orders for single or multiple copies can quickly be supplied.

The Cambridge Library Collection will bring back to life books of enduring scholarly value (including out-of-copyright works originally issued by other publishers) across a wide range of disciplines in the humanities and social sciences and in science and technology.

A Voyage Round the World

Performed by Order of His Most Christian Majesty, in the Years 1766–1769

Lewis de Bougainville
Translated by J.R. Forster

CAMBRIDGE UNIVERSITY PRESS

Cambridge, New York, Melbourne, Madrid, Cape Town,
Singapore, São Paolo, Delhi, Tokyo, Mexico City

Published in the United States of America by Cambridge University Press, New York

www.cambridge.org
Information on this title: www.cambridge.org/9781108031875

This edition first published 1772
This digitally printed version 2011

ISBN 978-1-108-03187-5 Paperback

A

VOYAGE

ROUND THE

WORLD.

Performed by Order of

HIS MOST CHRISTIAN MAJESTY,

In the Years 1766, 1767, 1768, and 1769.

BY

LEWIS DE BOUGAINVILLE,

Colonel of Foot, and Commodore of the Expedition, in the Frigate La Boudeuſe, and the Store-ſhip L'Etoile.

Tranſlated from the French

By JOHN REINHOLD FORSTER, F. A. S.

LONDON,

Printed for J. NOURSE, Bookſeller to HIS MAJESTY, in the Strand; and T. DAVIES, Bookſeller to the Royal Academy, in Ruſſel-ſtreet, Covent-garden.

MDCCLXXII.

TO

JAMES WEST, Esq.

High Steward of St. Alban's, Recorder of Pool,

AND

PRESIDENT of the ROYAL SOCIETY.

SIR,

I Beg leave to offer you the Tranſlation of a Work written by a learned, intelligent, and judicious Traveller, which abounds with remarkable events and curious obſervations; equally inſtructive to future navigators, and intereſting to ſcience in general, and Geography in particular.

THE place you occupy with great honour in the Royal Society, the zeal with which you promote and countenance whatever has a tendency towards the advancement

ment of Science, and the remarkable kindneſs and favour you always have treated me with, encourage me to prefix your name to this publication.

ACCEPT then, Sir, this public acknowledgement of the deep ſenſe of gratitude and attachment your benevolence has raiſed, with the ſincereſt wiſhes for your health, proſperity, and the enjoyment of every intellectual and moral pleaſure. Believe me to be, with the trueſt eſteem,

SIR,

Your moſt obliged,

and obedient

humble ſervant,

JOHN REINHOLD FORSTER.

THE

TRANSLATOR's PREFACE.

THE present translation of Mr. de Bougainville's Voyage round the World merits, in more than one respect, the attention of the public.

Circumnavigations of the globe have been of late the universal topics of all companies: every one takes upon him to be a competent judge in matters which very few understand, mostly for want of good and authentic information: this work will enable the reader to judge with greater precision of the vague discourses held on this subject.

Nautical advices and observations are always interesting, from whatever quarter they may happen to come, provided they are communicated by a man of known abilities; and nobody, we think, will question those of Mr. de Bougainville.

The superiority of the British discoveries in the great ocean, between America and Asia, cannot be ascer-

aſcertained, unleſs by an authentic account of the diſcoveries of the rival nation; who, after a great exertion, and the advantage of being ſupplied by the Spaniards with all the neceſſaries at a great diſtance from home, before they entered the South Pacific Ocean, however diſcovered very little; and what they diſcovered, had partly been ſeen by Engliſh navigators, or ſome Spaniſh ones of older date; ſo that the honour of the greateſt diſcoveries made within two centuries, in thoſe remote ſeas, is entirely reſerved to the Britiſh nation, and their ſpirit and perſeverance in conducting this great and intereſting event.

The envious and ſcandalous behaviour of the Portugueſe viceroy, at Rio de Janeiro, towards our philoſophers, which will for ever brand that mean barbarian with indelible ignominy, is confirmed by a ſimilar act of deſpotic barbariſm towards another nation, related in this work.

The French, who are ſo remarkable for the gravings with which they ornament their principal publications, will find, that the charts joined to this tranſlation, though reduced to a ſixteenth part of the ſurface of the originals, are, however, infinitely ſuperior to them in point of neatneſs, convenience, and

accuracy.

accuracy. Without being leſs uſeful, we have connected, in our charts, the whole run of their ſhips, from the beginning of their diſcoveries to Batavia. The chart of the Magellanic Straits is of the ſame ſize, and upon the ſame ſcale as in the original, but more accurate; and the names by which the Engliſh call the ſeveral points of land, the bays and the reaches, are all added to the French names. The omiſſion of the charts of Rio de la Plata, and of the Falkland Iſles, is by no means an imperfection; becauſe, very lately, two charts have been publiſhed in England, one equally good of the firſt, and a better one of the latter; it would therefore be needleſs to multiply the identical charts, or to give the public ſome imperfect ones.

Though Mr. de Bougainville is a man of undoubted veracity and abilities, he has, however, in a few inſtances, been miſled by falſe reports, or prejudiced in favour of his nation: we have, in ſome additional notes, corrected as far as it was in our power theſe miſtakes, and impartially vindicated the Britiſh nation, where we thought the author had been unjuſtly partial; for the love of one's country is, in our opinion, very conſiſtent with common juſtice and good breeding; qualities which never ſhould be wanting in a philoſopher.

Our author endeavours to make it highly probable, that the ſpice-trade, which has hitherto been the great ſource of the grandeur and wealth of the Dutch Eaſt India Company, will ſoon be divided among them, the French, and the Engliſh. We have reaſon to believe the French to be in a fair way of getting the ſpices in their plantations, as Mr. de Poivre has actually planted at Iſle de France ſome hundreds of clove and nutmeg-trees. Every true patriot will join in the wiſh, that our Engliſh Eaſt India Company, prompted by a noble zeal for the improvement of natural hiſtory, and every other uſeful branch of knowledge, might ſend a ſet of men properly acquainted with mathematics, natural hiſtory, phyſic, and other branches of literature, to their vaſt poſſeſſions in the Indies, and every other place where their navigations extend, and enable them to collect all kinds of uſeful and curious informations; to gather foſſils, plants, ſeeds, and animals, peculiar to theſe regions; to obſerve the manners, cuſtoms, learning, and religion of the various nations of the Eaſt; to deſcribe their agriculture, manufactures, and commerce; to purchaſe

Hebrew,

Hebrew, Persian, Braminic manuscripts, and such as are written in the various characters, dialects, and languages of the different nations; to make observations on the climate and constitution of the various countries; the heat and moisture of the air, the salubrity and noxiousness of the place, the remedies usual in the diseases of hot countries, and various other subjects. A plan of this nature, once set on foot in a judicious manner, would not only do honour to the East India Company, but it must at the same time become a means of discovering many new and useful branches of trade and commerce; and there is likewise the highest probability, that some unsearched island, with which the Eastern Seas abound, might produce the various spices, which would greatly add to the rich returns of the Indian cargoes, and amply repay the expences caused by such an expedition.

Mr. de Bougainville's work abounds in marine phrases, which makes the translation of it very difficult, even to a native; but a foreigner, and a man unacquainted with nautical affairs, must be under still greater difficulties: we should have been under this predicament, had it not been for the kind assiſtance of two worthy friends, who not only enabled us to do juſtice to the original, but alſo to make the whole intelligible to men converſant with navigation: it is therefore no more than juſtice to acknowledge this favour publicly *.

* We have thought proper to omit M. Pereire's diſcourſe on the nature of the language of Taiti, as being a very trifling performance, founded on the imperfect vocabulary, and defective pronunciation of Aotourou.

INTRODUCTION.

I THINK it would be of uſe to give, at the head of my relation, an account of all the voyages that ever were performed round the world, and of the different diſcoveries which have hitherto been made in the South Sea or Pacific Ocean.

Ferdinand Magalhaens, a Portugueſe, commanding five Spaniſh ſhips, left Seville in 1519, diſcovered the ſtraits which bear his name, and through them he came into the Pacific Ocean, where he firſt diſcovered two little deſart iſles, on the ſouth ſide of the Line, afterwards the Ladrones, and laſt of all the Philippines. His ſhip, called la Victoria, was the only one out of the five that returned to Spain by the Cape of Good Hope: On her return ſhe was carried on ſhore at Seville, and ſet up as a monument of this expedition, which was the boldeſt that had hitherto been undertaken by men. Thus it was for the firſt time phyſically demonſtrated, that the earth was of a ſpherical figure, and its circumference aſcertained.

Sir Francis Drake, an Engliſhman, ſet ſail from Plymouth, with five ſhips, the 15th of September, 1577, and returned thither with only one, the 3d of Novem-

 ber,

ber, 1580. He was the second that sailed round the world. Queen Elizabeth dined on board his ship, called the Pelican, which was afterwards preserved in a dock at Deptford, with a very honourable inscription on the main-mast. The discoveries attributed to Drake are very precarious. The charts of the South Seas contain a coast which is placed below the polar circle, some isles to the north of the Line, and likewise New Albion to the north.

Sir Thomas Cavendish, an Englishman, left Plymouth the 21st of July, 1586, with three ships, and returned with two on the 9th of September, 1588. This voyage, which was the third round the world, was productive of no new discoveries.

Oliver Van Noort, a Dutchman, sailed from Rotterdam the 2d of July, 1598, with four ships, passed through the straits of Magalhaens, sailed along the western coasts of America, from whence he went to the Ladrones, the Philippines, the Moluccas, the Cape of Good Hope, and returned to Rotterdam with one ship the 26th of August, 1601. He made no discoveries in the South Seas.

George Spilberg, a Dutchman, sailed from Zeeland the 8th of August, 1614, with six ships; he lost two ships before he came to the straits of Magalhaens, passed through them, attacked several places on the coasts of

Peru

Peru and Mexico; from whence, without discovering any thing on his course, he sailed to the Ladrones and Moluccas. Two of his ships re-entered the ports of Holland, on the first of July, 1617.

James Lemaire and William Cornelius Schouten immortalized their names much about the same time. They sailed from the Texel the 14th of June, 1615, with the ships Concord and Horn, discovered the straits that bear the name of Lemaire, and were the first that ever entered the South Seas by doubling Cape Horn. In that ocean they discovered the Isle of Dogs, in 15° 15′ south latitude, and about 142° west longitude from Paris; the Isle without Bottom (*Zonder Grond*) in 15° south latitude, one hundred leagues westward: Water Island in 14° 40 south latitude, and fifteen leagues more to the west; at twenty leagues westward of this, Fly Island, in 16° 10 south latitude; and between 173° and 175° west longitude from Paris, two isles, which they called Cocos and Traitor's; fifty leagues more westward, the Isle of Hope; next the Isle of Horn, in 14° 56′ south latitude, and about 179° east longitude from Paris; they then coasted New Guinea, passed between its western extremity and the Isle of Gilolo, and arrived at Batavia in October 1616. George Spilberg stopped them there, and they were sent to Europe, on board the East-India company's ships; Lemaire died of a sickness at at the Isle of Mauritius; Schouten returned to his country; the Concord and Horn came back in two years and ten days.

James l'Hermite, a Dutchman, commanding a fleet of eleven ships, sailed in 1623, with the scheme of making the conquest of Peru; he got into the South Seas round Cape Horn, and harrassed the Spanish coasts, from whence he went to the Ladrones, and thence to Batavia, without making any discoveries in the South Seas. He died, after clearing the straits of Sonda; and his ship, almost the only one of the whole fleet, arrived in the Texel the 9th of July, 1626.

In 1683, Cowley, an Englishman, sailed from Virginia, doubled Cape Horn, made several attacks upon the Spanish coasts, came to the Ladrones, and returned to England by the Cape of Good Hope, where he arrived on the 12th of October, 1686. This navigator has made no discoveries in the South Seas; he pretends to have found out the Isle of Pepis in the North Sea *, in 47° southern latitude, about eighty leagues from the coast of Patagonia; I have sought it three times, and the English twice, without finding it.

* North Sea signifies here the Atlantic Ocean, and is put in opposition to South Sea; the former taking in the ocean on this side the Magellanic straits, the latter that which is west of them. The appellation, though somewhat improper, by calling the sea about the south pole the North Sea, is however sometimes employed by some writers. F.

Woodes Rogers, an Englishman, left Bristol the 2d of August, 1708, doubled Cape Horn, attacked the Spanish coast up to California, from whence he took the same course which had already been taken several times before him, went to the Ladrones, Moluccas, Batavia, and doubling the Cape of Good Hope, he arrived in the Downs the first of October, 1711.

Ten years after, Roggewein, a Dutchman, left the Texel, with three ships; he came into the South Seas round Cape Horn, sought for Davis's Land without finding it; discovered to the south of the Tropic of Capricorn, an isle which he called Easter Island, the latitude of which is uncertain; then, between 15° and 16° south latitude, the Pernicious Isles, where he lost one of his ships; afterwards, much about the same latitude, the isles Aurora, Vesper, the Labyrinth composed of six islands, and Recreation Island, where he touched at. He next discovered three isles in 12° south, which he called the Bauman's Isles; and lastly, in 11° south, the Isles of Tienhoven and Groningen; then sailing along New Guinea and Papua, he came at length to Batavia, where his ships were confiscated. Admiral Roggewein returned to Holland, on board a Dutch India-man, and arrived in the Texel the 11th of July, 1723, six hundred and eighty days after his departure from the same port.

The taste for great navigations seemed entirely extinct, when, in 1741, Admiral Anson made a voyage round the world, the excellent account of which is in every body's hands, and has made no new improvement in geography.

After this voyage of Lord Anson's, there was no considerable one undertaken for above twenty years. The spirit of discovery seems to have been but lately revived. Commodore Byron sailed from the Downs the 20th of June, 1764, passed through the straits of Magalhaens; discovered some isles in the South Sea, sailing almost due north west, arrived at Batavia the 28th of November, 1765, at the Cape the 24th of February, 1766, and in the Downs the 9th of May, having been out upon this voyage six hundred and forty-eight days.

Two months after commodore Byron's return, captain Wallace sailed from England, with the Dolphin and Swallow sloops; he went through the straits of Magalhaens, and as he entered the South Seas, he was separated from the Swallow, commanded by captain Carteret; he discovered an isle in about 18°, some time in August, 1767: he sailed up to the Line, passed near Papua, arrived at Batavia in January, 1768, touched at the Cape of Good Hope, and returned to England in May the same year.

His

His companion Carteret, after having suffered many misfortunes in the South Sea, and lost almost all his crew, came to Macassar in March 1768, to Batavia the 15th of September, and to the Cape of Good Hope towards the end of December. It will appear in the sequel, that I overtook him on the 18th of February, 1769, in 11° north latitude. He arrived in England in June.

It appears, that of these thirteen voyages which have been made round the world *, none belongs to the French

* Dom Pernetty, in his Dissertation upon America, speaks of a voyage round the world, in 1719, by captain Shelvock; I have no knowledge of this voyage. Note of Mr. de B.

As M. de Bougainville's list of circumnavigators is very imperfect, we will endeavour to give a more compleat one in few words.

1. Fernando Magalhaens, 1519.
2. Sir Francis Drake sailed from Plymouth the 15th of November, 1577, but was obliged to put back on account of a storm; after which, he set sail again the 13th of December, and returned the 16th of September, 1580.
3. Sir Thomas Cavendish, 1586—88.
4. Simon de Cordes, a Dutchman, sailed in 1598—1600.
5. Oliver Van Noort sailed the 13th of September, 1598, and returned the 22d of August, 1601.
6. George Spielbergen, a German in the Dutch service, 1614—1617.
7. William Cornelius Schouten with Jacob Le Maire, 1615—1617.
8. Jacob l'Hermite with John Hugo Schapenham sailed from Goeree, in the province of Holland, the 29th of April, 1623, and arrived in the Texel the 9th of July, 1626.
9. Henry Brouwer, a Dutchman, in 1643.
10. Cowley, in 1683—1686.
11. William Dampier, an Englishman, sailed in 1689, and returned 1691. He has been omitted by M. de Bougainville in the list of circumnavigators, because he did not go round the world in one and the same ship.
12. Beauchesne Gouin, in 1699.
13. Edward Cooke, an Englishman, made the voyage in the years 1708 and 1711.

14. Woodes

French nation, and that only six of them have been made with the spirit of discovery; viz. those of Magalhaens, Drake, Le Maire, Roggewein, Byron, and Wallace; the other navigators, who had no other view than to enrich themselves by their attacks upon the Spaniards, followed the known tracks, without increasing the knowledge of geography.

In 1714, a Frenchman, called la Barbinais le Gentil, sailed, on board a private merchant ship, in order to carry on an illicit trade, upon the coast of Chili and Peru. From thence he went to China, where, after staying some time in various factories, he embarked in another ship than that which had brought him, and returned to Europe, having indeed gone in person round

14. Woodes Rogers, an Englishman, sailed from Bristol, June 15th, 1708, and returned 1711.

15. Clipperton and Shelvocke, two Englishmen, sailed the 13th of February, 1719, and returned in 1722; the former to Galway in Ireland, in the beginning of June, the latter to London, on the first of August.

16. Roggewein, a Mecklenburger, in the Dutch service, sailed the 16th of July, 1721, and returned the 11th of July, 1723.

17. Lord Anson, 1740—1744.

18. Commodore Byron, 1764—1766.

19. The Dolphin and Swallow sloops. The first 1766—1768. The second 1766—1769.

20. M. de Bougainville, 1766—1769.

21. The Endeavour sloop, captain Cooke, which sailed in August, 1768, to observe the transit of Venus, came to Batavia the latter end of 1770, and returned to England in July 1771.

From this list, it appears that the English have undertaken the greatest number of voyages, with a design to make discoveries, unattended by that selfishness with which most of the Dutch voyages were entered upon, merely with a view to promote the knowledge of geography, to make navigation more safe, and likewise to throw further lights on the study of nature. F.

the world, though that cannot be conſidered as a circumnavigation by the French nation *.

Let us now ſpeak of thoſe who going out either from Europe, or from the weſtern coaſts of South-America, or from the Eaſt-Indies, have made diſcoveries in the South Seas, without ſailing round the world.

It appears that one Paulmier de Gonneville, a Frenchman, was the firſt who diſcovered any thing that way, in 1503 and 1504. The countries which he viſited are not known; he brought however with him a native of one of them, whom the government did not ſend back, for which reaſon, Gonneville, thinking himſelf perſonally engaged, gave him his heireſs in marriage.

Alfonzo de Salazar, a Spaniard, diſcovered in 1525 the Iſle of St. Bartholomew, in 14° north latitude, and 158° eaſt longitude from Paris.

Alvaro de Saavedra, left one of the ports of Mexico in 1526, diſcovered, between 9° and 10° north, a heap of iſles, which he called the King's Iſles, much about the ſame longitude with the Iſle St. Bartholomew; he then went to the Philippines, and to the Moluccas, and

* The author is very ſolicitous to exclude le Gentil de la Barbinais from the honour of being the firſt circumnavigator of the French nation, in order to ſecure it to himſelf; though it is a real circumnavigation. The famous Italian, Giovan Franciſco Gemelli Carreri, cannot with propriety be called a circumnavigator, though he made the tour of the globe in the years 1693—1698, for he landed in Mexico, and croſſed America by land, and went again to the Manillas by ſea, and from thence to China and Europe on board of other ſhips. F.

on his return to Mexico, he was the firſt that had any knowledge of New Guinea and Papua. He diſcovered likewiſe, in twelve degrees north, about eighty leagues eaſt of the King's Iſles, a chain of low iſlands, which he called Iſlas de los Barbudos.

Diego Hurtado and Hernando de Grijalva, who ſailed from Mexico in 1533, to ſearch the South Seas, diſcovered only one iſle, ſituated in 20° 30′ north latitude, and about 100° weſt longitude from Paris; they called it St. Thomas Iſland.

Juan Gaëtan ſailed from Mexico in 1542, and likewiſe kept to the north of the æquator. He there diſcovered, between 20° and 9° in various longitudes, ſeveral iſles; viz. Rocca Partida, the Coral Iſles, the Garden Iſles, the Sailor Iſles, the Iſle of Areziſa, and at laſt he touched at New Guinea, or rather, according to his report, at the iſles that were afterwards called New Britain; but Dampier had not yet diſcovered the paſſage which bears his name.

The following voyage is more famous than all the preceding ones.

Alvaro de Mendoça and Mindana, leaving Peru in 1567, diſcovered thoſe celebrated iſles, which obtained the name of Solomon's Iſlands, on account of their riches; but ſuppoſing that the accounts we have of the riches of theſe iſles be not fabulous, yet their ſituation

is

is not known, and they have been ſought for ſince without any ſucceſs. It appears only, that they are on the ſouth ſide of the Line, between 8° and 12°. The Iſle Iſabella, and the land of Guadalcanal, which thoſe voyages mention, are not better known.

In 1595, Alvaro de Mindaña, the companion of Mendoça, in the preceding voyage, ſailed again from Peru, with four ſhips, in ſearch of the Solomon's Iſles: he had with him Fernando de Quiros, who afterwards became celebrated by his own diſcoveries. Mindaña diſcovered, between 9° and 11° ſouth latitude, about 108° weſt from Paris, the iſles of San Pedro, Magdalena, Dominica, St. Chriſtina, all which he called *las Marqueſas de Mendoça*, in honour of Donna Iſabella de Mendoça, who made the voyage with him: about twenty-four degrees more to the weſtward, he diſcovered the Iſle of San Bernardo; almoſt two hundred leagues to the weſt of that, the Solitary Iſle; and laſtly, the Iſle of Santa Cruz, ſituated nearly in 140° eaſt longitude from Paris. The fleet ſailed from thence to the Ladrones, and laſtly to the Philippines, where general Mindana did not arrive, nor did any one know ſince what became of him.

Fernando de Quiros, the companion of the unhappy Mindana, brought Donna Iſabella back to Peru. He ſailed from thence again with two ſhips, on the 21ſt of December, 1605, and ſteered his courſe almoſt weſt-ſouth-

ſouth-weſt. He diſcovered at firſt a little iſle, in about 25° ſouth latitude, and about 124° weſt longitude from Paris; then, between 18° and 19° ſouth, ſeven or eight low, and almoſt inundated iſlands, which bear his name, and in 13° ſouth lat. about 157° weſt from Paris, the iſle which he called Iſle of Beautiful People. Afterwards he ſought in vain for the Iſle of Santa Cruz, which he had ſeen on his firſt voyage, but diſcovered, in 13° ſouth lat. and near 176° eaſt longitude from Paris, the Iſle of Taumaco; likewiſe, about a hundred leagues weſt of that iſle, in 15° ſouth lat. a great continent, which he called *Tierra auſtral del Eſpiritù Santo* and which has been differently placed by the ſeveral geographers. There he ceaſed to go weſtward, and ſailed towards Mexico, where he arrived at the end of the year 1606, having again unſucceſsfully ſought the Iſle of Santa Cruz.

Abel Taſman ſailed from Batavia the 14th of Auguſt, 1642, diſcovered land in 42° ſouth latitude, and about 155° eaſt longitude from Paris, which he called Van Diemen's land: he ſailed from thence to the eaſtward, and in about 160° of our eaſt longitude, he diſcovered New Zeeland, in 42° 10′ ſouth. He coaſted it till to 34° ſouth lat. from whence he ſailed N. E. and diſcovered, in 22° 35′ ſouth lat. and nearly 174° eaſt of Paris, the Iſles of Pylſtaart, Amſterdam, and Rotterdam.

He

He did not extend his reſearches any farther, and returned to Batavia, ſailing between New Guinea and Gilolo.

The general name of New Holland has been given to a great extent of continent, or chain of iſlands, reaching from 6° to 34° ſouth lat. between 105° and 140° eaſt longitude from Paris. It was reaſonable to give it the name of New Holland, becauſe the different parts of it have chiefly been diſcovered by Dutch navigators. The firſt land which was found in theſe parts, was called the Land of Eendraght, from the perſon * that diſcovered it in 1616, in 24° and 25° ſouth latitude. In 1618, another part of this coaſt, ſituated nearly in 15° ſouth, was diſcovered by Zeachen, who gave it the name of Arnhem and Diemen; though this is not the ſame with that which Taſman called Diemen's land afterwards. In 1619, Jan van Edels gave his name to a ſouthern part of New Holland. Another part, ſituated between 30° and 33°, received the name of Leuwen. Peter van Nuitz communicated his name in 1627 to a coaſt which makes as it were a continuation of Leuwen's land to the weſtward. William de Witts called a part of the weſtern coaſt, near the tropic of Capricorn, after his own name, though it ſhould have born that of captain Viane, a Dutchman, who paid dear for the diſ-

* Not from the diſcoverer, but from the ſhip Eendraght (Concord).

diſcovery of this coaſt in 1628, by the loſs of his ſhip, and of all his riches.

In the ſame year 1628, Peter Carpenter, a Dutchman, diſcovered the great Gulph of Carpentaria, between 10° and 20° ſouth latitude, and the Dutch have often ſince ſent ſhips to reconnoitre that coaſt.

Dampier, an Engliſhman, ſetting out from the great Timor Iſle, made his firſt voyage in 1687, along the coaſts of New Holland; and touched between the land of Arnhem and of Diemen: this ſhort expedition was productive of no diſcovery. In 1699 he left England, with an expreſs intention of viſiting all that region, concerning which, the Dutch would not publiſh the accounts they had of it. He ſailed along the weſtern coaſt of it, from 28° to 15°. He ſaw the land of Eendraght, and of De Witt, and conjectured that there might exiſt a paſſage to the ſouth of Carpentaria. He then returned to Timor, from whence he went out again, examined the Iſles of Papua, coaſted New Guinea, diſcovered the paſſage that bears his name, called a great iſle which forms this paſſage or ſtrait on the eaſt ſide, New Britain, and ſailed back to Timor along New Guinea. This is the ſame Dampier who between 1683 and 1691, partly as a free-booter or privateer, and partly as a trader, ſailed round the world, by changing his ſhips.

This

This is the ſhort abſtract of the ſeveral voyages round the world, and of the various diſcoveries made in that vaſt Pacific Ocean before our departure from France* Before I begin the narrative of the expedition, with which I was charged, I muſt beg leave to mention, that this relation ought not to be looked upon as a work of amuſement; it has chiefly been written for ſeamen. Beſides, this long navigation round the globe does not offer ſuch ſtriking and intereſting ſcenes to the polite world, as a voyage made in time of war. Happy, if by being uſed to compoſition, I could have learnt to counterbalance the dulneſs of the ſubject by elegance of ſtile! But, though I was acquainted with the ſciences from my very youth, when the leſſons which M. d'Alembert was ſo kind to give me, enabled me to offer to the indulgent public, a work upon geometry, yet I am now far from the ſanctuary of ſcience and learning; the rambling and ſavage life I have led for theſe twelve years paſt, have had too great an effect upon my ideas and my ſtile. One does not become a good writer in the woods of Canada, or on the ſeas, and I have loſt

* The miſtakes and omiſſions of our author in regard to theſe navigators of the South Seas, who did not ſail round the world, are various and multifarious; but it would take up too much time to point them all out; and as there is a very complete liſt of all the navigators of the Pacific Ocean, in the *Hiſtorical Collection of the ſeveral Voyages and Diſcoveries in the South Pacific Ocean*, publiſhed by Alex. Dalrymple, Eſq. we refer our readers to it. F.

a brother, whofe productions were admired by the public, and who might have affifted me in that refpect.

Laftly, I neither quote nor contradict any body, and much lefs do I pretend to eftablifh or to overthrow any hypothefis; and fuppofing that the great differences which I have remarked in the various countries where I have touched at, had not been able to prevent my embracing that fpirit of fyftem-making, fo peculiar in our prefent age, and however fo incompatible with true philofophy, how could I have expected that my whim, whatever appearance of probability I could give it, fhould meet with fuccefs in the world? I am a voyager and a feaman; that is, a liar and a ftupid fellow, in the eyes of that clafs of indolent haughty writers, who in their clofets reafon in *infinitum* on the world and its inhabitants, and with an air of fuperiority, confine nature within the limits of their own invention. This way of proceeding appears very fingular and inconceivable, on the part of perfons who have obferved nothing themfelves, and only write and reafon upon the obfervations which they have borrowed from thofe fame travellers in whom they deny the faculty of feeing and thinking.

I fhall conclude this preliminary difcourfe by doing juftice to the zeal, courage, and unwearied patience of

 the

the officers and crew of my two ſhips *. It has not been neceſſary to animate them by any extraordinary incitement, ſuch as the Engliſh thought it neceſſary to grant to the crew of commodore Byron. Their conſtancy has ſtood the teſt of the moſt critical ſituations, and their good will has not one moment abated. But the French nation is capable of conquering the greateſt difficulties, and nothing is impoſſible to their efforts, as often as ſhe will think herſelf equal at leaſt to any nation in the world †.

* The officers on board the frigate la Boudeuſe, were M. de Bougainville, captain of the ſhip; Duclos Guyot, captain of a fire-ſhip; chevalier de Bournand, chevalier d'Oraiſon, chevalier du Bouchage, under-lieutenants (*enſeignes de vaiſſeau*); chevalier de Suzannet, chevalier de Kué, midſhipmen acting as officers; le Corre, ſuper-cargo (*officier-marchand*); Saint-Germain, ſhip's clerk; la Veze, the chaplain; la Porte, ſurgeon.

The officers of the ſtore-ſhip l'Etoile, conſiſted of M. M. Chenard de la Giraudais, captain of a fire-ſhip; Caro, lieutenant in an India-man; Donat, Landais, Fontainé, and Lavary-le-Roi, *officiers marchands*; Michaud, ſhip's-clerk; Vivez, ſurgeon.

There were likewiſe M. M. de Commerçon, a phyſician; Verron, an aſtronomer, and de Romainville, an engineer.

† It would be improper to derogate from the merit of any nation, unleſs that ſame nation intends to obtain it by deſtroying the character of another. Had Mr. de Bougainville beſtowed ſome encomiums upon the zeal and courage of the officers under his command, it would be thought that he were willing to do them juſtice: but ſince he, without the leaſt neceſſity, caſts a reflection upon the Engliſh officers in commodore Byron's expedition, it is no more but juſtice to retort the argument. It is an undeniable proof of the badneſs of the conſtitution, and of the arbitrary government of a country, when a ſet of worthy men, who have braved the moſt imminent danger, with an undaunted courage, for the welfare of their fellow-citizens, remain without any reward whatſoever, except that philoſophical one, the conſciouſneſs of good and laudable actions. But it is likewiſe the beſt proof of the happineſs of the government and conſtitution of a country, when merit and virtue is rewarded. Theſe propoſitions are ſo evidently founded on truth, that they want no further confirmation: and every true Engliſhman will congratulate himſelf on the

the happineſs to live under a government which thinks it a neceſſary duty to reward zeal, courage, and virtue, in a ſet of men who go through their duty with ſpirit and chearfulneſs; and what honour muſt not redound on an adminiſtration which forces, even a rival nation, to give an honourable teſtimony to its attention in juſtly and conſpicuouſly rewarding merit in its fellow-citizens, at the ſame time that theſe rivals endeavour to quiet the uneaſy minds of their poor diſſatisfied officers, with a vain and empty compliment. F.

A VOY-

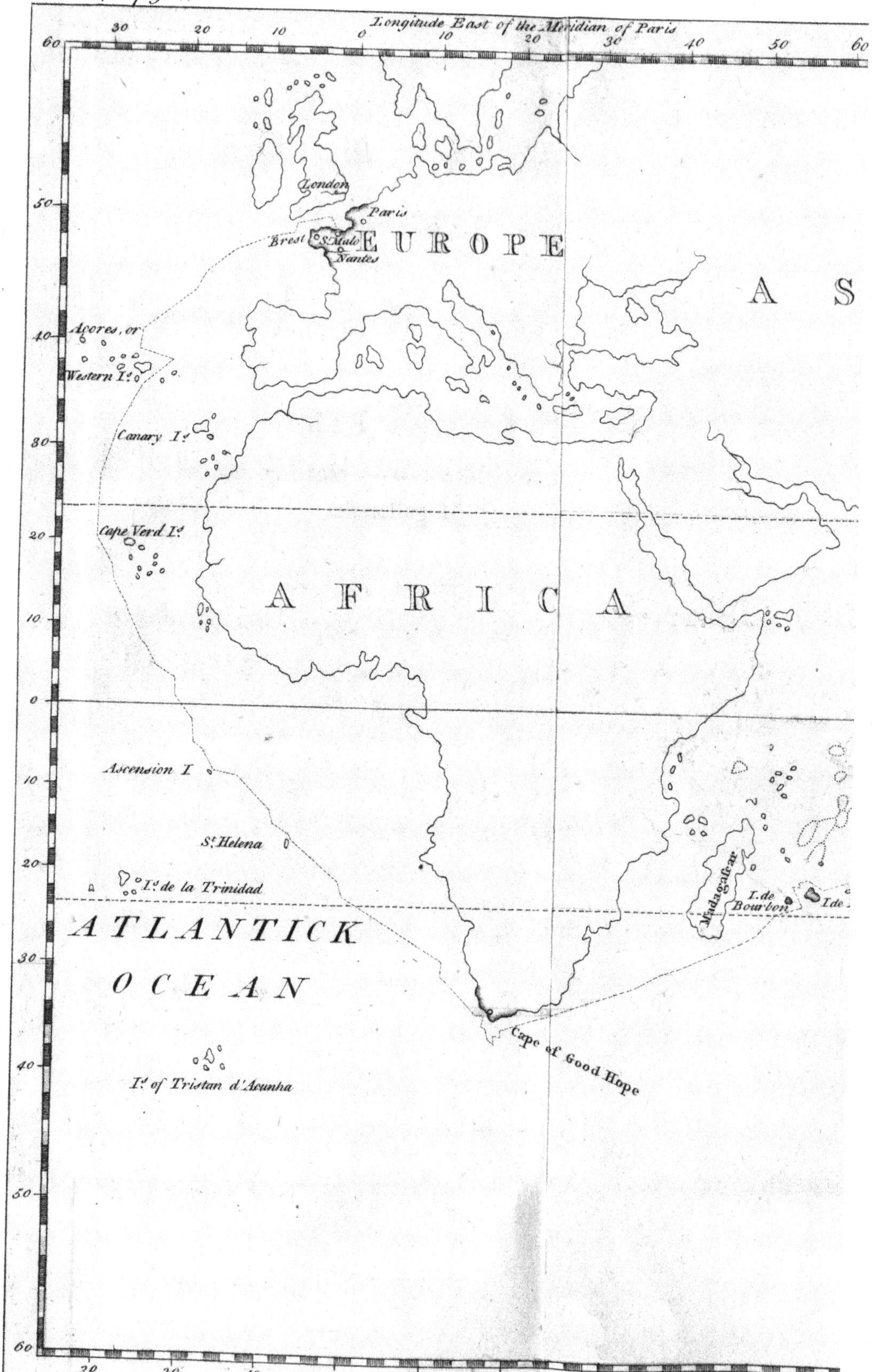

Pl. I. face page 1.

Longitude East of the Meridian of Paris

London

Paris

Brest

St. Malo

Nantes

EUROPE

AS

Açores, or Western Is.

Canary Is.

Cape Verd Is.

AFRICA

Ascension I.

St. Helena

Is. de la Trinidad

ATLANTICK OCEAN

Madagascar

I. de Bourbon

I de

Cape of Good Hope

Is. of Tristan d'Acunha

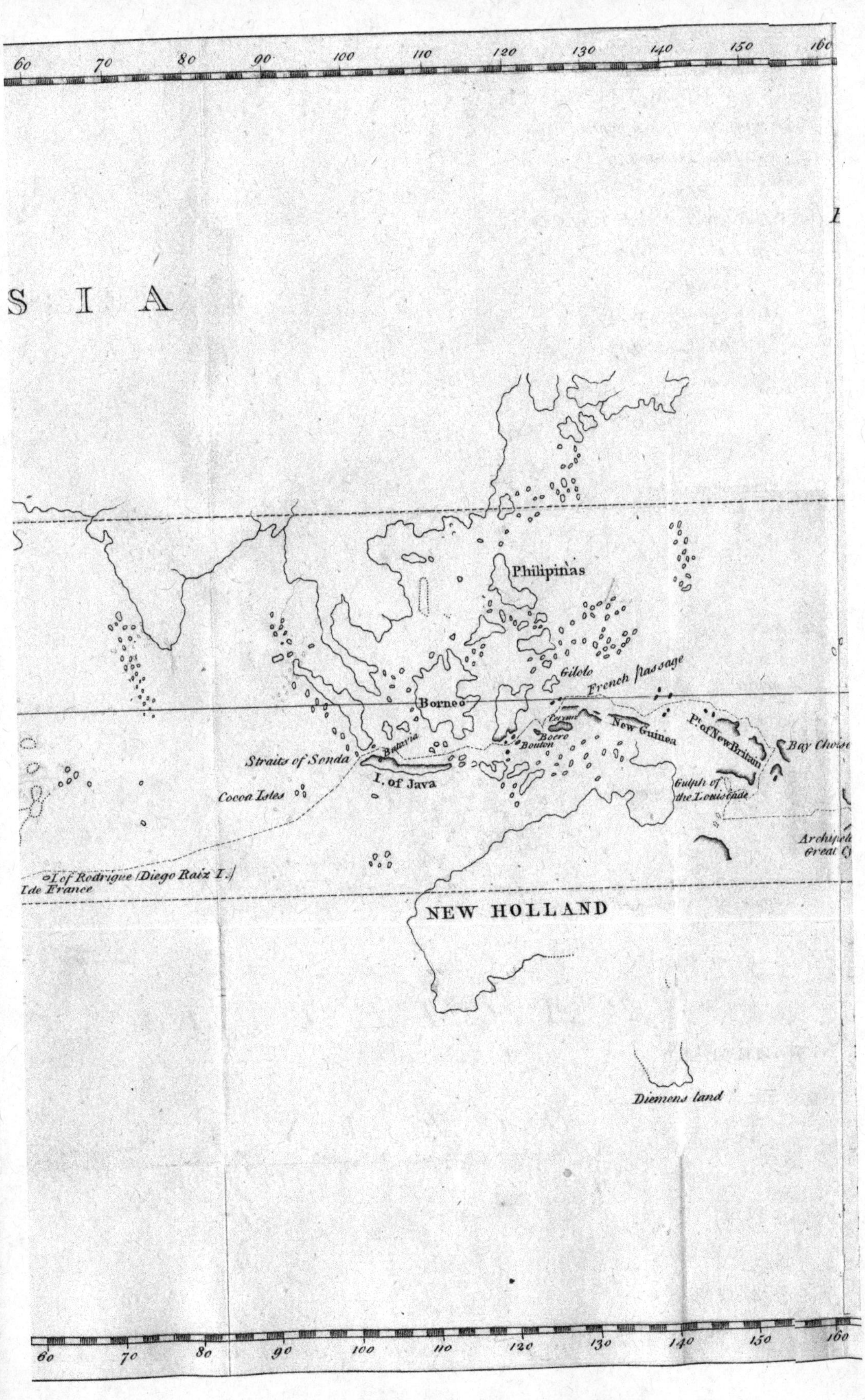
60
70
80
90
100
110
120
130
140
150
160
S I A
Philipinas
Gilolo
French passage
Borneo
New Guinea
Pt. of New Britain
Straits of Sonda
Batavia
I. of Java
Cocoa Isles
Ceram
Boero
Bouton
Gulph of the Louisiade
I. of Rodrigue (Diego Raiz I.)
I. de France
NEW HOLLAND
Diemens land
60
70
80
90
100
110
120
130
140
150
160

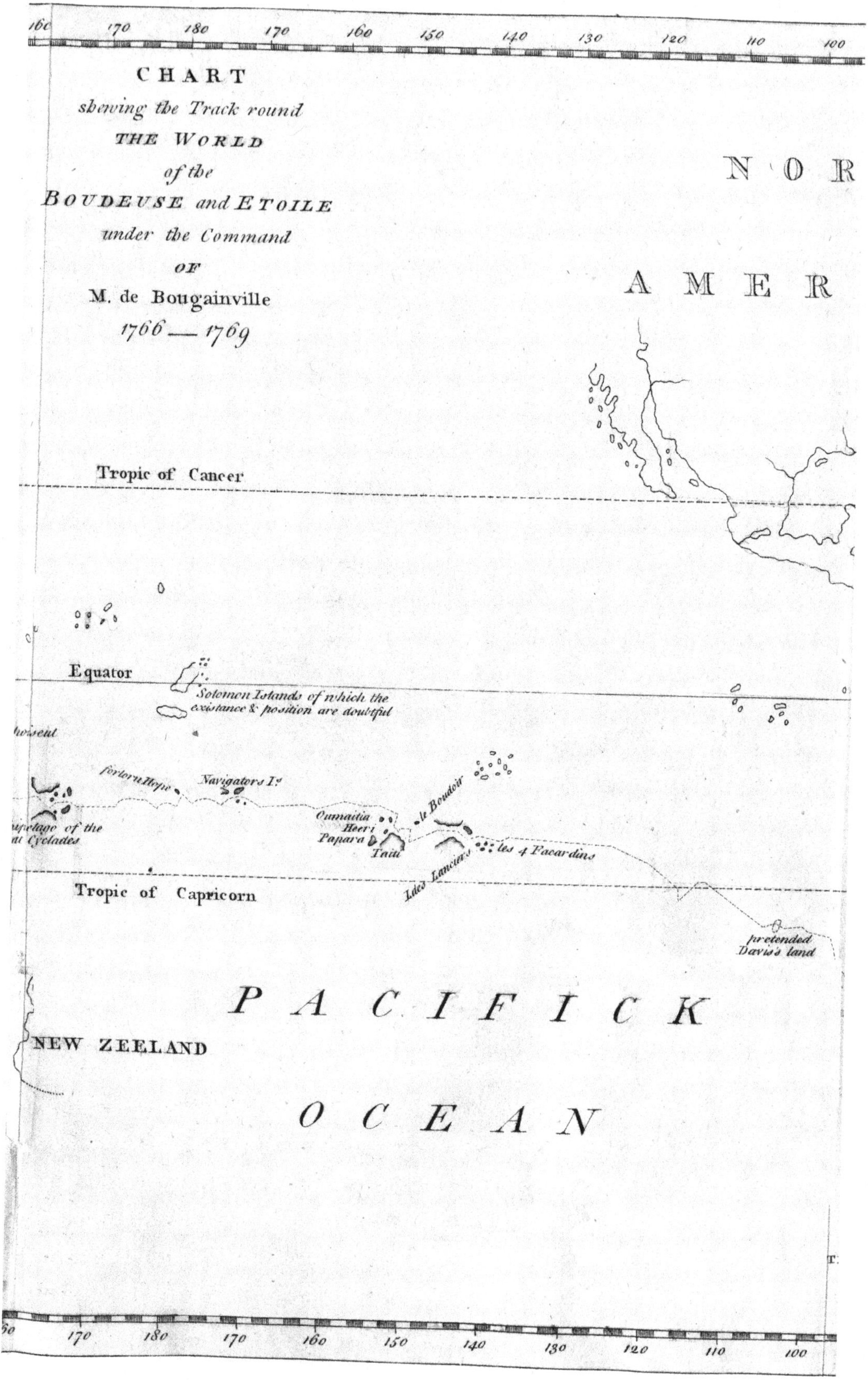

CHART
shewing the Track round
THE WORLD
of the
BOUDEUSE *and* ETOILE
under the Command
OF
M. de Bougainville
1766 — 1769

160 170 180 170 160 150 140 130 120 110 100

N O R

A M E R

Tropic of Cancer

Equator

Solomon Islands of which the existance & position are doubtful

hoiseul

forlorn Hope

Navigators Is.

ipelago of the
at Cyclades

Oumaitia
Heeri
Papara
Taiti

le Boudoir

les 4 Facardins

Ides Lanciers

Tropic of Capricorn

pretended Davis's land

P A C I F I C K

NEW ZEELAND

O C E A N

T:

60 170 180 170 160 150 140 130 120 110 100

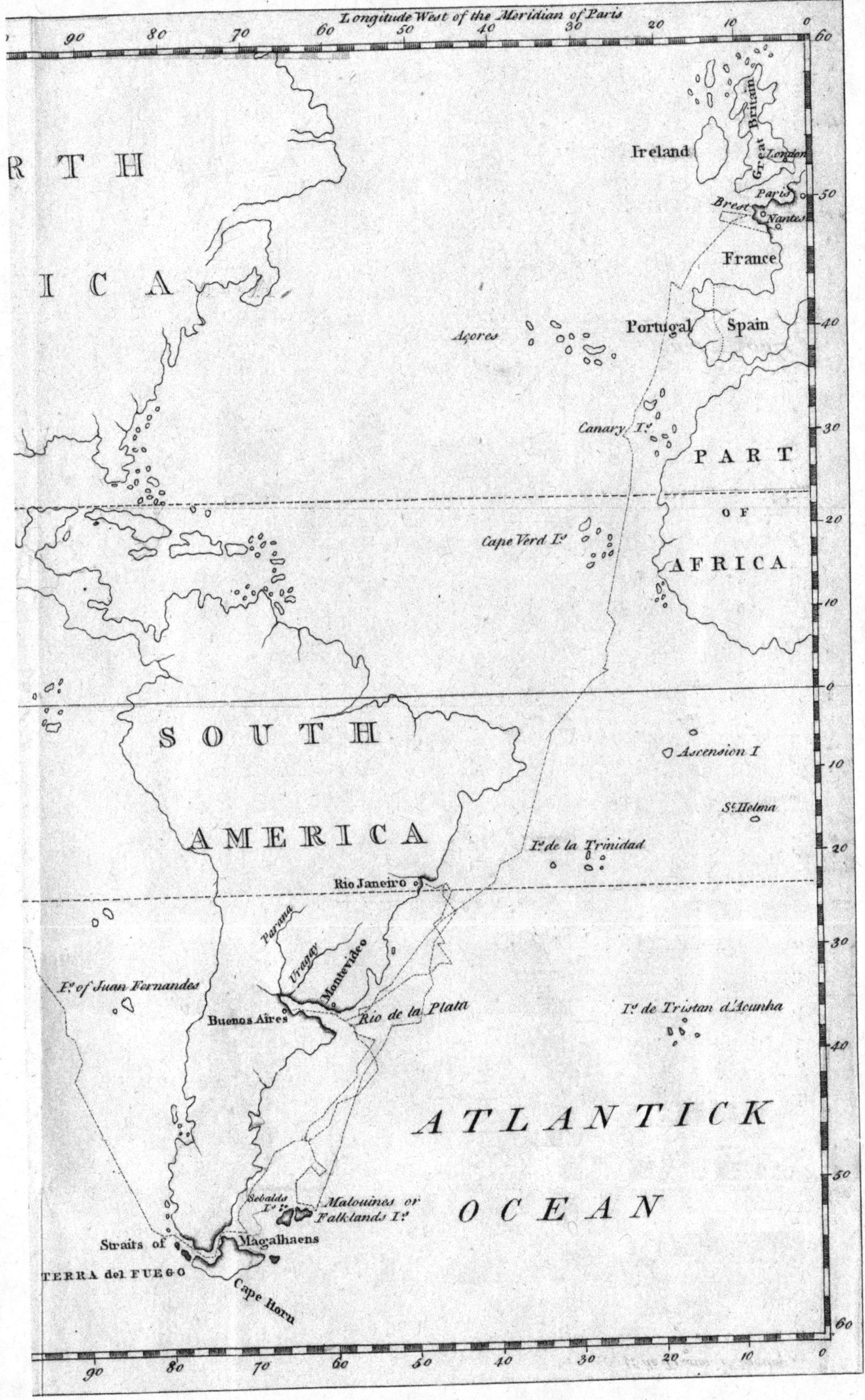

Longitude West of the Meridian of Paris
RTH
ICA
Ireland
Great Britain
London
Paris
Brest
Nantes
France
Portugal
Spain
Açores
Canary Is.
PART OF AFRICA
Cape Verd Is.
SOUTH AMERICA
Ascension I
St. Helena
Is. de la Trinidad
Rio Janeiro
Parana
Uragay
Montevideo
Is. of Juan Fernandes
Buenos Aires
Rio de la Plata
Is. de Tristan d'Acunha
ATLANTICK OCEAN
Sebalds Is.
Malouines or Falklands Is.
Straits of Magalhaens
TERRA del FUEGO
Cape Horn

A VOYAGE ROUND THE WORLD.

PART the FIRST.

Departure from France —— clearing the Straits of Magalhaens.

CHAP. I.

Departure of the Boudeuse from Nantes; puts in at Brest; run from Brest to Montevideo; junction with the Spanish frigates, intended for taking possession of the Malouines, or Falkland's islands.

IN February 1764, France began to make a settlement on the Isles Malouines. Spain reclaimed these isles as belonging to the continent of South America; and her right to them having been acknowledged by the king, I received orders to deliver our settlement to the Spaniards, and to proceed to the East Indies by crossing the South Seas between the Tropics. For this expedition I received the command

Object of t[he] voyage. 1766, November

mand

mand of the frigate la Boudeuſe, of twenty-ſix twelve-pounders, and I was to be joined at the Malouines by the ſtore-ſhip * l'Etoile, which was intended to bring me the proviſions neceſſary for a voyage of ſuch a length, and to follow me during the whole expedition. Several circumſtances retarded the junction of this ſtore-veſſel, and conſequently made my whole voyage near eight months longer than it would otherwiſe have been.

In the beginning of November, 1766, I went to Nantes, where the Boudeuſe had juſt been built, and where M. Duclos Guyot, a captain of a fireſhip, my ſecond officer, was fitting her out. The 5th of this month we came down from Painbeuf to Mindin, to finiſh the equipment of her; and on the 15th we ſailed

Departure rom Nantes.

from this road for the river de la Plata. There I was to find the two Spaniſh frigates, called la Eſmeralda and la Liebre, that had left Ferrol the 17th of October, and whoſe commander was ordered to receive the *Iſles Malouines*, or Falkland's iſlands, in the name of his Catholic majeſty.

quall of ind.

The 17th in the morning we ſuffered a ſudden guſt of wind from W. S. W. to N. W. it grew more violent in the night, which we paſſed under our bare poles, with our main-yards lowered, the clue of the fore-ſail,

* La flûte.

under

under which we tried before, having been carried away. The 18th, at four in the morning, our fore-top-maſt broke about the middle of its height; the main-top-maſt reſiſted till eight o'clock, when it broke in the cap, and carried away the head of the main-maſt. This laſt event made it impoſſible for us to continue our voyage, and I determined to put into Breſt, where we arrived the 21ſt of November.

Putting in at Breſt.

This ſquall of wind, and the confuſion it had occaſioned, gave me room to make the following obſervations upon the ſtate and qualities of the frigate which I commanded.

1. The prodigious tumbling home of her top-timbers, leaving too little opening to the angles which the ſhrouds make with the maſts, the latter were not ſufficiently ſupported.

2. The preceding fault became of more conſequence by the nature of the ballaſt, which we had been obliged to take in, on account of the prodigious quantity of proviſions we had ſtowed. Forty tuns of ballaſt, diſtributed on both ſides of the kelſon, and at a ſhort diſtance from it, and a dozen twelve-pounders placed at the bottom of the pump-well (we had only fourteen upon deck) added a conſiderable weight, which being much below the center of gravity, and almoſt entirely reſted upon the kelſon,

put the mafts in danger, if there had been any rolling.

Thefe reflections induced me to get the exceffive height of our mafts fhortened, and to exchange the cannon, which were twelve-pounders, for eight-pounders. Befides the diminution of near twenty ton weight, both in the hold and upon deck, gained by exchanging the artillery, the narrow make of the frigate alone was fufficient to render it neceffary. She wanted about two feet of the beam which fuch frigates have as are intended to carry twelve pounders.

Notwithftanding thefe alterations, which I was allowed to make, I could not help obferving that my fhip was not fit for navigating in the feas round Cape Horn. I had found, during the fquall of wind, that fhe made water from all her upper-works, which might expofe part of my bifcuit to be fpoiled by the water getting into the ftore-rooms in bad weather; an inconvenience, the confequences of which we fhould not be able to remedy during the voyage. I therefore afked leave to fend the Boudeufe back to France from the Falkland's iflands, under the command of the chevalier Bournand, lieutenant of a fhip, and to continue the voyage with the ftore-fhip l'Etoile alone, if the long winter nights fhould prevent my paffing the Straits of

Magal-

Magalhaens *. I obtained this permission, and the 4th of December, our masts being repaired, the artillery exchanged, and the frigate entirely caulked in her upper-works, we went out of the port and anchored in the road, where we continued a whole day, in order to embark the powder, and to set up the shrouds

December. Departure from Brest.

The 5th at noon we got under sail in the road of Brest. I was obliged to cut my cable, because the fresh east-wind and the ebb prevented my tacking about, as I was apprehensive of falling off too near the shore. I had eleven commissioned officers, and three volunteers; and the crew consisted of two hundred sailors, warrant-officers, soldiers, boys, and servants. The prince of Nassau Sieghen had got leave from the king to go upon this expedition. At four o'clock in the afternoon, the middle of the isle of Ushant bore N. by E. and from thence I took my departure.

Description of the Salvages.

During the first days, we had the wind pretty constant from W. N. W. to W. S. W. and S. W, very fresh. The 17th, afternoon, we got sight of the Salvages; the 18th, of the Isle of Palma; and the 19th, of the Isle of Ferro. What is called the Salvages, is a little isle of about a league in extent from E. to W. it is low in the

* Though the name of this circumnavigator is frequently spelled Magellan, it is, however, right to spell proper names as they are written in their original language; according to this rule we shall always write Magalhaens. F.

middle,

middle, and at each end a little hillock; a chain of rocks, ſome of which appear above water, extend to the weſtward about two leagues off the iſland; there are likewiſe ſome breakers on the eaſt-ſide, but they are not far from the ſhore.

Error in the calculation of the courſe.

The ſight of theſe rocks convinced us of a great error in our reckoning; but I would not make a computation before I had ſeen the Canaries, whoſe poſition is exactly determined. The ſight of the Iſle of Ferro gave me with certainty the correction which I was deſirous to make. The 19th, at noon, I took the latitude, and comparing it with the bearings of the Iſle of Ferro taken that ſame hour, I found a difference of four degrees and ſeven minutes, which I was more to the eaſtward, than by my reckoning. This error is frequent in croſſing from Cape Finiſterre to the Canaries, and I had found it on other voyages, as the currents oppoſite the ſtraits of Gibraltar ſet to the eaſtward with great rapidity.

oſition of he Salvages ectified.

I had, at the ſame time, an opportunity of remarking, that the Salvages are improperly placed on M. de Bellin's Chart. Indeed, when we got ſight of them the 17th, after noon, the longitude which their bearings gave us differed from our calculation by three degrees ſeventeen minutes to the eaſtward. However, this ſame difference appeared the 19th of four degrees ſeven mi-

4 nutes,

nutes, by correcting our place, according to the bearings of the isle of Ferro, whose longitude has been determined by astronomical observations. It must be observed, that during the two days which passed between our getting sight of the Salvages and of Ferro, we sailed with a fair wind; and consequently there can be very little miscalculation in that part of the course. Besides, the 18th, we set the Isle of Palma, bearing S. W. by W. corrected; and, according to M. Bellin, it was to bear S. W. I concluded, from these two observations, that M. Bellin has placed the Isle of Salvages about 32 more to the W. than it really is.

I therefore took a fresh departure the 19th of December at noon. We met with no remarkable occurrences on our voyage, till we came to the Rio de la Plata; our course furnished us only with the following observations, which may be interesting to navigators.

1767. January. Nautical [ob]servations

1. The 6th and 7th of January 1767, being between 1° 40′ and 0° 38′ north latitude; and about 28° longitude, we saw many birds, which induced me to believe, that we were near the rock of Penedo San Pedro; though M. Bellin does not mark it on his chart.

Passing of the line.

2. The 8th of January, in the afternoon, we passed the line between 27° and 28° of longitude.

Remark o[n] the variations.

3. Since the 2d of January we could no longer observe the variations; and I only reckoned them by the charts of

of William Mountain and James Obſon. The 11th, at ſun-ſet, we obſerved 3° 17′ of N. W. variation; and the 14th, in the morning, I obſerved again 10′ of N. W. variation with an azimuth-compaſs, the ſhip then being in 10° 30′ or 40′ S. latitude, and about 33° 20′ W. longitude, from Paris. Therefore it is certain, that, if my eſtimated longitude is exact, and I verified it as ſuch at the land-fall *, the line of no variation is ſtill further advanced to the weſtward ſince the obſervation of Mountain and Obſon; and it ſeems the progreſs of this line weſtward is pretty uniform. Indeed, upon the ſame degree of latitude, where Mountain and Obſon found 12° or 13° of difference in the ſpace of forty-four years, I have found a little more than 6° after an interval of 22 years. This progreſſion deſerves to be confirmed by a chain of obſervations. The diſcovery of the law by which theſe changes happen that are obſerved in the declination of the magnetic needle, beſides furniſhing us with a method of finding out the longitude at ſea, might perhaps lead us to the cauſes of this variation, and perhaps even to that of the magnetic power.

[C]auſes of the [v]ariations [f]ound in go[in]g to the [B]raſils.

4. About the line we have almoſt always obſerved very great variations on the north-ſide, though it is more common to obſerve them on the ſouth-ſide. We had an opportunity of gueſſing at the cauſe of it, the

* Land-fall, *atterage*, the firſt land a ſhip makes after a ſea-voyage. See Falconer's Marine Dictionary. F.

18th

18th of January paſſing over a bank with young fiſh, which extended beyond the reach of our ſight, from S. W. one quarter W. to N. E. one quarter E. upon a line of reddiſh white, about two fathoms broad. Our meeting with it, taught us that ſince ſome days the currents ſet in to the N. E. one quarter E. for all fiſh ſpawn upon the coaſts, whence the currents detach the fry and carry them into the open ſea. On obſerving theſe variations N. of which I have ſpoken, I did not infer from thence, that it was neceſſary there ſhould be variations weſtward together with them; likewiſe the 29th of January, in the evening, when we ſaw land, I had calculated at noon that it was ten or twelve leagues off, which gave riſe to the following obſervations.

It has long ago been a complaint among navigators, and ſtill continues, that the charts, and eſpecially thoſe of M. Bellin, lay down the coaſts of Braſil too much to the eaſtward. They ground this complaint upon their having got ſight of theſe coaſts in their ſeveral voyages, when they thought themſelves at leaſt eighty or a hundred leagues off. They add, that they have ſeveral times obſerved on theſe coaſts, that the currents had carried them S. W. and they rather chooſe to tax the charts and aſtronomical obſervations as erroneous, than suspect their ships reckoning subject to mistakes.

Upon the like reasonings we might have concluded the contrary on our course to Rio de la Plata, if by chance we had not discovered the reason of the variatiations N. which we met with. It was evident that the bank with the fry of fish, that we met with the 29th, was subject to the direction of a current; and its distance from the coast proved, that the current had already existed several days. It was therefore the cause of constant errors in our course; and the currents which navigators have often found to set in to the S W. on these shores, are subject to variations, and sometimes take contrary directions.

This observation being well confirmed, and our course being nearly S. W. were my authorities for correcting our mistakes as to the distances, making them agree with the observations of the latitude, and not to correct the points of the compass. By this method I got sight of the land, almost the same moment when I expected to see it by my calculation. Those amongst us, who always reckoned our course to the westward, according to the ship's journals, being contented to correct the difference of latitude by the observations at noon, expected to be close to the shore, according

to

to their calculation, long before we had so much as got sight of it: but can this give them reason to conclude, that the coast of the Brasils is much more westward than Mr. Bellin has laid it down?

Observations on the currents.

In general it seems, that in this part the currents vary, and sometimes set to the N. E. but more frequently to S. W. One glance at the bearings and position of the coast is sufficient to prove that they can only follow one or the other of these directions; and it is always easy to distinguish which of the two then takes place by the differences north or south, which the latitude gives. To these currents we may impute the frequent errors of which navigators complain; and I am of opinion Mr. Bellin has laid down the coasts of the Brasils with exactness. I believe it the more readily, as the longitude of Rio Janeiro has been determined by Messrs. Godin, and the Abbé de la Caille, who met there in 1751; and as some observations of the longitude have likewise been made at Fernambuco and Buenos Ayres. These three points being determined, there can be no considerable error in regard to the longitude of the eastern coasts of America, from 8° to 35° S. latitude; and this has been confirmed to us by experience.

Entry in Rio de la Plata.

Since the 27th of January we found ground, and on the 29th, in the evening, we saw the land, though we could not take the bearings, as night was coming

 on,

on, and the ſhore very low. The night was dark, with rain and thunder. We lay-to under our reefed top-ſails, the head towards the offing. On the 30th, by break of day, we perceived the mountains of Maldonado: it was then eaſily diſcovered that the land we ſaw the evening before, was the iſle of Lobos. However, as our latitude, when we arrived, was 35° 16′ 20″ we muſt have taken it for cape Santa Maria, which Mr. Bellin places in 35° 15′, though its true latitude is 34° 55′; I take notice of this falſe poſition, becauſe it might prove dangerous. A ſhip ſailing in 35° 15′ S. latitude, and expecting to find cape Santa Maria, might run the riſk of getting upon the Engliſh Bank without having ſeen any land. However, the ſoundings would caution them againſt the approaching danger; for, near the ſand, you find no more than ſix or ſeven fathoms of water. The French Bank, or Sand, which is no more than a prolongation of cape San Antonio, would be more dangerous; juſt before you come to the northern point of it, you find from twelve to fourteen fathoms of water.

Neceſſary correction in M. Bellin's chart.

[A]nchoring-[p]lace at the [M]aldonados.

The Maldonados are the firſt high lands one ſees on the north-ſide after entering the Rio de la Plata, and almoſt the only ones till you come to Montevideo. Eaſt of theſe mountains there is an anchorage upon a very low coaſt; it is a creek ſheltered by a little iſland.

The

The Spaniards have a little town at the Maldonados, with a garrifon. In its neighbourhood is a poor gold mine, that has been worked thefe few years; in it they likewife find pretty tranfparent ftones. About two leagues inland is a town newly built, and entirely peopled with Portugueze deferters; it is called Pueblo Nuevo.

Anchoring at Montevideo

The 31ft, at eleven in the morning, we anchored in Montevideo bay, having four fathom water, with a black, foft, muddy bottom. We had paffed the night between the 30th and 31ft in nine fathoms, the fame bottom, five or fix leagues eaft of the ifle of Flores. The two Spanifh frigates, which were to take poffeffion of the Ifles Malouines (Falkland's Ifland) had lain in the road a whole month. Their commander, Don Philip Ruis Puente, captain of a man of war, was appointed governor of thofe iflands; we went together to Buenos Ayres, in order to concert the neceffary meafures with the governor-general, for the ceffion of the fettlement, which I was to deliver up to the Spaniards. We did not make a long ftay there, and I returned to Montevideo on the 16th of February.

February.

Journey from Buenos Ayres to Montevideo.

The prince of Naffau went with me, and as a contrary wind prevented our returning in a fchooner, we landed oppofite Buenos Ayres, above the colony of San Sacramento, and made this tour by land. We croffed thofe thoſe immenſe plains, in which travellers are guided by the eye, taking care not to miſs the fords in the rivers, and driving before themſelves thirty or forty horſes, among which they muſt take ſome with nooſes, in order to have relays, when thoſe on which they ride are fatigued. We lived upon meat which was almoſt raw; and paſſed the nights in huts made of leather, in which our ſleep was conſtantly interrupted by the howlings of tygers that lurk around them. I ſhall never forget in what manner we croſſed the river St. Lucia, which is very deep, rapid, and wider than the Seine oppoſite the Hoſpital of Invalids at Paris. You get into a narrow, long canoe, one of whoſe ſides is half as high again as the other; two horſes are then forced into the water, one on the ſtarboard, and the other on the larboard ſide of the canoe, and the maſter of the ferry, being quite naked, (which, though a very wiſe precaution, is inſufcient to encourage paſſengers that cannot ſwim) holds up the horſes heads as well as he can above the water, obliging them to ſwim over the river, and to draw the canoe, if they be ſtrong enough for it.

Don Ruis arrived at Montevideo a few days after us. There arrived at the ſame time two boats laden, one with wood and refreſhments, the other with biſcuit and flour, which we took on board, in place of

 that

that which had been consumed on our voyage from Brest. The Spanish frigates being likewise ready. we prepared to leave Rio de la Plata.

CHAP. II.

Account of the establishment of the Spaniards in Rio de la Plata.

1767.

Incertainty concerning the source this river.

RIO de la Plata, or the river of Plate, does not go by that same name from its source. It is said to spring from the lake Xaragès, near 16° 30′ south, under the name of Paraguai, which it communicates to the immense extent of land it passes through. In about 27° it joins with the river Parana, whose name it takes, together with its waters. It then runs due south to lat. 34°; where it receives the river Uraguai, and directs its course eastward, by the name of la Plata, which it keeps to the sea.

The Jesuit geographers, who were the first that attributed the origin of this great river to the lake of Xaragès, have been mistaken, and other writers have followed their mistake in this particular. The existence of this lake, which has been in vain sought for, is now acknowledged to be fabulous. The marquis of Valdelirais and Don George Menezès, having been appointed,

pointed, the one by Spain and the other by Portugal, for ſettling the limits between the poſſeſſions of theſe two powers in this country, ſeveral Spaniſh and Portugueſe officers went through the whole of this portion of America, from 1751 till 1755. Part of the Spaniards went up the river Paraguai, expecting by this means to come into the lake of Xaragès; the Portugueſe on their part, ſetting out from Maragoſſo, a ſettlement of theirs upon the inner boundaries of the Braſils, in about 12° ſouth latitude, embarked on a river called Caourou, which the ſame maps of the Jeſuits marked, as falling into the lake of Xaragès. They were both much ſurpriſed at meeting in the river Paraguai, in 14° S. latitude, without having ſeen any lake. They proved, that what had been taken for a lake, was a great extent of very low grounds, which, during a certain ſeaſon, are covered by the inundations of the river.

[S]ources of the [R]iver Plata.

The Paraguai, or Rio de la Plata, ariſes between 5° and 6° S. latitude nearly in the middle between the two oceans, and in the ſame mountains whence the Madera comes, which empties itſelf into the river of Amazons. The Parana and Uraguai ariſe both in the Braſils; the Uraguai in the captainſhip of St. Vincent; the Parana near the Atlantic ocean, in the mountains that lie to the E. N. E. of Rio Janeiro, whence it takes

takes its courſe to the weſtward, and afterwards turns ſouth.

Date of the firſt ſettlements of the Spaniarde there.

The abbé Prevoſt has given the hiſtory of the diſcovery of the Rio de la Plata, and of the obſtacles the Spaniards met with, in forming the firſt ſettlements they made there. It appears from his account that Diaz de Solis firſt entered this river in 1515, and gave his name to it, which it bore till 1526, when Sebaſtian Cabot changed it to that of la Plata, or of Silver, on account of the quantity of that metal he found among the natives there. Cabot built the fort of Eſpiritu Santo, upon the river Tercero, thirty leagues above the junction of the Paraguai and Uraguai; but this ſettlement was deſtroyed almoſt as ſoon as it was conſtructed.

Don Pedro de Mendoza, great cup-bearer to the emperor, was then ſent to the river of Plate in 1535. He laid the firſt foundations of Buenos Ayres, under bad auſpices, on the right hand ſhore of the river, ſome leagues below its junction with the Uraguai, and his whole expedition was a chain of unfortunate events, that did not even end at his death.

The inhabitants of Buenos Ayres, being continually interrupted by the Indians, and conſtantly oppreſſed by famine, were obliged to leave the place and to retire to Aſſumption. This town, now the capital of Paraguai, was founded by ſome Spaniards, attendants of Men-

Mendoza, upon the weſtern ſhore of the river, three hundred leagues from its mouth, and was in a very ſhort ſpace of time conſiderably enlarged. At length Don Pedro Ortiz de Zarata, governor of Paraguay, rebuilt Buenos Ayres in 1580, on the ſame ſpot where the unhappy Mendoza had formerly laid it out, and fixed his reſidence there: the town became the ſtaple to which European ſhips reſorted, and by degrees the capital of all theſe tracts, the ſee of a biſhop, and the reſidence of a governor-general.

Situation of the town of Buenos Ayres.

Buenos Ayres is ſituated in 34° 35 ſouth latitude, its longitude is 61° 5 weſt from Paris, according to the aſtronomical obſervations of father Feuillée. It is built regular, and much larger than the number of its inhabitants would require, which do not exceed twenty thouſand, whites, negroes, and meſtizos. The way of building the houſes gives the town this great extent; for, if we except the convents, public buildings, and five or ſix private manſions, they are all very low, and have no more than a ground-floor, with vaſt court-yards, and moſt of them a garden. The citadel, which includes the governor's palace, is ſituated upon the ſhore of the river, and forms one of the ſides of the great ſquare, oppoſite to which the town-hall is ſituated; the cathedral and epiſcopal palace occupy the two other sides of the square, in which a public market is daily held.

This town wants a harbour.

There is no harbour at Buenos Ayres, nor so much as a mole, to facilitate the landing of boats. The ships can only come within three leagues of the town; there they unload their goods into boats, which enter a little river, named Rio Chuelo, from whence the merchandizes are brought in carts to the town, which is about a quarter of a league from the landing-place. The ships which want careening, or take their lading at Buenos Ayres, go to la Encenada de Baragon, a kind of port about nine or ten leagues E. S. E. of this town.

Religious establishmen[ts].

Buenos Ayres contains many religious communities of both sexes. A great number of holidays are yearly celebrated by processions and fireworks. The monks have given the title of Majordomes or Stewards of the founders of their orders, and of the holy Virgin, to the principal ladies in this town. This post gives them the exclusive charge of ornamenting the church, dressing the statue of the tutelar saint, and wearing the habit of the order. It is a singular sight for a stranger to see ladies of all ages in the churches of St. Francis and St. Dominique assist in officiating, and wear the habit of those holy institutors.

The Jesuits have offered a much more austere mode of sanctification than the former to the pious ladies. Ad-

 joining

joining to their convent, they had a houſe, called Caſa de los Exercicios de las Mugeres, i. e. the Houſe for the Exerciſes of Women. Married and unmarried women, without the conſent of their huſbands or parents, went to be ſanctified there by a retreat of twelve days. They were lodged and boarded at the expence of the community. No man was admitted into this ſanctuary, unleſs he wore the habit of St. Ignatius; even ſervant-maids were not allowed to attend their miſtreſſes thither. The exerciſes practiſed in this holy place were meditation, prayer, catechetical inſtructions, confeſſion, and flagellation. They ſhewed us the walls of the chapel, yet ſtained with the blood, which, as they told us, was diſperſed by the rods wherewith penitence armed the hands of theſe Magdalens.

All men are brothers, and religion makes no diſtinction in regard to their colour. There are ſacred ceremonies for the ſlaves, and the Dominicans have eſtabliſhed a religious community of negroes. They have their chapels, maſſes, holidays, and decent burials, and all this coſts every negro that belongs to the community only four reals a year. This community of negroes acknowledges St. Benedict of Palermo, and the Virgin, as their patrons, perhaps on account of theſe words of ſcripture; "Nigra ſum, ſed formoſa filia Jeruſalem." On the holidays of theſe tutelary ſaints, they

chuſe

chuſe two kings, one to repreſent the king of Spain, the other the Portugueze monarch, and each of them chooſes a queen. Two bands, armed and well dreſſed, form a proceſſion, and follow the kings, marching with the croſs, banners, and a band of muſic. They ſing, dance, repreſent battles between the two parties, and repeat litanies. This feſtivity laſts from morning till night, and the ſight of it is diverting.

Environs of Buenos Ayr[e]s and their pr[o]ductions.

The environs of Buenos Ayres are well cultivated. Moſt of the inhabitants of that city have their country-houſes there, called Quintas, furniſhing all the neceſſaries of life in abundance. I except wine, which they get from Spain, or from Mandoza, a vineyard about two hundred leagues from Buenos Ayres. The cultivated environs of this city do not extend very far; for at the diſtance of only three leagues from the city, there are immenſe fields, left to an innumerable multitude of horſes and black cattle. One ſcarce meets with a few ſcattered huts, on croſſing this vaſt country, erected not ſo much with a view of cultivating the ſoil, as rather to ſecure the property of the ground, or of the cattle upon it to their ſeveral owners. Travellers, who croſs this plain, find no accommodations, and are obliged to ſleep in the ſame carts they travel in, and which are the only kind of carriages made uſe of on long journeys here. Thoſe who travel on horſeback are often expoſed to lie in the fields, without any covering.

Abundance of cattle.

The country is a continued plain, without other foreſts than thoſe of fruit trees. It is ſituated in the happieſt climate, and would be one of the moſt fertile in the world in all kinds of productions, if it were cultivated. The ſmall quantity of wheat and maize which is ſown there, multiplies by far more than in our beſt fields in France. Notwithſtanding theſe natural advantages, almoſt the whole country lies neglected, as well in the neighbourhood of the Spaniſh ſettlements, as at the greateſt diſtance from them; or, if by chance you meet with any improvements, they are generally made by negro-ſlaves. Horſes and horned cattle are in ſuch great abundance in theſe plains, that thoſe who drive the oxen before the carts, are on horſeback; and the inhabitants, or travellers, when preſſed by hunger, kill an ox, take what they intend to eat of it, and leave the reſt as a prey to wild dogs and tygers*, which are the only dangerous animals in this country.

The dogs were originally brought from Europe: the eaſe with which they are able to get their livelihood in the open fields, has induced them to leave the habitations,

* It is now certain, that the animal, here called tyger, is the Couguara or Brown (tyger) Cat, of Penn. Syn. quad. p. 179. a very large animal, and very fierce in hot countries. F.

tions, and they have encreaſed their ſpecies innumerably. They often join in packs to attack a wild bull, and even a man on horſeback, when they are preſſed by hunger. The tygers are not numerous, except in woody parts, which are only to be found on the banks of rivulets. The inhabitants of theſe countries are known to be very dexterous in uſing nooſes; and it is fact, that ſome Spaniards do not fear to throw a nooſe, even upon a tyger; though it is equally certain that ſome of them unfortunately became the prey of theſe ravenous creatures. At Montevideo, I ſaw a ſpecies of tyger-cat, whoſe hairs were pretty long, and of a whitiſh grey. The animal is very low upon its legs, about five feet long, fierce, and very ſcarce.

Scarcity of wood; me[ans] of remedyi[ng] it.

Wood is very dear at Buenos Ayres, and at Montevideo. In the neighbourhood of theſe places, are only ſome little ſhrubs, hardly fit for fuel. All timber for building houſes, and conſtructing and refitting the veſſels that navigate in the river, comes from Paraguai in rafts. It would, however, be eaſy to get all the timber for conſtructing the greateſt ſhips from the upper parts of the country. From Montegrande, where they have the fineſt wood, it might be tranſported in ſingle round ſtems, through the river Ybicui, into the Uraguai, and from the Salto-Chico of the Uraguai,

ſome

ſome veſſels made on purpoſe for this uſe, might bring it to ſuch places upon the river, where docks were built.

Account of the natives of this country.

The Indians, who inhabit this part of America, north and ſouth of the river de la Plata, are of that race called by the Spaniards Indios bravos.—They are middle-ſized, very ugly, and afflicted with the itch. They are of a deep tawny colour, which they blacken ſtill more, by continually rubbing themſelves with greaſe. They have no other dreſs than a great cloak of roe-deer ſkins, hanging down to their heels, in which they wrap themſelves up. Theſe ſkins are very well dreſſed; they turn the hairy ſide inwards, and paint the outſide with various colours. The diſtinguiſhing mark of their cacique is a band or ſtrap of leather, which is tied round his forehead; it is formed into a diadem or crown, and adorned with plates of copper. Their arms are bows and arrows; and they likewiſe make uſe of nooſes and of balls*. Theſe Indians are always on horſeback, and have no fixed habitations, at leaſt not near the Spaniſh ſettlements. Sometimes they come with their

* Theſe balls are two round ſtones, of the ſize of a two-pound ball, both enchaſed in a ſtrap of leather, and faſtened to the extremities of a thong, ſix or ſeven feet long. The Indians, when on horſeback, uſe this weapon as a ſling, and often hit the animal they are purſuing, at the diſtance of three hundred yards.

wives

wives to buy brandy of the Spaniards; and they do not ceaſe to drink of it, till they are ſo drunk as not to be able to ſtir. In order to get ſtrong liquors, they ſell their arms, furs, and horſes; and having diſpoſed of all they are poſſeſſed of, they ſeize the horſes they can meet with near the habitations, and make off. Sometimes they come in bodies of two or three hundred men, to carry off the cattle from the lands of the Spaniards, or to attack the caravans of travellers. They plunder and murder, or carry them into ſlavery. This evil cannot be remedied: for, how is it poſſible to conquer a nomadic nation, in an immenſe uncultivated country, where it would be difficult even to find them: beſides, theſe Indians are brave and inured to hardſhips; and thoſe times exiſt no longer, when one Spaniard could put a thouſand Indians to flight.

Race of ro[bbers, ſettle[d] on the nort[h] ſide of the river.

A ſet of robbers united into a body, a few years ago, on the north ſide of the river, and may become more dangerous to the Spaniards than they are at preſent, if efficacious meaſures are not taken to deſtroy them. Some malefactors eſcaped from the hands of juſtice, retired to the north of the Maldonadoes; ſome deſerters joined them; their numbers encreaſed inſenſibly; they took wives from among the Indians, and founded a race of men who live upon robberies. They make inroads, and carry off the cattle in the Spaniſh poſſeſſions, which they conduct to the boundaries of the Braſils, where they barter it with the Pauliſts *, againſt arms and clothes. Unhappy are the travellers that fall into their hands. They are now, it is ſaid, upwards of ſix hundred in number, have left their firſt habitation, and are retired much further to the north-weſt.

Extent of the government de la Plata.

The governor-general of the province de la Plata reſides, as I have already mentioned, at Buenos Ayres. In all matters which do not concern the marine, he is reckoned dependent upon the viceroy of Peru; but the great diſtance between them almoſt annuls this dependency, and it only exiſts in regard to the ſilver, which he is obliged to get out of the mines of Potoſi; this, however, will no longer be brought over in ſhapeleſs pieces, as a mint has been eſtabliſhed this year at Potoſi. The particular governments of Tucuman and Paraguai (the principal ſettlements of which are Santa-Fé, Corrientes, Salta, Tujus, Cordoua, Mendoza, and Aſſumption) are dependent, together with the famous miſſions of the Jeſuits, upon the governor-general of la Plata. This vaſt province contains, in a word, all the poſſeſſions of the Spaniards, eaſt of the Cordilleras, from the river of Amazons to the ſtraits of Magalhaens. It is true, there is no ſettlement ſouth of Buenos Ayres; and nothing

* The Pauliſts are another race of robbers, who left Braſil, and formed a republic, towards the end of the ſixteenth century.

but

but the neceſſity of providing themſelves with ſalt, induces the Spaniards to penetrate into thoſe parts. For this purpoſe a convoy of two hundred carts, eſcorted by three hundred men, ſets out every year from Buenos Ayres, and goes to the latitude of forty degrees, to load the ſalt in lakes near the ſea, where it is naturally formed. Formerly the Spaniards uſed to ſend ſchooners to the bay of St. Julian, to fetch ſalt.

I ſhall ſpeak of the miſſions in Paraguay when I come to the ſecond voyage, which ſome circumſtances obliged us to make again into the river of la Plata; I ſhall then enter into the account of the expulſion of the Jeſuits, of which we were witneſſes.

The commerce of the province de la Plata is leſs profitable than any in Spaniſh America; this province produces neither gold nor ſilver, and its inhabitants are not numerous enough to be able to get at all the other riches which the ſoil produces and contains The commerce of Buenos Ayres itſelf is not in the ſame ſtate it was in about ten years ago; it is fallen off conſiderably, ſince the trade by land is no longer permitted; that is, ſince it has been prohibited to carry European goods by land from Buenos Ayres to Peru and Chili; ſo that the only objects of the commerce with theſe two provinces are, at preſent, cotton, mules, and

 maté,

maté, or the Paraguay-herb* The money and interest of the merchants at Lima have obtained this order, against which those of Buenos Ayres have complained. The law-suit is carried on at Madrid, and I know not how or when it will be determined. However, Buenos Ayres is a very rich place: I have seen a register-ship sail from thence, with a million of dollars on board; and if all the inhabitants of this country could get rid of their leather or skins in Europe, that article alone would

* Maté, or Paraguay-tea, or South-sea-tea, are pounded dry leaves of a plant growing in South America, and chiefly in Paraguay. The Jesuits, when in possession of the interior parts of the provinces of Paraguay, got by a manoeuvre similar to that of the Dutch, in regard to the spice-trade, the exclusive commerce of this commodity. They cultivated this plant in enclosures, upon the rivers Uragui and Parana, and wherever it grew wild, it was destroyed; and after the space of nineteen years they became the sole masters of this trade, which was very lucrative; for as this plant is thought to be an excellent restorative, and a good paregoric, and therefore of indispensible necessity to the workmen in the famous Peruvian mines, it is carried constantly to Peru and Chili; the whole consumption of it being yearly upon an average of 160,000 arrobas, of 25 pounds Spanish weight each; and the price is, at a medium, thirty-six piasters per arroba, so that this plant was worth to the Jesuits 5,760,000 piastres per ann. the tenth part of which sum must be deducted out of the whole, for instruments of agriculture, the erection and repairing of buildings necessary for manufacturing this plant, feeding and cloathing of about 300,000 Indians and Negroes: so that still above five millions of piastres were the clear yearly profit of the pious fathers. These cunning men sold these leaves in powder on purpose that no botanist might get a sight of them, and thus be enabled to find out the plant to which the leaves belong, in case some plants should have escaped their selfish destruction of them. Some writers call this plant Maté, which is, I believe, the name of the vessel it is drank out of. Others call it Caa, and make this the generic name of it, and its species are Caa-cuys, Caa-mini, and Caa-guaz, the last of which is the coarsest sort prepared, with the stalks left to it, for which reason it is likewise called Yerva de Palos; but the Caa-mini or Yerva de Caamini is the best sort and sold dearer; the Caa-cuys will not keep so long as the other two sorts. This plant is thought to be the Ilex Cassine, Linn. Sp. pl. p. 181. or the Dahoon-holly. Forster's Flora Americ. Septentr. p. 7. and Catesby car. i. t. 31. F.

would ſuffice to enrich them. Before the laſt war, they carried on a prodigious contraband-trade with the colony of Santo Sacramento, a place in the poſſeſſion of the Portugueſe, upon the left ſide of the river, almoſt directly oppoſite Buenos Ayres. But this place is now ſo much ſurrounded by the new works, erected by the Spaniards, that it is impoſſible to carry on any illicit trade with it, unleſs by connivance; even the Portugueſe, who inhabit the place, are obliged to get their ſubſiſtence by ſea from the Braſils. In ſhort, this ſtation bears the ſame relation to Spain here, as Gibraltar does in Europe; with this difference only, that the former belongs to the Portugueſe, and the latter to the Engliſh.

Colony of Santo Sacramento.

Account of the town of Montevideo

The town of Montevideo has been ſettled forty years ago, is ſituated on the north ſide of the river, thirty leagues above its mouth, and built on a peninſula, which lies convenient to ſecure from the eaſt wind, a bay of about two leagues deep, and one league wide at its entrance. At the weſtern point of this iſle, is a ſingle high mountain, which ſerves as a look out, and has given a name to the town; the other lands, which ſurround it, are very low. That ſide which looks towards a plain, is defended by a citadel. Several batteries guard the ſide towards the ſea and the harbour. There is a battery upon a very little iſle, in the bottom

bottom of the bay, called Isle au Francois, or French-Island. The anchorage at Montevideo is safe, though sometimes molested by pamperos, which are storms from the south-west, accompanied by violent tempests. There is no great depth of water in the whole bay; and one may moor in three, four, or five fathoms of water in a very soft mud, where the biggest merchant-ships run a-ground, without receiving any damage; but sharp-built ships easily break their backs, and are lost. The tides do not come in regular; according as the wind is, the water is high or low. It is necessary to be cautious, in regard to a chain of rocks that extends some cables-length off the east point of the bay; the sea forms breakers upon them, and the people of this country call them la Punta de las Carretas.

Anchorage in this bay.

It is an excellent place to put in at for refreshments.

Montevideo has a governor of its own, who is immediately under the orders of the governor-general of the province. The country round this town is almost entirely uncultivated, and furnishes neither wheat nor maize; they must get flour, biscuit, and other provisions for the ships from Buenos Ayres. In the gardens belonging to the town, and to the adjoining houses, they cultivate scarce any legumes; there is, however, plenty of melons, calabashes, figs, peaches, apples, and quinces. Cattle are as abundant there as in any other part of this country; which, together with the wholeſomeneſs of the air, makes Montevideo an excellent place to put in at for the crew; only good meaſures muſt be taken to prevent deſertion. Every thing invites the ſailor thither; it being a country, where the firſt reflection which ſtrikes him, on ſetting his feet on ſhore, is, that they live there almoſt without working. Indeed, how is it poſſible to reſiſt the compariſon of ſpending one's days in idleneſs and tranquility, in a happy climate, or of languiſhing under the weight of a conſtantly laborious life, and of accelerating the misfortunes of an indigent old age, by the toils of the ſea?

CHAP. III.

Departure from Montevideo; navigation to the Malouines; delivery of them into the hands of the Spaniards; historical digression on the subject of these islands.

1767. February.

Departure from Montevideo.

THE 28th of February, 1767, we weighed from Montevideo, in company with two Spanish frigates, and a tartane laden with cattle. I agreed with Don Ruis, that whilst we were in the river, he should lead the way; but that as soon as we were got out to sea, I was to conduct the squadron. However, to obviate the dangers in case of a separation, I gave each of the frigates a pilot, acquainted with the coasts of the Malouines. In the afternoon we were obliged to come to an anchor, as a fog prevented our seeing either the main land, or the isle of Flores. The next morning we had contrary wind; however, I expected that we should have weighed, as the strong currents in the river favoured us; but seeing the day almost at an end, without any signal being given by the Spanish commodore, I sent an officer to tell him, that having had a sight of the isle of Flores, I found myself too near the English sand-bank, and that I advised we should weigh the next day, whether the wind was fair or not. Don Ruis answer-

anſwered, that he was in the hands of the pilot of the river, who would not weigh the anchor till we had a ſettled fair wind. The officer then informed him from me, that I ſhould ſail by day-break; and that I would wait for him, by plying to windward, or by anchoring more to the north, unleſs the tides or the violence of the wind ſhould ſeparate us againſt my will.

The tartane had not caſt anchor the laſt night; and we loſt ſight of her, and never ſaw her again. She returned to Montevideo three weeks after, without fulfilling its intended expedition. The night was ſtormy; the pamperos blew very violently, and made us drag our anchor; however, we caſt another anchor, and that fixed us. By day-break we ſaw the Spaniſh ſhips, with their top-maſts handed, main-yards lowered, and had dragged their anchors much further than ourſelves. The wind was ſtill contrary and violent, the ſea very high, and it was nine o'clock before we could proceed under our main-ſails; at noon we loſt ſight of the Spaniards, who remained at anchor, and the third of March in the evening we were got out of the river.

Storm in th[e] river.

1767. March.

During our voyage to the Malouines, we had variable winds from N. W. to S. W. almoſt always ſtormy weather and high ſeas: we were obliged to try under our main-ſail on the 16th, having ſuffered ſome damage. Since the 17th in the afternoon, when

Voyage f[rom] Montevid[eo] to the Ma[l]louines.

when we came into ſoundings, the weather was very foggy. The 19th, not ſeeing the land, though the horizon was clear, and I was eaſt of the Sebald's iſles by my reckoning, I was afraid I had gone beyond the Malouines, and therefore reſolved to ſail weſtward; the wind, which is a rare circumſtance, favoured my reſolution. I proceeded very faſt in twenty-four hours, and having then found the ſoundings off the coaſt of Patagonia, I was ſure as to my poſition, and ſo proceeded again very confidently to the eaſtward. Indeed, the 21ſt, at four o'clock in the afternoon, we diſcovered the Sebald's iſles, remaining in N. E. ¼ E. eight or ten leagues diſtant, and ſoon after we ſaw the coaſt of the Malouines. I could have ſpared myſelf all the trouble I had been in, if I had in time ſailed cloſe-hauled, in order to approach the coaſt of America, and ſo find the iſlands by their latitude.

Fault committed in the direction of this courſe.

The 23d in the evening we entered and anchored in the great bay, where the two Spaniſh frigates likewiſe came to an anchor on the 24th. They had ſuffered greatly during their courſe; the ſtorm on the 16th having obliged them to bear away; and the commodore-ſhip, having ſhipped a ſea, which carried away her quarter-badges, broke through the windows of the great cabbin, and poured a great quantity of water into her. Almoſt all the cattle they took on board at Montevideo for the colony, died through the badneſs of the weather. The twenty-fifth the three veſſels came into port, and moored.

The Spaniards take poſſeſſion of our ſettlement at the Malouines. April.

The firſt of April I delivered our ſettlement to the Spaniards, who took poſſeſſion of it, by planting the Spaniſh colours, which were ſaluted at ſun-riſing and ſun-ſetting from the ſhore and from the ſhips. I read the king's letter to the French inhabitants of this infant colony, by which his majeſty permits their remaining under the government of his moſt catholic majeſty. Some families profited of this permiſſion; the reſt, with the garriſon, embarked on board the Spaniſh frigates, which ſailed for Montevideo the 27th in the morning*.

Hiſtorical details concerning the Malouines.

Some hiſtorical remarks concerning theſe iſles, will, I hope, not be deemed unneceſſary.

Americo Veſpucci diſcovers them.

It appears to me, that the firſt diſcovery of them may be attributed to the celebrated Americo Veſpucci, who, in the third voyage for the diſcovery of America,

* When I delivered the ſettlement to the Spaniards, all the expences, whatſoever, which it had coſt till the firſt of April 1767, amounted to 603,000 livres, including the intereſt of five per cent. on the ſums expended ſince the firſt equipment. France having acknowledged the catholic king's right to the Malouines, he, by a principle of the law of nations, owed no reimburſement to theſe coſts. However, as his majeſty took all the ſhips, boats, goods, arms, ammunition, and proviſions that belonged to our ſettlement, he being equally juſt and generous, deſired that we ſhould be reimburſed for what we had laid out; and the above ſum was remitted to us by his treaſurers; part at Paris, and the reſt at Buenos Ayres.

rica, ſailed along the northern coaſts of them in 1502. It is true, he did not know whether it belonged to an iſle, or whether it was part of the continent; but it is eaſy to conclude, from the courſe he took, from the latitudes he came to, and from the very deſcription he gives of the coaſts, that it is that of the Malouines. I ſhall aſſert with equal right, that Beauchesne Gouin, returning from the South Seas in 1700, anchored on the eaſt ſide of the Malouines, thinking he was at the Sebald's iſles.

French and English navigators viſit them after him.

His account ſays, that after diſcovering the iſle to which he gave his own name, he anchored on the eaſt ſide of the moſt eaſterly of Sebald's iſles. I muſt firſt of all obſerve, that the Malouines, being in the middle between the Sebald's iſles and the iſle of Beauchesne, have a conſiderable extent, and that he muſt have neceſſarily fallen in with the coaſt of the Malouines, as is impoſſible not to ſee them, when at anchor eaſtward of the Sebald's iſles. Beſides, Beauchesne ſaw a ſingle iſle of an immenſe extent; and it was not till after he had cleared it, that he perceived two other little ones: he paſſed through a moiſt country, filled with marſhes and freſh-water lakes, covered with wild-geeſe, teals, ducks, and ſnipes; he ſaw no woods there; all this agrees prodigiouſly well with the Malouines. Sebald's iſles, on the contrary, are four little

tle rocky iſles, where William Dampier, in 1683, attempted in vain to water, and could not find a good anchoring-ground.

Be this as it will, the Malouines have been but little known before our days—Moſt of the relations report them as iſles covered with woods. Richard Hawkins, who came near the northern coaſt of them, which he called Hawkins's Maiden-land, and who pretty well deſcribed them, aſſerts that they were inhabited, and pretends to have ſeen fires there. At the beginning of this century, the St. Louis, a ſhip from St. Malo, anchored on the ſouth-eaſt ſide, in a bad bay, under the ſhelter of ſome little iſles, called the iſles of Anican, after the name of the privateer; but he only ſtayed to water there, and continued his courſe, without caring to ſurvey them.

The Frenc ſettle there

However, their happy poſition, to ſerve as a place of refreſhment or ſhelter to ſhips going to the South-Seas, ſtruck the navigators of all nations. In the beginning of the year 1763, the court of France reſolved to form a ſettlement in theſe iſles. I propoſed to government, that I would eſtabliſh it at my own expence, aſſiſted by Meſſrs. de Nerville and d'Arboulin, one my couſin-german, the other my uncle. I immediately got the Eagle of twenty guns, and the Sphinx of twelve, conſtructed and furniſhed with proper neceſſaries

ſaries for ſuch an expedition, by the care of M. Duclos Guyot, now my ſecond. I embarked ſeveral Acadian families, a laborious intelligent ſet of people, who ought to be dear to France, on account of the inviolable attachment they have ſhewn, as honeſt but unfortunate citizens.

The 15th of September I ſailed from St. Malo. M. de Nerville was on board the Eagle with me. After touching twice, once at the iſle of St. Catharine, on the coaſt of the Braſils, and once at Montevideo, where we took in many horſes and horned cattle, we made the land of Sebald's iſles the 31ſt of January, 1764. I ſailed into a great bay, formed by the coaſt of the Malouines, between its N. W. point, and Sebald's iſles; but not finding a good anchoring ground, ſailed along the north coaſt; and, coming to the eaſtern extremity of theſe iſles, I entered a great bay on the third of February, which ſeemed very convenient to me, for forming the firſt ſettlement.

Account of the manner in which it was made.

The ſame illuſion which made Hawkins, Woods Rogers, and others, believe that theſe iſles were covered with wood, acted likewiſe upon my fellow voyagers. We were ſurpriſed, when we landed, to ſee that what we took for woods as we ſailed along the coaſt, was nothing but buſhes of a tall ruſh, ſtanding very cloſe together. The bottom of its ſtalks being dried, got

the

the colour of a dead leaf to the height of about five feet; and from thence ſprings a tuft of ruſhes, which crown this ſtalk; ſo that at a diſtance theſe ſtalks together have the appearance of a wood of middling height. Theſe ruſhes only grow near the ſea ſide, and on little iſles; the mountains on the main land are, in ſome parts, covered all over with heath, which are eaſily miſtaken for buſhes.

In the various excurſions, which I immediately ordered, and partly made in the iſland myſelf, we did not find any kind of wood; nor could we diſcover that theſe parts had been frequented by any nation.

I only fo md, and in great quantity too, an exceeding good turf, which might ſupply the defect of wood, both for fuel, and for the forge; and I paſſed through immenſe plains, every where interſected by little rivulets, with very good water. Nature offered no other ſubſiſtence for men than fiſh and ſeveral ſorts of land and water fowl. It was very ſingular, on our arrival, to ſee all the animals, which had hitherto been the only inhabitants of the iſland, come near us without fear, and ſhew no other emotions than thoſe which curioſity inſpires at the ſight of an unknown object. The birds ſuffered themſelves to be taken with the hand, and ſome would come and ſettle upon people that ſtood ſtill;

ſtill, ſo true it is, that man does not bear a characteriſtic mark of ferocity, which mere inſtinct is capable of pointing out to theſe weak animals, the being that lives upon their blood. This confidence was not of long duration with them; for they ſoon learnt to miſtruſt their moſt cruel enemies.

First year. The 17th of March, I fixed upon the place of the new colony, which at firſt was only compoſed of twenty ſeven perſons, among whom were five women, and three children. We ſet to work immediately to build them huts covered with ruſhes, to conſtruct a magazine, and a little fort, in the middle of which a ſmall obeliſk was erected. The king's effigy adorned one of its ſides, and under its foundations we buried ſome coins, together with a medal, on one ſide of which was graved the date of the undertaking, and on the other the figure of the king, with theſe words for the exergue, "Tibi ſerviat ultima Thule." *

* The inſcription on this medal was as follows.

Settlement of the Iſles Malouines, ſituated in 51° 30′ of S. latitude, 60° 50′ W. long. from the meridian of Paris, by the Eagle frigate, captain P. Duclos Guyot, captain of a fire ſhip, and the ſphinx ſloop; captain F. Chenard de la Giraudais, lieutenant of a frigate, equipped by Louis Antoine de Bougainville, colonel of infantry, captain of a ſhip, chief of the expedition, G. de Nerville, captain of infantry, and P. d' Arboulin, poſt-maſter general of France: conſtruction of a fort, and an obeliſk, decorated with a medallion of his majeſty Louis XV. after the plans of A. L'Huillier, engineer and geographer of the field and army, ſerving on this expedition; during the adminiſtration of E′. de Choiſeul, duke of Stainville, in February, 1764.

And the exergue. Conamur tenues grandia.

How-

However, to encourage the colonists, and encrease their reliance on speedy assistance, which I promised them, M. de Nerville consented to remain at their head, and to share the risks to which this weak settlement was exposed, at the extremity of the globe, where it was at that time the only one in such a high southern latitude. The fifth of April, 1764, I solemnly took possession of the isles in the king's name, and the eighth I sailed for France.

The fifth of January, 1765, I saw my colonists again, and found them healthy and content. After landing what I had brought to their assistance, I went into the straits of Magalhaens, to get a cargo of timber, palisadoes and young trees, and I began a navigation, which is become necessary to the colony. Then I found the ships of commodore Byron, who, after surveying the Malouines for the first time, passed the straits, in order to get into the South-seas. When I left the Malouines the 27th of April following, the colony consisted of twenty-four persons, including the officers. Second year

In 1765 we sent back the Eagle to the Malouines, and the king sent the Etoile, one of his store ships, with her. These two vessels, after landing the provisions and new colonists, sailed together to take in wood in the straits of Magalhaens. The settlement now began to get a kind of form. The governor and the ordonnateur * lodged in very convenient houſes built of ſtone, and the other inhabitants lived in houſes of which the walls were made of ſods. There were three magazines, both for the public ſtores, and thoſe of private perſons. The wood out of the ſtraits had ſerved to build ſeveral veſſels, and to conſtruct ſchooners for the purpoſe of ſurveying the coaſt. The Eagle returned to France from this laſt voyage, with a cargo of train oil and ſeals-ſkins, tanned in the iſland. Several trials had been made towards cultivation, which gave no reaſon to deſpair of ſucceſs, as the greateſt part of the corn brought from Europe was eaſily naturalized to the country. The encreaſe of the cattle could be depended upon, and the number of inhabitants amounted then to about one hundred and fifty.

However, as I have juſt mentioned, commodore Byron came in January, 1765, to ſurvey the Malouines. He touched to the weſtward of our ſettlement, in a port which we had already named Port de la Croiſade, and he took poſſeſſion of theſe iſlands for the crown of England, without leaving a ſingle inhabitant there It was not before 1766, that the Engliſh ſent a colony to ſettle in Port de la Croiſade, which they had named

* An officer who has the care of the ſtores.

Port

Port Egmont; and captain Macbride, of the Jason frigate, came to our settlement the same year, in the beginning of December. He pretended that these parts belonged to his Britannic majesty, threatened to land by force, if he should be any longer refused that liberty, visited the governor, and sailed away again the same day.

Such was the state of the Malouines, when we put them into the hands of the Spaniards, whose prior right was thus inforced by that which we possessed by making the first settlement *. The account of the productions of these isles, and the animals which are to be found there, will furnish matter for the following chapter, and are the result of the observations of M. de Nerville, during a residence of three years. I believed

* The author has on purpose omitted to mention, that the English are the first discoverers of these isles. Captain Davis, in the expedition of 1592, under the command of Sir Thomas Cavendish, saw them; and so did Sir Richard Hawkins two years after in 1594, and called them Hawkins's Maiden Land. In the year 1598 they were seen by the Dutchman Sebald de Waert, and called Sebald's isles, and with that name they were put in all Dutch charts. Dampier discovered them likewise in 1683, but suspected they had no water. Strong gave these isles, in the year 1689, the name of Falkland-Islands, which was adopted by the celebrated astronomer Halley, and is now become of universal use in all our maps and charts. The privateers in the times of the wars of king Willam and queen Mary frequently saw these isles, and no sooner than in 1699-1700 they were seen for the first time by a Frenchman called Beauchesne Gouin. It is pretty evident from this account, that the English have an undoubted prior claim to these barren rocks and marshes, situated in a cold climate, subject to the severest rigours of winter, without the benefit of woods to alleviate them; and on which, was it not for the wretched fuel of turf, all the French, English, and Spanish settlements would have been starved with cold. F.

it was so much more proper to enter upon this detail, as M. de Commerçon has not been at the Malouines, and as their natural history is in some regards im portant *.

CHAP. IV.

Detail of the natural history of the Isles Malouines.

A Country which has been but lately inhabited always offers interesting objects, even to those who are little versed in natural history; and though their remarks may not be looked upon as authorities, yet they may satisfy, in part, the curiosity of the investigators of the system of nature.

First aspect they bear.

The first time we landed upon these isles, no inviting objects came in sight, and, excepting the beauty of the port in which we lay, we knew not what could prevail upon us to stay on this apparently barren ground: the horizon terminated by bald mountains, the land lacerated by the sea, which seems to claim the empire over it; the fields bearing a dead aspect, for want of inhabitants; no woods to comfort those who

* The work which I now publish was already finished, when the History of a Voyage to the Malouines, by Dom Pernetty, appeared, otherwise I should have omitted the following accounts.

who intended to be the firſt ſettlers; a vaſt ſilence, now and then interrupted by the howls of marine monſters; and, laſtly, the ſad uniformity which reigned throughout; all theſe were diſcouraging objects, which ſeemed that in ſuch dreary places nature would refuſe aſſiſtance to the efforts of man. But time and experience taught us, that labour and conſtancy would not be without ſucceſs even there. The reſources with which nature preſented us, were immenſe bays, ſheltered from the violence of the winds by mountains, which poured forth caſcades and rivulets; meadows covered with rich paſtures, proper for the food of numerous flocks; lakes and pools to water them; no conteſts concerning the property of the place; no fierce, or poiſonous, or importune animals to be dreaded; an innumerable quantity of the moſt uſeful amphibia; birds and fiſh of the beſt taſte; a combuſtible ſubſtance to ſupply the defect of wood; plants known to be ſpecifics againſt the diſeaſes common to ſea-faring men; a healthy and continually temperate climate, much more fit to make men healthy and robuſt, than thoſe enchanting countries, where abundance itſelf becomes noxious, and heat cauſes a total inactivity. Theſe advantages ſoon expunged the impreſſions which the firſt appearance had made, and juſtified the attempt.

To this we may add, that the Engliſh in their relation of Port-Egmont, have not ſcrupled to ſay, that the countries adjacent furniſhed every thing neceſſary for a good ſettlement. Their taſte for natural hiſtory will, without doubt, engage them to make and to publiſh enquiries which will rectify theſe.

Geographical poſition of the Malouiaes. The Malouines are ſituated between 51° and 52° 30′ S. lat. and 65° 30′ W. long. from Paris; and between 80 and 90 leagues diſtant from the coaſt of America or Patagonia, and from the entrance of the ſtraits of Magalhaens.

The map wich we give of theſe iſlands, has certainly not a geographical accuracy, which muſt have been the work of many years. It may, however, ſerve to indicate nearly the extent of theſe iſles from eaſt to weſt, and from north to ſouth; the poſition of the coaſts, along which our ſhips have ſailed; the figure and depth of the great bays, and the direction of the principal mountains *.

Of the harbours. The harbours, which we have examined, are both extenſive and ſecure; a tough ground, and iſlands happily ſituated to break the fury of the waves, contribute to make them ſafe and eaſily defenſible; they have lit-

* As M. de Bougainville's map of the Malouines or Falkland's iſles, is a mere inaccurate out-line; we refer our readers to the more exact plans of theſe iſlands, publiſhed in England. F.

tle

tle creeks, in which the ſmalleſt veſſels can retire. The rivulets come down into the ſea; ſo that nothing can be more eaſy, than to take in the proviſion of freſh water.

The tides are ſubject to all the emotions of the ſea, which ſurrounds the iſles, and have never riſen at ſettled periods, which could have been calculated. It has only been obſerved, that, juſt before high-water, they have three determinate variations; the ſea, at that time, in leſs than a quarter of an hour, riſes and falls thrice, as if ſhaken up and down; and this motion is more violent during the ſolſtices, the equinoxes, and the full moons. Tides.

The winds are generally variable; but ſtill thoſe between north and weſt, and between ſouth and weſt, are more prevalent than the others. In winter, when the winds are between north and weſt, the weather is foggy and rainy; if between weſt and ſouth, they bring ſnow, hail, and hoar froſt; if from between ſouth and eaſt, they are leſs attended with miſts, but violent, though not quite ſo much as the ſummer winds, which blow between ſouth-weſt and north-weſt: theſe latter, which clear the ſky and dry the ſoil, do not begin to blow till the ſun appears above the horizon; they encreaſe as that luminary riſes; are at the greateſt height when he croſſes the meridian; and loſe their force when Winds.

when he goes to disappear behind the mountains. Besides being regulated by the sun's motion, they are likewise subject to be governed by the tides, which encrease their force, and sometimes alter their direction. Almost all the nights throughout the year are calm, fair, and star-light, especially in summer. The snow, which is brought by the south-west winds in winter, is inconsiderable; it lies about two months upon the tops of the highest mountains; and a day or two, at most, upon the surface of the other grounds. The rivers do not freeze, and the ice of lakes and pools has not been able to bear men upwards of twenty-four hours together. The hoar-frosts in spring and autumn do no damage to the plants, and at sun-rising are converted into dew. In summer, thunder is seldom heard; and, upon the whole, we felt neither great cold, nor great heat; and the distinction of seasons appeared almost insensible. In such a climate, where the revolutions of the seasons affect by no means the constitution, it is natural that men should be strong and healthy; and this has been experienced during a stay of three years.

Water. The few mineral substances found at the Malouines, are a proof of the goodness of the water, which is every where conveniently situated; no noxious plants infect the places where it runs through; its bed is generally gravel or sand, and sometimes turf, which give it a little yellowiſh hue, without diminiſhing its goodneſs and lightneſs.

All the plains have much more depth of ſoil than is neceſſary for the plough to go in. The ſoil is ſo much interwoven with roots of plants, to the depth of near twelve inches, that it was neceſſary, before it was poſſible to proceed to cultivation, to take off this cruſt or layer; and to cut it, that it might be dried and burnt. It is known, that this proceſs is excellent to make the ground better, and we made uſe of it. Below this firſt layer, is a black mould, never leſs than eight or ten inches deep, and frequently much deeper; the next is the yellow, or original virgin-ſoil, whoſe depth is undeterminate. It reſts upon ſtrata of ſlate and ſtones; among which no calcareous ones have ever been found; as the trial has been made with aquafortis. It ſeems, that the iſles are without ſtones of this kind. Journeys have been undertaken to the very tops of the mountains, in order to find ſome; but they have never procured any other than a kind of quartz, and a ſandſtone, not friable; which produced ſparks, and even a kind of phoſphoreſcent light, accompanied with a ſmell of brimſtone. Stones proper for building are not wanting; for moſt of the coaſts are formed of them. There are ſtrata of a very hard and ſmall grained ſtone; and likewiſe other ſtrata, more or leſs ſloping, which Soil.

consist of slates, and of a kind of stone containing particles of talc. There are likewise stones, which divide into shivers; and on them we observed impressions of a kind of fossil shells, unknown in these seas; we made grind-stones of it to sharpen our tools. The stone taken out of the quarries was yellowish, and not yet come to a sufficient degree of hardness, as it could be cut with a knife; but it hardened in the air. Clay, sand, and earth, fit for making potters-ware and bricks, were easily found.

Turf and its qualities.

The turf, which is generally to be met with above the clay, goes up a great way in the country. From whatever point one sets out, one could not go a league without meeting with considerable strata of it, always easy to be distinguished by the inequalities in the ground, by which some of its sides were discovered. It continually is formed from the remains of roots and plants in marshy places; which are always known by a sharp-pointed kind of rushes. This turf being taken in a bay, near our habitation, where it shews a surface of twelve feet high to the open air, gets a sufficient degree of dryness there. This was what we made use of; its smell was not disagreeable; it burnt well, and its cinders, or embers, were superior to those of sea-coals; because, by blowing them, it was as easy to light a candle as with burning coals; it was sufficient

for

for all the works of the forge, excepting the joining of great pieces.

Plants.

All the ſea-ſhores, and the inner parts of the iſles are covered with a kind of gladiolus, or rather a ſpecies of gramen. It is of an excellent green, and is above ſix feet high, and ſerves for a retreat to ſeals and ſea-lions: on our journies it ſheltered us, as it did them. By its aſſiſtance we could take up our quarters in a moment. Its bent and united ſtalks, formed a thatch or roof, and its dry leaves a pretty good bed. It was likewiſe with this plant that we covered our houſes; its ſtalk is ſweet, nouriſhing, and preferred to all other food by the cattle.

Next to this great plant, the heath, the ſhrubs, and the gum plant were the only objects that appeared in the fields. The other parts are covered by ſmall plants, which, in moiſt ground, are more green and more ſubſtantial. The ſhrubs were of great uſe to us as fuel, and they were afterwards kept for heating the ovens, together with the heath; the red fruit of the latter attracted a great quantity of game in the ſeaſon.

Reſinous gum-plant

The gum-plant, which is new and unknown in Europe, deſerves a more ample deſcription. It is of a bright green, and has nothing of the figure of a plant; one would ſooner take it to be an excreſcence of the earth of this colour; for it has neither ſtalk, branches, nor leaves---Its ſurface, which is convex, is of ſo cloſe a texture, that nothing can be introduced between it, without tearing it. The firſt thing we did, was to ſit down or ſtand upon it; it is not above a foot and a half high. It would bear us up as ſafely as a ſtone, without yielding under our weight. Its breadth is very diſproportionate to its height; and I have ſeen ſome of more than ſix feet in diameter, without being any higher than common. Its circumference is regular only in the ſmaller plants, which are generally hemiſpherical; but when they are grown up they are terminated by humps and cavities, without any regularity. In ſeveral parts of its ſurface, are drops of the ſize of peaſe, of a tough yellowiſh matter; which was at firſt called gum; but as it could not be diſſolved, except by ſpirituous ſolvents, it was named a roſin. Its ſmell is ſtrong, aromatic, and like that of turpentine. In order to know the inſide of this plant, we cut it cloſe to the ground, and turned it down. As we broke it, we ſaw that it comes from a ſtalk, whence an infinite number of concentric ſhoots ariſe, conſiſting of leaves like ſtars, enchaſed one within the other, by means of an axis common to all.

Theſe ſhoots are white within, except at a little diſtance of the ſurface, where the air colours them green. When they are broken, a milky juice comes out in

great

great abundance; which is more viscid than that of spurge*. The stalk abounds with the juice, as do the roots, which extend horizontally; and often at some distance send forth new shoots, so that you never find one of these plants alone. It seems to like the sides of hills; and it thrives well in any exposure. It was not before the third year that we endeavoured to know its flower and seeds, both of which are very small, because we had been disappointed in our attempts to bring it over to Europe. At last, however, some seeds were brought, in order to endeavour to get possession of so singular and new a plant, which might even prove useful in physic; as its rosin had already been successfully applied to slight wounds by several sailors. One thing deserves to be observed, namely, that this plant loses its rosin by the air alone, and the washing of the rains. How can we make this agree with its quality of dissolving in spirits alone? In this state it was amazingly light, and would burn like straw.

After this extraordinary plant, we met with one of approved utility; it forms a little shrub, and sometimes creeps under the plants, and along the coast. We accidentally tasted it, and found it had a spruce taste, which put us in mind of trying to make beer of it; we had brought a quantity of melasses and malt with Beer-plant.

*Euphorbia Linn. Tithymalus Tournef. F.

us:

us; the trials we made, anſwered beyond expectation; and the ſettlers being once inſtructed in the proceſs, never were in want of this liquor afterwards, which was anti ſcorbutic, by the nature of the plant; it was with good ſucceſs employed in baths, which were made for ſick perſons, who came from the ſea. Its leaves are ſmall and dentated, and of a bright green. When it is cruſhed between the fingers, it is reduced into a kind of meal, which is ſomewhat glutinous, and has an aromatic ſmell.

A kind of celery or wild parſley, in great quantities; abundance of ſorrel, water creſſes, and a kind of maiden-hair *, with undated leaves, furniſhed as much as could be required againſt the ſcurvy, together with the above plant.

Fruits. Two ſmall fruits, one of which is unknown, and looks like a mulberry, the other no bigger than a pea, and called lucet, on account of the ſimilarity it bears to that which is found in North-America, were the only ones which were to be had in autumn. Thoſe which grew upon the buſhes were good for nothing, excepting for children, who will eat the worſt of fruits, and for wild-fowl. The plant on which the fruit, which we called mulberry, grew, is creeping; its leaf reſembles that of the hornbeam; its branches are long, and it is propagated like the ſtrawberry.

* Ceterac Aſplenium, Linn. F.

The

The lucet is likewiſe a creeping plant, bearing the fruit all along its branches, which are beſet with little ſhining round leaves, of the colour of myrtle leaves; their fruits are white, and coloured red on that ſide which is turned towards the ſun; they have an aromatic taſte, and ſmell like orange-bloſſoms, as do the leaves, of which the infuſion drank with milk is very pleaſant to the taſte. This plant is hidden among the graſs, and prefers a wet ſoil: a prodigious quantity of it grows in the neighbourhood of lakes.

Among ſeveral other plants, which we found superfluous to examine, there were many flowers, but all without ſmell, one excepted, which is white, and has the ſmell of the tuberoſe. We likewiſe found a true violet, as yellow as a jonquil. It is worth notice that we have never found any bulbous-rooted plant. Another ſingularity is, that in the ſouthern part of the iſle we inhabited, beyond a chain of hills which divides it from eaſt to weſt, it appeared that there were hardly any of the reſinous gum-plants, and that in their ſtead we found abundance of another plant of the ſame form, but of a different green, wanting the ſolidity of the other, and not producing any roſin, but only fine yellow flowers in the proper ſeaſon. This plant, which was eaſily opened, conſiſted as the other, of ſhoots which all ſpring from the ſame ſtalk, and ter- Flowers.

minate at its ſurface. Coming back over the hills, we found a tall ſpecies of maiden hair; its leaves are not waved, but in the form of ſword blades. From the plant ariſe two principal ſtalks, which bear their ſeeds on the underſide, like the other ſpecies of maiden hair. There were likewiſe a great quantity of friable plants growing upon ſtones, they ſeemed to partake of the nature of ſtone, and of vegetables; they were thought to be ſpecies of lichen, but the aſcertaining whether they would be of uſe in dying, was put off to another time.

Sea plants. As to the ſubmarine plants, they were more inconvenient than of any uſe. The whole harbour is covered with ſea weeds, eſpecially near the ſhore, by which means the boats found it difficult to land; they are of no other ſervice than to break the force of the waters when the ſea runs very high. We hoped to make a good uſe of them by employing them for a manure. The tides brought us ſeveral ſpecies of coralines, which were very much varied, and of the fineſt colours; theſe, together with the ſpunges and ſhells, have deſerved places in the cabinets of the curious. All the ſpunges have the figure of plants, and are branched in ſo many different ways, that we could hardly believe them to be the work of marine inſects. Their texture is ſo compact, and their fibres ſo delicate, that it is inconceivable how theſe animals can lodge in them.

The

The coaſts of the Malouines have provided the collections in Europe with ſeveral new ſhells; the moſt curious of which, is that called *la poulette*. There are three ſorts of this bivalve; and among them the ſtriated one had never before been ſeen, except in the foſſil ſtate; this may prove the aſſertion, that the foſſil-ſhells, found much below the level of the ſea, are not luſus naturæ, and accidentally formed; but that they have really been inhabited by living animals, at the time when the land was covered by the water. Along with this ſhell, which is very common here, there are limpets *; eſteemed on account of their fine colours; whelks †, of ſeveral kinds; ſcallops ‡; great ſtriated and ſmooth muſcle-ſhells §, and the fineſt mother of pearl.

There is only a ſingle ſpecies of quadruped upon these iſlands; it is a medium between the wolf and the fox. The land and water-fowls are innumerable. The ſea-lions and ſeals are the only amphibia. All the coaſts abound with fiſh, moſt of them little known. The whales keep in the open ſea; ſome of them happen now and then to be ſtranded in the bays, and their remains are ſometimes ſeen there. Some other bones of an enormous ſize, a good way up in the country, whither the force of the waves could never Animals.

* Lepas Linn. † Buccinum Linn. ‡ Oſtreæ Pectines Linn. § Mya Linn. F.

 carry

carry them, prove that either the ſea is diminiſhed, or that the ſoil is encreaſed.

The wolf-fox, *(loup-renard)* thus called, on account of its digging a kennel under ground, and having a more buſhy tail than a wolf, lives upon the downs along the ſea-ſhore. It attacks the wild fowls; and makes its roads from one bay to another, with ſo much ſagacity, that they are always the ſhorteſt that can be deviſed; and, at our firſt landing on the iſle, we had almoſt no doubt of their being the paths of inhabitants. It ſeems this animal faſts during a time of the year; for it is then vaſtly lean. Its ſize and make is that of a common ſhepherd's dog; and it barks in the ſame manner, though not ſo loud. In what manner can it have been tranſported to theſe iſlands *?

The birds and fiſh have enemies, which endanger their tranquility. Theſe enemies of the birds are the above kind of wolf, which deſtroys many of their eggs and young ones; the eagles, hawks, falcons, and owls.

The fiſh are ſtill worſe uſed; without mentioning the whales, which feeding, as is well known, upon fry only,

* For a navigator, of Mr. Bougainville's experience and abilities, this query is very extraordinary; and, ſtill more ſo, for a man who has ſpent ſo many years in Canada, near the coaſts of Labrador; and who certainly muſt have read accounts from Greenland, where often land-animals, on large maſſes of ice fixed to the ſhore, and broke looſe by the ſea, are driven into the ocean; and again landed on the ſhores of countries, very diſtant from their native home. F.

ly, deſtroy prodigious numbers; they are likewiſe expoſed to the amphibious creatures, and to birds; ſome of which are always watching on the rocks, whilſt others conſtantly ſkim along the ſurface of the ſea.

It would require a great deal of time, and the eyes of an able naturaliſt, in order to deſcribe the following animals well. I ſhall here give the moſt eſſential obſervations, and extend them only to ſuch animals as were of ſome utility.

Among the web-footed birds, the ſwan is the firſt in order; it only differs from the European one by its neck; which is of a velvet black, and makes an admirable contraſt with the whiteneſs of the reſt of its body; its feet are fleſh-coloured. This kind of ſwan is likewiſe to be found in Rio de la Plata, and in the ſtraits of Magalhaens. Web-footed birds.

Four ſpecies of wild-geeſe made part of our greateſt riches. The firſt only feeds on dry land; and has, improperly, been called buſtard*. Its high legs ſerve to elevate it above the tall graſs, and its long neck to obſerve any danger. It walks and flies with great eaſe; and has not that diſagreeable cackling cry, peculiar to the reſt of its kind. The plumage of the male is white,

* In the northern parts of America is a kind of wild gooſe, which was called by the French, when in poſſeſſion of Canada, Outarde, or Buſtard; the Engliſh call it the Canada-gooſe; it has been repreſented by Cateſby, I. t. 92. Edward t. 151. and the PlanchesEnluminées, t. 346. Perhaps this may be the ſame ſpecies. F.

mixed

mixed with black and aſh-colour on the wings. The female is yellow; and its wings are adorned with changing colours; it generally lays ſix eggs. Its fleſh is wholeſome, nouriſhing, and palatable; it ſeldom happened that we had any ſcarcity of this kind of geeſe; for, beſides theſe which are bred in the iſle, they come in great flocks in autumn, with the eaſt wind, probably from ſome uninhabited country. The ſportſmen eaſily diſtinguiſh theſe new-comers, by the little fear they ſhew of men. The other three ſpecies are not ſo much in requeſt; for they feed on fiſh, and get a trainy taſte. Their figure is not ſo elegant as that of the firſt ſpecies; one of theſe kinds ſeldom riſes above the water, and is very noiſy. The colours of their feathers are chiefly white, black, yellow, and aſh-colour. All theſe ſpecies, and likewiſe the ſwan, have a ſoft down under the feathers; which is white or grey, and very thick.

Two kinds of ducks, and two of teals, frequent the ponds and rivers. The former are but little different from thoſe of our climate; ſome of thoſe which we killed, were quite black, and others quite white. As to the teals, the one has a blue bill, and is of the ſize of the ducks; the other is much leſs. Some of them had the feathers on the belly of a fleſh colour. Theſe ſpecies are in great plenty, and of an excellent taſte.

Here are two kinds of Divers, of a ſmall ſize. One of them has a grey back, and white belly; the feathers on the belly are ſo ſilky, ſhining, and cloſe, that we imagined theſe were the birds, of whoſe plumage the fine muffs are made: this ſpecies is here ſcarce*. The other, which is more common, is quite brown, but ſomewhat paler on the belly than on the back. The eyes of theſe creatures are like rubies. Their ſurpriſing livelineſs is heightened and ſet off ſtill more by the circle of white feathers that ſurrounds them; and has cauſed the name of Diver with Spectacles to be given to the bird. They breed two young ones at a time, which are probably too tender to ſuffer the coldneſs of the water, whilſt they have nothing but their down; for then the mother conveys them on her back†. Theſe two ſpecies have not webbed feet, as the other water-fowl; but their toes are ſeparate, with a ſtrong membrane on each ſide; in this manner, each toe reſembles a leaf, which is roundiſh towards the claw; and the lines, which run from the toe to the circumference of the membrane, together with its green-colour and thinneſs, increaſe the reſemblance.

* This bird, though the author calls it a Diver, ſeems, according to the deſcription of it, to be rather the Grebe; which is ſo plentiful on the lake of Geneva, whoſe beautiful ſkins are dreſt, and made into muffs and tippets. Br. Zool. 2. p. 396. 8vo. Ed. F.

† This ſpecies ſeems to be the white and duſky grebe. Br. Zool. 2. p. 397. and vol. 4. f. 17. F.

Two

Two ſpecies of birds, which were called by our people ſaw-bills *, I know not for what reaſon, only differed from each other in ſize, and ſometimes becauſe there were now and then ſome with brown bellies; whereas, the general colour of that part, in other birds of the kind, was white. The reſt of the feathers are of a very dark blueiſh-black; in conſequence of their ſhape, and the cloſe texture and ſilkineſs of their vent feathers, we muſt rank them with the divers, though I cannot be poſitive in this reſpect. They have a pointed bill, and the feet webbed without any ſeparation between the toes; the firſt toe, being the longeſt of the three, and the membrane which joins them, ending in nothing at the third toe, gives a very remarkable character. Their feet are fleſh-coloured †. Theſe birds deſtroy numbers of fiſh; they place themſelves upon the rocks, join together by numerous families, and lay their eggs there. As their fleſh is very good to eat, we killed two or three hundred of them at a time; and the abundance of their eggs offered another reſource to ſupply our wants. They were ſo little afraid of our ſportſmen, that it was ſufficient to go againſt them with no better arms than ſticks. Their enemy is a bird of prey, with webbed feet; meaſuring near ſeven feet from tip to tip, and having a

* Becs-ſcies.

† As far as we can gueſs, from this very imperfect deſcription, the birds here mentioned ſeem to be of the kind called Guillemot. Br. Zool. vol. 2. p. 410. and vol. 4. t. 20. F.

long

long and ſtrong bill, diſtinguiſhed by two tubes of the ſame ſubſtance as the bill itſelf, which are hollow throughout. This is the bird which the Spaniards call *Quebrantahueſſos* *.

A great quantity of mews, variouſly and prettily marked, of gulls and of terns, almoſt all of them grey, and living in families, come ſkimming along the water, and fall upon the fiſh with extraordinary quickneſs; they were ſo far of uſe to us, that they ſhewed us the proper ſeaſon of catching pilchards; they held them ſuſpended in the air for a moment only, and then preſently gave back entire, the fiſh they had ſwallowed juſt before. At other ſeaſons they feed upon a little fiſh, called *gradeau*, and ſome other ſmall fry. They lay their eggs in great quantities round the marſhes, on ſome green plants, pretty like the water lily ‡, and they were very wholeſome food.

We found three ſpecies of penguins: the firſt of them is remarkable on account of its ſhape, and the beauty of its plumage, and does not live in families as the

* The *Quebrantahueſſos* is a bird belonging to the genus called by Dr. Linnæus, *Procellaria*, or petrel; ſome of the ſailors call it Albatroſs, but then we muſt take care not to confound the common albatroſs, repreſented by Mr. Edwards, tab. 88, which is not this Quebrantahueſſos, but I believe the bird deſcribed by our author to be not yet well known by our ornithologiſts; and the imperfect account of Bougainville and Dom Pernetty are far from being ſatisfactory to natural hiſtorians. Our late great circumnavigators and philoſophers will probably oblige the literary world with a drawing and account of this bird. F.

‡ Nenuphars, Nymphæa Linn, F.

ſecond

ſecond ſpecies, which is the ſame with that deſcribed in Lord Anſon's Voyage *. The penguin of the firſt claſs is fond of ſolitude and retired places. It has a peculiar noble and magnificent appearance, having an eaſy gait, a long neck when ſinging or crying, a longer and more elegant bill than the ſecond ſort, the back of a more blueiſh caſt, the belly of a dazzling white, and a kind of palatine or necklace of a bright yellow, which comes down on both ſides of the head, as a boundary between the blue and the white, and joins on the belly ‡. We hoped to be able to bring one of them over to Europe. It was eaſily tamed ſo far as to follow and know the perſon that had the care of feeding it: fleſh, fiſh, and bread, were its food; but we perceived that this food was not ſufficient, and that it abſorbed the fatneſs of the bird; accordingly, when the bird was grown lean to a certain degree, it died. The third ſort of penguins live in great flocks or families like the ſecond; they inhabit the high cliffs, where we found the ſaw-bills (becs-ſcies), and they lay their eggs there. Their diſtinguiſhing characters are, the ſmallneſs of their ſize, their dark yellow colour, a tuft of gold-yellow feathers, which are ſhorter than thoſe of the

* The place referred to here in Lord Anſon's Voyage is book I. chap. vii. p 92. edit. 14th, in 8vo. 1769; but from thence, as well as from our author's account, it is impoſſible to determine which ſpecies of the penguin is meant. F.

‡ The firſt of theſe penguins ſeems to be that deſcribed by Mr. Pennant in the Philoſ. Tranſ. vol. lix. and repreſented in an accurate drawing. F.

egret,

egret *, and which they raiſe when provoked, and laſtly, ſome other feathers of the ſame colour, which ſtand in the place of eye-brows; our people called them hopping penguins, becauſe they chiefly advance by hopping and ſkipping. This ſpecies carries greater air of livelineſs in its countenance than the two others †.

Three ſpecies of petrels, (alcyons) which appear but ſeldom, did not forebode any tempeſts, as thoſe do which are ſeen at ſea. They are however the ſame birds, as our ſailors affirmed, and the leaſt ſpecies has all the characters of it. Though this may be the true alcyons ‡, yet ſo much is certain, that they build their neſts on ſhore, whence we have had their young ones covered only with down, but perfectly like their parents in other reſpects. The ſecond ſort only differs from them in ſize, being ſomewhat leſs than a pigeon. Theſe two ſpecies are black, with ſome white feathers on the belly §. The third ſort was at firſt

* Aigrette, a ſpecies of heron.

† This laſt ſpecies of penguin, or auk, ſeems to be the ſame with the alca cirrhata of Dr. Pallas, Spicileg. Zool. Faſc. v. p. 7. tab. i. & v. fig. 1—3. F.

‡ The author certainly has the noted fable of the antients in view, according to which, the alcyons had a ſwimming neſt, and brooded at ſea at a time in winter, when the weather was calm. The few calm days during which theſe birds were employed in brooding, were therefore called alcyonia. F.

§ The two petrels here mentioned ſeem to be the little, and the ſea-ſwallow or frigat; the firſt of which is deſcribed, Br. Zool, vol. ii. p. 434, and repreſented, vol. iv. t. 82. The ſecond, or ſwallow-petrel, or frigat, is to be met with in Rocheforty's Voyage, t. 135. Dr. Linnæus calls the firſt procellaria pelagica, the latter the fregatta, and, if I am not miſtaken, the third kind here mentioned, is, the fulmar, Br. Zool, vol. ii. p. 431. and vol. iv. t. 82. Dr. Linnæus's Procellaria glacialis. F.

called

called white-pigeon, on account of its feathers being all of that colour, and its bill being red: there is reaſon to ſuppoſe it is a true white alcyon, on account of its conformity with the other ſpecies.

Birds with cloven feet.

Three ſorts of eagles, of which the ſtrongeſt have a dirty white, and the others a black plumage, with yellow and white feet, attack the ſnipes and little birds; neither their ſize nor the ſtrength of their claws allowing them to fall upon others. A number of ſparrow hawks and falcons, together with ſome owls, are the other enemies of the fowl. Their plumage is rich, and much varied in colour.

The ſnipes are the ſame as the European ones; they do not fly irregularly when they riſe, and are eaſy to be ſhot. In the breeding ſeaſon they ſoar to a prodigious height; and after ſinging and diſcovering their neſt, which they form without precaution in the midſt of the fields, on ſpots where hardly any plants grow, they fall down upon it from the height they had riſen to before; at this ſeaſon they are poor; the beſt time for eating them is in autumn.

In ſummer we ſaw many curlews, which were not at all different from ours.

Throughout the whole year we ſaw a bird pretty like a curlew on the ſea-ſide; it was called a ſea-pie *, on

* The ſea-pie is ſometimes called oyſter-catcher, becauſe this bird forces the ſhells open with its bill, which are left bare on the ſhore, at the receſſion of the tide. Br. Zool, vol. iv. p. 376, Dr. Linnæus's Hæmatopus Oſtralegus. F.

on account of its black and white plumage; its other characteriſtics are, a bill of the colour of red coral, and white feet. It hardly ever leaves the rocks, which are dry at low water, and lives upon little ſhrimps. It makes a whiſtling noiſe, eaſy to be imitated, which proved uſeful to our ſportſmen, and pernicious to the bird.

Egrets are pretty common here; at firſt we took them for common herons, not knowing the value of their plumes. Theſe birds begin to feed towards night; they have a harſh barking noiſe, which we often took for the noiſe of the wolf we have mentioned before.

Two ſorts of ſtares or thruſhes came to us every autumn; a third ſpecies remained here conſtantly, it was called the red bird*; its belly is quite covered with feathers of a beautiful fiery red, eſpecially during winter; they might be collected, and would make very rich tippets. One of the two remaining ſpecies is yellow, with black ſpots on the belly, the other has the colour of our common thruſhes. I ſhall not give any particular account of an infinite number of little birds, that are pretty like thoſe ſeen in the maritime provinces of France.

* This ſeems to be the American red-breaſt, or turdus migratorius, Linn. and Kalm's Voyage, vol. ii. p. 90, where likewiſe a figure of it is given. F.

 The

Amphibious creatures.

The ſea-lions and ſeals are already known; theſe animals occupy the ſea-ſhore, and lodge, as I have before mentioned, among the tall plants, called gladioli *. They go up a league into the country in innumerable herds, in order to enjoy the freſh herbs, and to baſk in the ſun. It ſeems the ſea-lion deſcribed in Lord Anſon's Voyage ought, on account of its ſnout, to be looked upon as a kind of marine elephant, eſpecially as he has no mane; is of an amazing ſize, being ſometimes twenty-two feet long, and as there is another ſpecies much inferior in ſize, without any ſnout, and having a mane of longer hairs than thoſe on the reſt of the body, which therefore ſhould be conſidered as the true ſea-lion †. The ſeal (*loup marin*) has neither mane nor ſnout; thus all the three ſpecies are eaſily diſtinguiſhed. Under the hair of all theſe creatures, there is no ſuch down as is found in thoſe caught in North America and Rio de la Plata. Their greaſe or train oil, and their ſkins, might form a branch of commerce.

ish.

We have not found a great variety of ſpecies of fiſh. That ſort which we caught moſt frequently, we called

* Glayeuls.

† The animal here mentioned as the true ſea-lion exceeds the ſea-lion deſcribed in Lord Anſon's Voyage; for this is twenty-five feet long, and that in the iſle of Juan Fernandes only twenty. See Voyage aux Iſles Malouines, par Dom Pernetty. F.

 mullet,

mullet *, to which it bears ſome reſemblance. Some of them were three feet long, and our people dried them. The fiſh called gradeau is very common, and ſometimes found above a foot long. The ſardine only comes in the beginning of winter. The mullets being purſued by the ſeals, dig holes in the ſlimy ground, on the banks of the rivulets, where they take ſhelter, and we took them without difficulty, by taking off the layer of mud that covered their retreats. Beſides theſe ſpecies, a number of other very ſmall ones were taken with a hook and line, and among them was one which was called a tranſparent pike †. Its head is ſhaped like that of our pike, the body without ſcales, and perfectly diaphanous. There are likewiſe ſome congers on the rocks, and the white porpeſſe, called *la taupe*, or the mole, appears in the bays during the fine ſeaſon. If we had had time, and men enough to ſpare, for the fiſhery at ſea, we ſhould have found many other fiſh, and certainly ſome ſoals, of which a few have been found, thrown upon the ſands. Only a ſingle ſort of freſh water fiſh, without ſcales, has been taken; it is of a green colour, and of the ſize of a common trout ‡. It is true, we have made but few reſearches in this par-

* Muge ou mulet. † Brochet tranſparent.

‡ This kind of trout has been likewiſe mentioned in a pamphlet publiſhed laſt winter about the Falkland iſles. F.

ticular,

ticular, we had but little time; and other fiſh in abundance.

Cruſtaceous fiſh.

Here have been found only three ſmall ſorts of cruſtacea; viz. the cray-fiſh, which is red, even before it is boiled, and is properly a prawn; the crab, with blue feet, reſembling pretty much that called *tourelourou*, and a minute ſpecies of ſhrimp. Theſe three cruſtacea, and all muſcles, and other ſhell fiſh, were only picked up for curioſity's ſake, for they have not ſo good a taſte as thoſe in France.——This land ſeems to be entirely deprived of oyſters.

Laſtly, by way of forming a compariſon with ſome cultivated iſle in Europe, I ſhall quote what Puffendorf ſays of Ireland, which is ſituated nearly in the ſame latitude in the northern hemiſphere, as the Malouines in the ſouthern one, viz. "that this iſland is pleaſant "on account of the healthineſs and ſerenity of the air, "and becauſe heat and cold are never exceſſive there. "The land being well divided by lakes and rivers, offers "great plains, covered with excellent paſture, has no "venemous creatures, its lakes and rivers abound with "fiſh, &c." See the Univerſal Hiſtory.

CHAP.

CHAP. V.

Navigation from the Malouines to Rio-Janeiro; junction of the Boudeuse with the Etoile.---Hostilities of the Portuguese against the Spaniards. Revenues of the king of Portugal from Rio-Janeiro.

1767. June. Departure from the Malouines for Rio-Janeiro.

I WAITED, in vain, for the Etoile, at the Malouines; the months of March and April had passed, and that store-ship did not arrive. I could not attempt to traverse the Pacific Ocean with my frigate alone; as she had no more room than what would hold six months provision for the crew. I still waited for the store-ship, during May. Then seeing that I had only two months provisions, I left the Malouines the second of June, in order to go to Rio-Janeiro; which I had pointed out as a rendezvous to M. de la Giraudais, commander of the Etoile, in case some circumstances should prevent his coming to join me at the Malouines.

During this navigation, we had very fair weather. The 20th of June, in the afternoon, we saw the high head-lands of the Brasils; and, on the 21st, we discovered the entrance of Rio-Janeiro Along the coast we saw several fishing-boats. I ordered Portuguese colours to be hoisted, and fired a cannon: upon this signal one of the

the boats came on board, and I took a pilot to bring us into the road. He made us run along the coaſt, within half a league of the iſles which lie along it. We found many ſhoals every where. The coaſt is high, hilly, and woody; it is divided into little detached and perpendicular hillocks, which vary their proſpect. At half an hour paſt five, in the afternoon, we were got within the fort of Santa-Cruz; from whence we were hailed; and at the ſame time a Portugueze officer came on board, to aſk the reaſon of our entering into port. I ſent the chevalier Bournand with him, to inform the count d' Acunha, viceroy of the Braſils, of it, and to treat about the ſalute. At half an hour paſt ſeven, we anchored in the road, in eight fathoms water, and black muddy bottom.

[D]iſcuſſion [c]oncerning [t]he ſalute.

The chevalier de Bournand returned ſoon after; and told me, that, concerning the ſalute, the count d' Acunha had anſwered him, that if a perſon, meeting another in a ſtreet, took off his hat to him, he did not before inform himſelf, whether or no this civility would be returned; that if we ſaluted the place, he would conſider what he ſhould do. As this anſwer was not a ſufficient one, I did not ſalute. I heard at the ſame time, by means of a canoe, which M. de la Giraudais ſent to me, that he was in this port; that his departure from Rochefort, which ſhould have been in December, had

[Ju]nction with [th]e Etoile.

had been retarded till the beginning of February; that after three months ſailing, the water which his ſhip made, and the bad condition of her rigging, had forced him to put in at Montevideo, where he had received information concerning my voyage, by means of the Spaniſh frigates returning from the Malouines; and he had immediately ſet ſail for Rio-Janeiro, where he had been at anchor for ſix days.

This junction enabled me to continue my expedition; though the Etoile, bringing me upwards of fifteen months ſalt proviſions and liquor, had hardly for fifty days bread and legumes to give me. The want of theſe indiſpenſable proviſions, obliging me to return and get ſome in Rio de la Plata; as we found at Rio-Janeiro, neither biſcuit, nor wheat, nor flour.

Difficulties raiſed by the Portugueſe againſt a Spaniſh ſhip.

There were, at this time, two veſſels in this port which intereſted us; the one a French, and the other a Spaniſh one. The former, called l'Etoile du Matin, or the Morning Star, was the king's ſhip bound for India; which, on account of its ſmallneſs, could not undertake to double the Cape of Good Hope during winter; and, therefore, came hither to wait the return of the fair ſeaſon. The Spaniſh veſſel was a man of war, of ſeventy-four guns, named the Diligent, commanded by Don Franceſco de Medina. Having ſailed from the river of Plata, with a cargo of ſkins and piaſtres; a leak which his ſhip had

 ſprung,

ſprung, much below her water-line, had obliged him to bring her hither, in order to refit her for the voyage to Europe. He had been here eight months; and the refuſal of neceſſary aſſiſtance, and the difficulties which the viceroy laid in his way, had prevented his finiſhing the repair: accordingly, Don Franciſco ſent the ſame evening that I arrived, to beg for my carpenters and caulkers; and the next morning I ſent them to him from both the veſſels.

Aſſiſtance which we gave her.

The viceroy viſits us on board the frigate.

The 22d we went in a body to pay a viſit to the viceroy; he came and returned it on the 25th; and, when he left us, I ſaluted him with nineteen guns, which were returned from the ſhore. On this viſit, he offered us all the aſſiſtance in his power; and even granted me the leave I aſked, of buying a ſloop, which would have been very uſeful, during the courſe of my expedition; and, he added, that if there had been one belonging to the king of Portugal, he would have offered it me. He likewiſe aſſured me, that he would make the moſt exact enquiries, in order to diſcover thoſe, who, under the very windows of his palace, had murdered the chaplain of the Etoile, a few days before our arrival; and that he would proceed with them according to the utmoſt ſeverity of the law. He promiſed juſtice; but the law of nations was very ineffectually executed at this place.

How-

However, the viceroy's civilities towards us continued for ſeveral days: he even told us his intention of giving us a *petit ſouper*, or collation, by the water-ſide, in bowers of jaſmine and orange-trees; and he ordered a box to be prepared for us at the opera. We ſaw, in a tolerable handſome hall, the beſt works of Metaſtaſio represented by a band of mulattoes; and heard the divine compoſition of the great Italian maſters, executed by an orcheſtra, which was under the direction of a hump backed prieſt, in his canonicals.

The favour which we enjoyed, occaſioned great matter of aſtoniſhment to the Spaniards, and even to the people of the country; who told us, that their governor's proceedings would not be the ſame for a long time. Indeed, whether the aſſiſtance we gave the Spaniards, and our own connections with them diſpleaſed him, or whether he could no longer feign a conduct, ſo diametrically oppoſite to his natural temper, he ſoon became, in regard to us, what he had been to every body elſe.

The 28th of June, we heard that the Portugueſe had ſurpriſed and attacked the Spaniards at Rio-Grande; that they had driven them from a ſtation which they occupied on the left ſhore of that river; and that a Spaniſh ſhip, touching at the iſle of St. Catherine, had been detained there. They fitted out here, with great expedition,

Hoſtilities the Portugueze aga[illegible] the Spania[illegible]

dition, the San Sebaſtiano, of ſixty-four guns, built here; and a frigate, mounting forty guns, called Noſſa Senhora da Gracia. This laſt was deſtined, it was ſaid, to eſcort a convoy of troops and ammunition to Rio-Grande, and to the colony of Santo Sacramento. Theſe hoſtilities and preparations gave us reaſon to apprehend that the viceroy intended to ſtop the Diligent; which was careening upon the iſle das Cobras, and we accelerated her refitment as much as poſſible. She really was ready on the laſt day of June, and began to take in the ſkins, which were part of her lading; but on the ſixth of July, when ſhe wanted to take back her cannon, which, during the repair, had been depoſited on the iſle das Cobras, the viceroy forbade their being delivered; and declared, that he arreſted the ſhip, till he had received the orders of his court, on the ſubject of the hoſtilities committed at Rio-Grande. In vain did Don Medina take all the neceſſary ſteps on this occaſion; count d'Acunha would not ſo much as receive the letter, which the Spaniſh commander ſent him by an officer, from on board his ſhip.

1767. July.

[B]ad proceedings of the viceroy towards us.

We partook of the diſgrace of our allies. Having, upon the repeated leave of the viceroy, concluded the bargain for buying a ſnow, his excellency forbade the ſeller to deliver it to me. He likewiſe gave orders, that we ſhould not be allowed the neceſſary timber out of the

the royal dock-yards, for which we had already agreed: he then refused me the permission of lodging with my officers, (during the time that the frigate underwent some essential repairs) in a house near the town, offered me by its proprietor: and which commodore Byron had occupied in 1765, when he touched at this port. On this account, and likewise upon his refusing me the snow and the timber, I wanted to make some remonstrances to him. He did not give me time to do it; and, at the first words I uttered, he rose in a furious passion, and ordered me to go out; and being certainly piqued, that, in spite of his anger, I remained sitting with two officers, who accompanied me, he called his guards; but they, wiser than himself, did not come, and we retired; so that nobody seemed to have been disturbed. We were hardly gone, when the guards of his palace were doubled, and orders given to arrest all the French that should be found in the streets after sun-setting. He likewise sent word to the captain of the French ship of four guns, to go and anchor under the fort of Villagahon; and the next morning I got her towed there by my boats.

They dete[rmine] us t[o] leave Rio Janeiro.

From hence forward, I was intent upon my departure; especially as the inhabitants, with whom we had any intercourse of trade, must fear every thing from the viceroy. Two Portuguese officers became the victims of the

the civility they ſhewed us; the one was impriſoned in the citadel; the other exiled to Santa, a ſmall town between St. Catherine and Rio Grande. I made haſte to take in our water, to get the moſt neceſſary proviſions out of the Etoile, and to embark refreſhments. I had been forced to enlarge our tops; and the Spaniſh captain furniſhed me with the neceſſary timber for that purpoſe, which had been refuſed us out of the docks. I likewiſe got ſome planks, which we could not do without; and which were ſold to us ſecretly.

At laſt, on the 12th, every thing being ready, I ſent an officer to let the viceroy know, I ſhould weigh with the firſt fair wind. I adviſed M. d' Etcheveri, who commanded l'Etoile du Matin, (the Morning-Star) to ſtop at Rio-Janeiro as little as he could; and rather to employ the time that remained, till the favourable ſeaſon for doubling the Cape of Good Hope came on, in going to ſurvey the iſles of Triſtan d'Acunha, where he would find wood, water, and abundance of fiſh; and I gave him ſome memoirs I had concerning theſe iſles. I have ſince heard, that he has followed my advice.

During our ſtay at Rio-Janeiro, we enjoyed one of the ſprings, which are obvious in poetical deſcriptions; and the inhabitants teſtified, in the moſt genteel manner, the diſpleaſure which their viceroy's bad proceedings againſt us, gave them. We were ſorry, that it was not

6 in

in our power to stay any longer with them. The Brasils, and the capital in it, have been described by so many authors, that I could mention nothing, without tediously repeating what has been said before. Rio Janeiro has once been conquered by France; and is, of course, well known there. I will confine myself to give an account of the riches, of which that city is the staple *; and of the revenues which the king of Portugal gets from thence. I must previously mention, that M. de Commercon, an able naturalist, who came with us on board the Etoile, in order to go on the expedition, assured me, that this was the richest country in plants he had ever met with; and that it had supplied him with whole treasures in botany.

Account of the riches of Rio-Janeiro.

Rio-Janeiro is the emporium and principal staple of the rich produce of the Brasils. The mines, which are called *general*, are the nearest to the city; being about seventy-five leagues distant. They annually bring in to the king, for his fifth part, at least one hundred and twelve arobas of gold; in 1762 they brought in a hundred and nineteen. Under the government of the general mines, are comprehended those of Rio das Mortes, of Sabara, and of Sero-frio. The last place, besides gold, produces all the diamonds that come from the Brasils. They are in the bed of a river; which is led aside, in

* Debouché.

order

order afterwards to ſeparate the diamonds, topazes, chryſolites, and other ſtones of inferior goodneſs, from the pebbles, among which they ly.

Regulations for examining the mines.

Mines of diamonds.

All theſe ſtones, diamonds excepted, are not contraband: they belong to the poſſeſſors of the mines; but they are obliged to give a very exact account of the diamonds they find; and to put them into the hands of a ſurveyor *, whom the king appoints for this purpoſe. The ſurveyor immediately depoſits them in a little caſket, covered with plates of iron, and locked up by three locks. He has one of the keys, the viceroy the other, and the *Provador de Hazienda Reale* the third. This caſket is incloſed in another, on which are the ſeals of the three perſons above mentioned, and which contains the three keys to the firſt. The viceroy is not allowed to viſit its contents; he only places the whole in a third coffer, which he ſends to Liſbon, after putting his ſeal on it. It is opened in the king's preſence; he chooſes the diamonds which he likes out of it; and pays their price to the poſſeſſors of the mines, according to a tariff ſettled in their charter.

The poſſeſſors of the mines pay the value of a Spaniſh piaſtre or dollar per day to his Moſt Faithful Majeſty, for every ſlave ſent out to ſeek diamonds; the number of theſe ſlaves amounts to eight hundred. Of

* Intendant.

all

all the contraband trades, that of diamonds is moſt ſeverely puniſhed. If the ſmuggler is poor, he loſes his life; if his riches are ſufficient to ſatisfy what the law exacts, beſides the confiſcation of the diamonds, he is condemned to pay double their value, to be impriſoned for one year, and then exiled for life to the coaſt of Africa. Notwithſtanding this ſeverity, the ſmuggling trade with diamonds, even of the moſt beautiful kind, is very extenſive; ſo great is the hope and facility of hiding them, on account of the little room they take up.

All the gold which is got out of the mines cannot be ſent to Rio Janeiro, without being previouſly brought into the houſes, eſtabliſhed in each diſtrict, where the part belonging to the crown is taken. What belongs to private perſons is returned to them in wedges, with their weight, their number, and the king's arms ſtamped upon them. All this gold is aſſayed by a perſon appointed for that purpoſe, and on each wedge or ingot, the alloy of the gold is marked, that it may afterwards be eaſy to bring them all to the ſame alloy for the coinage. Gold-mines.

Theſe ingots belonging to private perſons are regiſtered in the office of *Praybuna*, thirty leagues from Rio Janeiro. At this place is a captain, a lieutenant, and fifty men: there the tax of one fifth part is paid,

 and

and further, a poll-tax of a *real* and a half per head, of men, cattle, and beaſts of burden. One half of the produce of this tax goes to the king, and the other is divided among the detachment, according to the rank. *As it is impoſſible to come back from the* mines without paſſing by this ſtation, the ſoldiers always ſtop the paſſengers, and ſearch them with the utmoſt rigour.

The private people are then obliged to bring all the ingots of gold which fall to their ſhare, to the mint at Rio Janeiro, where they get the value of it in caſh: this commonly conſiſts of demi-doubloons, worth eight Spaniſh dollars. Upon each demi-doubloon, the king gets a piaſtre or dollar for the alloy, and for the coinage. The mint at Rio Janeiro is one of the fineſt buildings exiſting. It is furniſhed with all the conveniences neceſſary towards working with the greateſt expedition. As the gold comes from the mines at the ſame time that the fleets come from Portugal, the coinage muſt be accelerated, and indeed they coin there with amazing quickneſs.

The arrival of theſe fleets, and eſpecially of that from Liſbon, renders the commerce of Rio Janeiro very flouriſhing. The fleet from Porto is laden only with wines, brandy, vinegar, victuals, and ſome coarſe cloths, manufactured in and about that town. As ſoon as the

 fleets

fleets arrive, all the goods they bring are conveyed to the custom-house, where they pay a duty of ten per cent to the king. It must be observed that the communication between the colony of Santo Sacramento and Buenos Ayres being entirely cut off at present, that duty must be considerably lessened; for the greater part of the most precious merchandizes which arrived from Europe were sent from Rio Janeiro to that colony, from whence they were smuggled through Buenos Ayres to Peru and Chili; and this contraband trade was worth a million and a half of piastres or dollars annually to the Portuguese. In short, the mines of the Brasils produce no silver, and all that which the Portuguese got, came from this smuggling trade. The negro trade was another immense object. The loss which the almost entire suppression of this branch of contraband trade occasions, cannot be calculated. This branch alone employed at least thirty coasting vessels between the Brasils and Rio de la Plata.

Revenues of the king of Portugal from Rio Janeiro.

Besides the old duty of ten per cent which is paid at the royal custom-house, there is another duty of two and a half per cent, laid on the goods as a free gift, on account of the unfortunate event which happened at Lisbon in 1755. This duty must be paid down at the custom house immediately, whereas for the

 tenth,

tenth, you may have a respite of six months, on giving good security.

The mines of S. Paolo and Parnagua pay the king four arrobas as his fifth, in common years. The most distant mines, which are those of Pracaton and Quiaba, depend upon the government * of Matagrosso. The fifth of these mines is not received at Rio Janeiro, but that of the mines of Goyas is. This government has likewise mines of diamonds, but it is forbidden to search in them.

All the expences of the king of Portugal at Rio Janeiro, for the payment of the troops and civil officers, the carrying on of the mines, keeping the public buildings in repair, and refitting of ships, amount to about six hundred thousand piastres. I do not speak of the expence he may be at in constructing ships of the line and frigates, which he has lately begun to do here.

* Capitainie.

A sum-

A ſummary account, and the amount of the ſeparate articles of the king's revenue, taken at a medium in Spaniſh dollars.

	Dollars.
One hundred and fifty arrobas of gold, of which in common years all the fifths amount to - - - - - - - - -	1,125,000
The duty on diamonds - - - - -	240,000
The duty on the coinage - - - -	400,000
Ten per cent. of the cuſtom-houſe - -	350,000
Two and a half per cent. free gift - -	87,000
Poll tax, ſale of employs, offices, and other products of the mines - - - - -	225,000
The duty on negroes - - - - - - -	110,000
The duty on train-oil, ſalt, ſoap, and the tenth on the victuals of the country -	130,000
Total in dollars or piaſters - - - - -	2,667,000

From whence, if you deduct the expences above mentioned, it will appear that the king of Portugal's revenues from Rio Janeiro, amount to upwards of ten millions of our money (livres *).

* Upwards of 450,000 pounds ſterling; at 4 s. and 6 d. per dollar.

CHAP.

CHAP. VI.

Departure from Rio Janeiro: second voyage to Montevideo; damage which the Etoile receives there.

1767. July. Departure from Rio Janeiro.

THE 14th of July we weighed from Rio Janeiro, but for want of wind we were obliged to come to an anchor again in the road. We sailed on the 15th, and two days after, the frigate being a much better sailer than the Etoile, I was obliged to unrig my top-gallant masts, as our lower masts required a careful management. The winds were variable, but brisk, and the sea very high. In the night between the 19th and 20th, we lost our main-top-sail, which was carried away on its clue-lines. The 25th there was an eclipse of the sun, visible to us. I had on board my ship M. Verron, a young astronomer, who came from France in the Etoile, with a view to try, during the voyage, some methods towards finding the longitude at sea.

Eclipse of the sun.

According to our estimation of the ships place, the moment of immersion, as calculated by the astronomer, was to be on the 25th, at four hours nineteen minutes in the evening. At four hours and six minutes, a cloud prevented our seeing the sun, and when we got sight of of him again, at four hours thirty-one minutes, about an inch and a half was already eclipſed. Clouds ſucceſſively paſſed over the ſun's diſk, and let us ſee him only at very ſhort intervals, ſo that we were not able to obſerve any of the phaſes of the eclipſe, and conſequently could not conclude our longitude from it. The ſun ſet to us before the moment of apparent conjunction, and we reckoned that that of immerſion had been at four hours twenty-three minutes.

Entrance into Rio de la Plata.

On the 26th we came into ſoundings; the 28th in the morning we diſcovered the Caſtilles. This part of the coaſt is pretty high, and is to be ſeen at ten or twelve leagues diſtance. We diſcovered the entrance to a bay, which probably is the harbour where the Spaniards have a fort, and where I have been told there is very bad anchorage. The 29th we entered Rio de la Plata, and ſaw the Maldonados. We advanced but little this day and the following. Almoſt the whole night between the 30th and 31ſt we were becalmed, and ſounded conſtantly. The current ſet to the north-weſtward, which was pretty near the ſituation of the iſle of Lobos. At half an hour paſt one after midnight, having ſounded thirty-three fathoms, I thought I was very near the iſle, and gave the ſignal for caſting anchor. At half paſt three we weighed, and ſaw the iſle of Lobos in N. E. about a league and a half diſtant. The wind

wind was S. and S. E. weak at firſt, but blew more freſh towards ſun-riſing, and we anchored in the bay of Montevideo the 31ſt in the afternoon. We had loſt much time on account of the Etoile; becauſe, beſides the advantage of our being better ſailers, that ſtore-ſhip, which at leaving Rio Janeiro made four inches of water every hour, after a few days ſail made ſeven inches in the ſame ſpace of time, which did not allow her to crowd her ſails.

Second time of touching at Montevideo.

News which we hear at this place.

We were hardly moored, when an officer came on board, being ſent by the governor of Montevideo, to compliment us on our arrival, and informed us that orders had been received from Spain to arreſt all the Jeſuits, and to ſeize their effects: that the ſhip which brought theſe diſpatches had carried away forty fathers of that community, deſtined for the miſſions: that the order had already been executed in the principal houſes without any difficulty or reſiſtance; and that, on the contrary, theſe fathers bore their diſgrace with reſignation and moderation. I ſhall ſoon enter into a more circumſtantial account of this great tranſaction, of which I have been able to obtain full information, by my long ſtay at Buenos Ayres, and the confidence with which the governor-general Don Franciſco Bukarely * honoured me.

* Buccarelli.

1767. August.

As we were to ſtay in Rio de la Plata till after the equinox, we took lodgings at Montevideo, where we ſettled our workmen, and made an hoſpital. This having been our firſt care, I went to Buenos Ayres, on the 11th of Auguſt, to accelerate our being furniſhed with the neceſſary proviſions, by the provider-general of the king of Spain; at the ſame price as he had agreed to deliver them to his Catholic Majeſty. I likewiſe wanted to have a conference with M. de Buccarelli, on the ſubject of what had happened at Rio-Janeiro; though I had already, by expreſs, ſent him the diſpatches from Don Franciſco de Madina. I found he had prudently reſolved to content himſelf with ſending an account of the hoſtilities of the viceroy of the Braſils to Europe, and not to make any repriſals. It would have been eaſy to him, to have taken the colony of Santo Sacramento in a few days; eſpecially as that place was in want of every neceſſary, and had not yet obtained, in November, the convoy of articles and ammunition that were preparing to be ſent thither, when we left Rio-Janeiro.

The governor-general made every thing as convenient as poſſible, towards quickly making up our wants. At the end of Auguſt, two ſchooners, laden with biſcuit and flour for us, ſailed for Montevideo; whither I likewiſe went to celebrate the day of St. Louis. I left the chevalier du Bouchage, an under lieutenant, at Buenos Ayres, in order to get the remainder of our provisions on board; and to take care of our affairs there till our departure; which, I hoped, would be towards the end of September. I could not foresee that an accident would detain us six weeks longer. In a hurricane, blowing hard at S. W. the San Fernando, a register-ship, which was at anchor near the Etoile, dragged her anchors, ran foul of the Etoile at night; and, at the first shock, broke her bowsprit level with the deck. Afterwards the knee and rails of her head were carried away; and it was lucky that they separated, notwithstanding the bad weather, and the obscurity of the night, without being more damaged.

Damage which the Etoile receives.

1767. September.

This accident greatly enlarged the leaks in the Etoile, which she had had from the beginning of her voyage. It now became absolutely necessary to unload this vessel, if not to heave her down *, in order to discover and stop this leak, which seemed to lie very low, and very forward. This operation could not be performed at Montevideo; where, besides, there was not timber sufficient to repair the masts; I therefore wrote to the chevalier du Bouchage, to represent our situation to the marquis de Buccarelli; and to obtain, that by his leave the Etoile might be allowed to come up the river, and to go into

* Virer en quille.

to

the Encenada de Baragan; I likewise gave him orders to send timber and the other materials, which we should want thither. The governor-general consented to our demands; and, the 7th of September, not being able to find any pilot, I went on board the Etoile, with the carpenters and caulkers of the Boudeuse, in order to sail the next morning, and undertake in person a navigation, which we were told was very hazardous. Two register-ships; the San-Fernando and the Carmen, provided with a pilot, were ready the same day, to sail for Montevideo to Encenada; and I intended to follow them; but the San-Fernando, which had got the pilot, named Philip, on board, weighed in the night, between the seventh and eighth, purely with a view of hiding his track from us; and left her companion in the same distress. However, we sailed on the eighth in the morning, preceded by our canoes; the Carman remaining to wait for a schooner to direct her route. In the evening we reached the San-Fernando, passed by her; and, on the tenth in the afternoon, we came to an anchor in the road of the Encenada: Philip, who was a bad pilot, and a wicked fellow, always steering in our water.

Navigation from Montevido to Baragan.

In this road I found the Venus frigate of twenty-six guns, and some merchant-ships; which were bound, together with her, to sail directly for Europe. I likewise found there la Esmeralda, and la Liebre; who were

 pre-

preparing to return to the Malouines, with provisions and ammunitions of all sorts; from whence they were to sail for the South Seas, in order to take in the Jesuits of Chili and Peru. There was likewise the xebeck * el Andaluz; which arrived from Ferrol, at the end of July, in company with another xebeck, named el Aventurero; but the latter was lost on the point of what is called the English-Sand; and the crew had time to save their lives. The Andaluz was preparing to carry presents and missionaries to the inhabitants of Tierra del Fuego; the king of Spain being desirous of testifying his gratitude to those people, for the services they rendered the Spaniards of the ship la Concepcion, which was lost on their coasts in 1765.

The Etoile oes to be eparied here.

I went on shore at Baragan, whither the chevalier du Bouchage had already sent part of the timber we wanted. He found it very difficult and expensive to collect it at Buenos Ayres, in the king's arsenal, and in some private timber-yards; the stores of both consisting of the timbers of such ships as were wrecked in the river. At Baragan we found no supplies; but, on the contrary, difficulties of many kinds; and every thing conspired to make all operations go on very slowly. The Encenada de Baragan is, indeed, merely a bad kind of bay, formed by the mouth of a little river,

* Chambekin.

which

which is about a quarter of a league broad; but the depth of water is only in the middle, in a narrow channel; which is conſtantly filling more and more; and, in which, only ſhips drawing no more than twelve feet water can enter. In all the other parts of the river, there is not ſix inches of water during the ebb; but as the tides are irregular in Rio de la Plata; and the water ſometimes high or low, for eight days together, according to the winds that blow, the landing of boats was connected with great difficulties. There are no magazines on ſhore; the houſes, or rather huts, are but few, made of ruſhes, covered over with leather, and built without any regularity, on a barren ſoil; and their inhabitants are hardly able to get their ſubſiſtance; all which cauſes ſtill more difficulties. The ſhips, which draw too much water to be able to enter this creek, muſt anchor at the point of Lara, a league and a half weſt. There they are expoſed to all the winds; but the ground being very good for anchoring, they may winter there, though labouring under many inconveniences.

I left M. de la Giraudais, at the point of Lara, to take care of what related to his ſhip; and I went to Buenos Ayres, from whence I ſent him a large ſchooner, by which he might heave down as ſoon as he came into the Encenada. For that purpoſe, it was neceſſary to unload part of the goods ſhe had on board; and M. de Bucca- 1767. October

 relli

relli gave us leave to deposit them on board the Esmeralda and the Liebre. The 8th of October the Etoile was able to go into port; and it appeared, that her repair would not take so much time as was at first expected. Indeed, they had hardly begun to unload her, when her leak diminished considerably; and she did not leak at all when she drew only eight feet of water forward. After taking up some planks of her sheathing, they saw that the seam of her entrance was entirely without oakum for the length of four feet and a half, from the depth of eight feet of her draught upwards. They discovered likewise two auger holes, into which they had not put the bolts. All these faults and damages being quickly repaired, new railing put on the head, a new bowsprit made and rigged, and the ship being new caulked all over, she returned to the point of Lara on the 21st, where she took in her lading again, from on board the Spanish frigates. In that road she likewise stowed the wood, flour, biscuit, and different provisions I sent her.

Departure of several vessels for Europe, and arrival of others.

From thence, the Venus and four other vessels laden with leather, sailed for Cadiz, at the end of September, having on board two hundred and fifty Jesuits, and the French families from the Malouines, seven excepted, who having no room in these ships, were obliged to wait for another opportunity. The marquiss of Buc-

carelli

carelli tranſported them to Buenos Ayres, where he provided them with ſubſiſtence and lodgings. At the ſame time we got intelligence of the arrival of the Diamante, a regiſter ſhip, bound for Buenos Ayres, and of the San-miguel, another regiſter ſhip, bound for Lima. The ſituation of the laſt ſhip was very diſtreſſing: after ſtruggling with the winds at Cape Horn during forty-five days, thirty-nine men of her crew being dead, and the others attacked by the ſcurvy, and a ſea carrying away her rudder, ſhe was obliged to bear away for this river, and arrived at the port of Maldonados ſeven months after leaving Cadiz, having no more than three ſailors and a few officers that were able to do duty. At the requeſt of the Spaniards we ſent an officer with ſome ſailors to bring her into the port of Montevideo. On the fifth of October the Spaniſh frigate la Aguila arrived there, having left Ferrol in March. She touched at the iſle of St. Catherine, and the Portugueſe had arreſted her there at the ſame time that they ſtopped the Diligent at Rio Janeiro.

CHAP.

CHAP. VII.

Accounts of the miſſions in Paraguay, and the expulſion of the Jeſuits from that province.

WHILST we carried on our preparations for leaving Rio de la Plata, the marquiſs of Buccarelli made ſome on his part to go on the Uraguai. The Jeſuits had already been arreſted in all the other provinces of his department; and this governor-general intended to execute the orders of his catholic majeſty, in perſon, in the miſſions. It depended upon the firſt ſteps that were taken, either to make the people conſent to the alterations that were going to be made, or to plunge them again into their former ſtate of barbariſm. But before I give an account of what I have ſeen of the cataſtrophe of this ſingular government, I muſt ſpeak ſomething of its origin, progreſs, and form. I ſhall ſpeak of it *ſine irâ & ſtudio, quorum cauſas procul habeo.*

[S]tate of the [es]tabliſhment [o]f the miſ[si]ons.

In 1580 the Jeſuits were firſt admitted into theſe fertile regions, where they have afterwards, in the reign of Philip the third, founded the famous miſſions, which in Europe go by the name of Paraguay, and in America, with more propriety, by that of Uraguay, from

from the river of that name, on which they are situated. They were always divided into colonies, which at first were weak and few, but by gradual progress have been encreased to the number of thirty-seven, viz. twenty-nine on the right side of the Uraguay, and eight on the left side, each of them governed by two Jesuits, in the habit of the order. Two motives, which sovereigns are allowed to combine, if they do not hurt each other, namely, religion and interest, made the Spanish monarch desirous of the conversion of the Indians; by making them catholics, they became civilized, and he obtained possession of a vast and abundant country; this was opening a new source of riches for the metropolis, and at the same time making proselytes to the true Deity. The Jesuits undertook to fulfil these projects; but they represented, that in order to facilitate the success of so difficult an enterprize, it was necessary they should be independent of the governors of the province, and that even no Spaniard should be allowed to come into the country.

Conditions agreed on between the court of Spain and the Jesuits.

The motive on which this demand was grounded, was, the fear lest the vices of the Europeans should diminish the ardour of their proselytes, or even remove them farther from Christianity; and likewise lest the Spanish haughtiness should render a yoke, already too heavy, insupportable to them. The court of Spain, approving

proving of theſe reaſons, ordered that the miſſionaries ſhould not be controuled by the governor's authority, and that they ſhould get ſixty thouſand piaſtres a year from the royal treaſure, for the expences of cultivation, on condition that as the colonies ſhould be formed, and the lands be cultivated, the Indians ſhould annually pay a piaſtre per head to the king, from the age of eighteen to ſixty. It was likewiſe ſtipulated, that the miſſionaries ſhould teach the Indians the Spaniſh language; but this clauſe it ſeems has not been executed.

Zeal and ſucceſs of the miſſionaries.

The Jeſuits entered upon this carrier with the courage of martyrs, and the patience of angels. Both theſe qualifications were requiſite to attract, retain, and uſe to obedience and labour, a race of ſavage, inconſtant men, who were attached to their indolence and independence. The obſtacles were infinite, the difficulties encreaſed at each ſtep; but zeal got the better of every thing, and the kindneſs of the miſſionaries at laſt brought theſe wild, diffident inhabitants of the woods, to their feet. They collected them into fixed habitations, gave them laws, introduced uſeful and polite arts among them; and, in ſhort, of a barbarous nation, without civilized manners, and without religious principles, thev made a good-natured well governed people, who ſtrictly obſerved the Chriſtian ceremonies. Theſe Indians, charmed with the perſuaſive eloquence of their apoſtles, willingly

 obeyed

obeyed a ſet of men, who, they ſaw would ſacrifice themſelves for their happineſs; accordingly, when they wanted to form an idea of the king of Spain, they repreſented him to themſelves in the habit of the order of St. Ignatius.

Revolt of the Indians againſt the Spaniards.

However, there was a momentary revolt againſt his authority in the year 1757. The catholic king had exchanged the colonies on the left ſhore of the Uraguay againſt the colony of Santo Sacramento with the Portugueſe. The deſire of deſtroying the ſmuggling trade, which we have mentioned ſeveral times, had engaged the court of Madrid to this exchange. Thus the Uraguay became the boundary of the reſpective poſſeſſions of the two crowns. The Indians of the colonies, which had been ceded, were tranſported to the right hand ſhore, and they made them amends in money for their loſt labour and tranſpoſition. But theſe men, accuſtomed to their habitations, could not bear the thought of being obliged to leave the grounds, which were highly cultivated, in order to clear new ones. They took up arms: for long ago they had been allowed the uſe of them, to defend themſelves from the incurſions of the Pauliſts, a band of robbers, deſcended from Braſilians, and who had formed themſelves into a republic towards the end of the ſixteenth century. They revolted without any Jeſuits ever heading them. It is however

Cauſes of their diſcontent.

 ſaid,

ſaid, they were really kept in the revolted villages, to exerciſe their ſacerdotal functions.

They take up arms and are defeated.

The governor-general of the province de la Plata, Don Joſeph Andonaighi, marched againſt the rebels, and was followed by Don Joachim de Viana, governor of Montevideo. He defeated them in a battle, wherein upwards of two thouſand Indians were ſlain. He then proceeded to conquer the country; and Don Joachim ſeeing what terror their firſt defeat had ſpread amongſt them, reſolved to ſubdue them entirely with ſix hundred men. He attacked the firſt colony, took poſſeſſion of it without meeting any reſiſtance; and that being taken, all the others ſubmitted.

The diſturbances are appeaſed.

At this time the court of Spain recalled Don Joſeph Andonaighi, and Don Pedro Cevallos arrived at Buenos Ayres to replace him. Viana received orders at the ſame time to leave the miſſions, and bring back his troops. The intended exchange was now no longer thought of, and the Portugueſe, who had marched againſt the Indians with the Spaniards, returned with them likewiſe. At the time of this expedition, the noiſe was ſpread in Europe of the election of king Nicholas, an Indian, whom indeed the rebels ſet up as a phantom of royalty.

Don Joachim de Viana told me, that when he received orders to leave the miſſions, a great number of Indians,

The Indians appear disgusted with the administration of th Jesuits.

Indians, diſcontented with the life they led, were willing to follow him. He oppoſed it, but could not hinder ſeven families from accompanying him; he ſettled them at the Maldonados, where, at preſent, they are patterns of induſtry and labour. I was ſurpriſed at what he told me concerning this diſcontent of the Indians. How is it poſſible to make it agree with all I had read of the manner in which they are governed? I ſhould have quoted the laws of the miſſions as a pattern of an adminiſtration inſtituted with a view to diſtribute happineſs and wiſdom among men.

Indeed, if one caſts a general view at a diſtance upon this magic government, founded by ſpiritual arms only, and united only by the charms of perſuaſion, what inſtitution can be more honourable to human nature? It is a ſociety which inhabits a fertile land, in a happy climate, of which, all the members are laborious, and none works for himſelf; the produce of the common cultivation is faithfully conveyed into public ſtore-houſes, from whence every one receives what he wants for his nouriſhment, dreſs, and houſe-keeping; the man who is in full vigour, feeds, by his labour, the new-born infant; and when time has conſumed his ſtrength, his fellow-citizens render him the ſame ſervices which he did them before. The private houſes are convenient, the public buildings fine; the worſhip uniform

uniform and ſcrupulouſly attended: this happy people knows neither the diſtinction of rank, nor of nobility, and is equally ſheltered againſt ſuper-abundance and wants.

The great diſtance and the illuſion of perſpective made the miſſions bear this aſpect in my eyes, and muſt have appeared the ſame to every one elſe. But the theory is widely different from the execution of this plan of government. Of this I was convinced by the following accounts, which above a hundred ocular witneſſes have unanimouſly given me.

Accounts of the interior government.

The extent of country in which the miſſions are ſituated, contains about two hundred leagues north and ſouth, and about one hundred and fifty eaſt and weſt, and the number of inhabitants is about three hundred thouſand; the immenſe foreſts afford wood of all ſorts; the vaſt paſtures there, contain at leaſt two millions of cattle; fine rivers enliven the interior parts of this country, and promote circulation and commerce throughout it. This is the ſituation of the country, but the queſtion now is, how did the people live there? The country was, as has been told, divided into pariſhes, and each pariſh was directed by two Jeſuits, of which, one was rector, and the other his curate. The whole expence for the maintenance of the colonies was but ſmall, the Indians being fed, dreſſed, and lodged, by the

the labour of their own hands; the greatest costs were those of keeping the churches in repair, all which were built and adorned magnificently. The other products of the ground, and all the cattle, belonged to the Jesuits, who, on their part, sent for the instruments of various trades, for glass, knives, needles, images, chaplets of beads, gun-powder and muskets. Their annual revenues consisted in cotton, tallow, leather, honey, and above all, in *maté*, a plant better known by the name of Paraguay tea, or South-Sea tea, of which that company had the exclusive commerce, and of which likewise the consumption is immense in the Spanish possessions in America, where it is used instead of tea.

The Indians shewed so servile a submission to their rectors, that not only both men and women suffered the punishment of flagellation, after the manner of the college, for public offences, but they likewise came of themselves to sollicit this chastisement for mental faults. In every parish the fathers annually elected *corrégidors*, and their assistants, to take care of the minutiæ of the government. The ceremony of their election was performed on new year's day, with great pomp, in the court before the church, and was announced by ringing of bells, and the playing of a band of music. The newly elected persons came to the feet of the father rector to receive the marks of their dignity, which however,

however did not exempt them from being whipped like the others. Their greateſt diſtinction was that of wearing habits, whereas, a ſhirt of cotton ſtuff was the only dreſs of the other Indians of both ſexes. The feaſts of the pariſh, and that of the rector, were likewiſe celebrated by public rejoicings, and even by comedies, which probably reſembled thoſe ancient pieces of ours, called *myſtéres* or myſteries.

The rector lived in a great houſe near the church; adjoining to it were two buildings, in one of which were the ſchools for muſic, painting, ſculpture, and architecture; and likewiſe, work-houſes of different trades; Italy furniſhed them with maſters to teach the arts, and the Indians, it is ſaid, learn with facility: the other building contained a great number of young girls at work in ſeveral occupations, under the inſpection of old women: this was named the *guatiguaſu*, or the ſeminary. The apartment of the rector communicated internally with theſe two buildings.

This rector got up at five o'clock in the morning, employed an hour in holy meditation, and ſaid his maſs at half paſt ſix o'clock; they kiſſed his hands at ſeven o'clock, and then he publicly diſtributed an ounce of *maté* to every family. After maſs, the rector breakfaſted, ſaid his breviary, conferred with the corregidors, four of whom were his miniſters, and viſited the ſeminary,

the

the ſchools, and the work-ſhops. Whenever he went out, it was on horſeback, and attended by a great retinue; he dined alone with his curate at eleven of the clock, then chatted till noon, and after that, made a *ſieſta* till two in the afternoon; he kept cloſe in his interior apartments till it was prayer time, after which, he continued in converſation till ſeven in the evening; then the rector ſupped, and at eight he was ſuppoſed to be gone to bed.

From eight of the clock in the morning, the time of the people was taken up either in cultivating the ground, or in their work-ſhops, and the corregidors took care to ſee them employ their time well; the women ſpun cotton; they got a quantity of it every Monday, which they were obliged to bring back converted into ſpun yarn at the end of the week; at half an hour paſt five in the evening they came together to ſay the prayers of their roſary, and to kiſs the hands of their rector once more, then came on the diſtribution of an ounce of *maté* and four pounds of beef for each family, which was ſuppoſed to conſiſt of eight perſons; at the ſame time they likewiſe got ſome maize. On Sundays they did no work; the divine worſhip took up more time; they were after that allowed to amuſe themſelves with plays as dull as the reſt of their whole life.

Consequences drawn from it.

From this exact detail it appears that the Indians had in some manner no property, and that they were subject to a miserable, tedious uniformity of labour and repose. This tiresomeness, which may with great reason be called deadly or extreme, is sufficient to explain what has been told to us, that they quitted life without regret, and died without having ever lived or enjoyed life. When once they fell sick, it seldom happened that they recovered, and being then asked whether they were sorry to be obliged to die, they answered, no; and spoke it as people whose real sentiments coincide with their words. We can no longer be surprised, that when the Spaniards penetrated into the missions, this great people, which was governed like a convent, should shew an ardent desire of forcing the walls which confined them. The Jesuits represented the Indians, upon the whole, as men incapable of attaining a higher degree of knowledge than that of children; but the life they led, prevented these grown children from having the liveliness of little ones.

Expulsion of the Jesuits from the province of Plata.

The society were occupied with the care of extending their missions, when the unfortunate events happened in Europe, which overturned the work of so many years, and of so unwearied patience in the new world. The court of Spain having resolved upon the expulsion of the Jesuits, was desirous that this might be done

done at the ſame time throughout all its vaſt dominions. Cevallos was recalled from Buenos Ayres, and Don Franciſco Buccarelli appointed to ſucceed him. He ſet out, being inſtructed in the buſineſs which he was intended for, and with orders to defer the execution of it till he received freſh orders, which would ſoon be ſent him. The king's confeſſor, the count d'Aranda, and ſome miniſters, were the only perſons to whom this ſecret affair was entruſted. Buccarelli made his entry at Buenos Ayres in the beginning of 1767.

Meaſures taken at the court of Spain for thi[s] purpoſe.

When Don Pedro de Cevallos was arrived in Spain, a packet was diſpatched to the marquis of Buccarelli, with orders both for that province, and for Chili, whither he was to ſend them over land. This veſſel arrived in Rio de la Plata in June, 1767, and the governor inſtantly diſpatched two officers, one to Peru, and the other to Chili, with the diſpatches from court, directed to them. He then ſent his orders into the various parts of his province, where there were any Jeſuits, viz. to Cordoua, Mendoza, Corrientes, Santa-Fé, Salta, Montevideo, and Paraguay. As he feared, that among the commanders of theſe ſeveral places, ſome might not act with the diſpatch, ſecrecy, and exactneſs which the court required, he enjoined, by ſending his orders to them, that they ſhould not open them till on a certain day, which he had fixed for the execution, and to do it only

Meaſures taken by th[e] governor-g[e]neral of th[e] province.

 in

in the preſence of ſome perſons, whom he named, and who ſerved in the higheſt eccleſiaſtical and civil offices, at the above mentioned places. Cordoua, above all, intereſted his attention. In that province was the principal houſe of the Jeſuits, and the general reſidence of their provincial. There they prepared and inſtructed in the Indian language and cuſtoms, thoſe who were deſtined to go to the miſſions, and to become heads of colonies; there their moſt important papers were expected to be found. M. de Buccarelli reſolved to ſend an officer of truſt there, whom he appointed the king's lieutenant of that place, and on whom, under this pretext, he ſent a detachment of ſoldiers to attend.

It now remained to provide for the execution of the king's orders in the miſſions, and this was the moſt critical point. It was dubious whether the Indians would ſuffer the Jeſuits to be arreſted in the midſt of the colonies, and this violent ſtep muſt at all events have been ſupported by a numerous body of troops. Beſides this, it was neceſſary, before they thought of removing the Jeſuits, to have another form of government ready to ſubſtitute in their ſtead, and by that means to prevent confuſion and anarchy. The governor reſolved to temporize, and was contented at that time to write to the miſſions, that a corregidor and a cacique from each colony ſhould be ſent to him immediately, in order to communicate the king's letters to them. He dispatched this order with the greatest quickness, that the Indians might already be on the road, and beyond the missions, before the news of the expulsion of the Jesuits could reach thither. By this he had two aims in view; the one, that of getting hostages of the fidelity of the colonies, when the Jesuits would be taken from thence; the other, that of gaining the affection of the principal Indians, by the good treatment he intended for them at Buenos Ayres, and of instructing them in the new situation upon which they would enter; for, as soon as the restraint would be taken away, they were to enjoy the same privileges, and have the same property as the king's other subjects.

The secret near being divulged b[y] an unforese[en] accident.

Every measure was concerted with the greatest secrecy, and though people wondered that a vessel should arrive from Spain without any other letters than those for the general, yet they were very far from suspecting the cause of it. The moment of the general execution was fixed to the day when all the couriers were supposed to have arrived at their different destinations, and the governor waited for that moment with impatience, when the arrival of the two xebecs * of the king from Cadiz, the Andaluz and the Adventurero, was near making all these precautions useless. The governor-

* Chambekins.

general

general had ordered the governor of Montevideo, that in caſe any veſſels ſhould arrive from Europe, he ſhould not allow them to ſpeak with any perſon whatſoever, before he had ſent him word of it; but one of the two xebecs being in the forlorn ſituation we have before mentioned, at the entrance of the river, it was very neceſſary to ſave the crew of it, and give her all the aſſiſtance which her ſituation required.

The two xebecs had ſailed from Spain, after the Jeſuits had been arreſted there, and this piece of news could by no means be prevented from ſpreading. An officer of theſe ſhips was immediately ſent to M. de Buccarelli, and arrived at Buenos Ayres the 9th of July, at ten in the evening. The governor did not loſe time, he inſtantly diſpatched orders to all the commanders of the places, to open their former packets of diſpatches, and execute their contents with the utmoſt celerity. At two of the clock after midnight, all the couriers were gone, and the two houſes of the Jeſuits at Buenos Ayres inveſted, to the great aſtoniſhment of thoſe fathers, who thought they were dreaming, when rouſed from their ſleep in order to be impriſoned, and to have their papers ſeized. The next morning an order was publiſhed in the town, which forbade, by pain of death, to keep up any intercourſe with the Jeſuits, and five merchants were arreſted, who intended, it is ſaid, to ſend advices to them at Cordoua.

Conduct of the governor-general.

The

The Jesuits are arrested in all the Spanish towns.

The king's orders were executed with the same facility in all the towns. The Jesuits were surprised every where, without having the least notice, and their papers were seized. They were immediately sent from their houses, guarded by detachments of soldiers, who were ordered to fire upon those that should endeavour to escape. But there was no occasion to come to this extremity. They shewed the greatest resignation, humbling themselves under the hand that smote them, and acknowledging, as they said, that their sins had deserved the punishment which God inflicted on them. The Jesuits of Cordoua, in number above a hundred, arrived towards the end of August, at the Encenada, whither those from Corrientes, Buenos Ayres, and Montevideo, came soon after. They were immediately embarked, and the first convoy sailed, as I have already said, at the end of September. The others, during that time, were on the road to Buenos Ayres, where they should wait for another opportunity.

Arrival of the caciques and corregidors at Buenos Ayres from the missions.

On the 13th of September arrived all the corregidors, and a cacique of each colony, with some Indians of their retinue. They had left the missions before any one guessed at the reason of their journey there. The news which they received of it on the road had made some impression on them, but did not prevent their continuing the journey. The only instruction which the rectors

rectors gave their dear proselytes at parting, was, to believe nothing of what the governor-general should tell them: "Prepare, my children," did every one tell them, "to hear many untruths." At their arrival, they were immediately sent to the governor, where I was present at their reception. They entered on horseback to the number of a hundred and twenty, and formed a crescent in two lines; a Spaniard understanding the language of the *Guaranis*, served them as an interpreter. The governor appeared in a balcony; he told them, that they were welcome; that they should go to rest themselves, and that he would send them notice of the day which he should fix in order to let them know the king's intentions. He added, in general, that he was come to release them from slavery, and put them in possession of their property, which they had not hitherto enjoyed. They answered by a general cry, lifting up their right hands to heaven, and wishing all prosperity to the king and governor. They did not seem discontented, but it was easy to discover more surprize than joy in their countenance. On leaving the governor's palace, they were brought to one of the houses of the Jesuits, where they were lodged, fed, and kept at the king's expence. The governor, when he sent for them, expresly mentioned the famous Cacique Nicholas, but they wrote him word, that his great age and his infirmities did not allow him to come out.

They appear before the governor-general.

At

At my departure from Buenos Ayres, the Indians had not yet been called to an audience of the general. He was willing to give them time to learn ſomething of the language, and to become acquainted with the Spaniſh cuſtoms. I have been ſeveral times to ſee them. They appeared to me of an indolent temper, and ſeemed to have that ſtupid air ſo common in creatures caught in a trap. Some of them were pointed out to me as very intelligent, but as they ſpoke no other language but that of the Guaranis, I was not able to make any eſtimate of the degree of their knowledge; I only heard a cacique play upon the violin, who, I was told, was a great muſician; he played a ſonata, and I thought I heard the ſtrained ſounds of a ſerinette. Soon after the arrival of theſe Indians at Buenos Ayres, the news of the expulſion of the Jeſuits having reached the miſſions, the marquis de Bucarelli received a letter from the provincial, who was there at that time, in which he aſſured him of his ſubmiſſion, and of that of all the colonies to the king's orders.

Extent of the miſſions

Theſe miſſions of the Guaranis and Tapes, upon the Uraguay, were not the only ones which the Jeſuits founded in South America. Somewhat more northward they had collected and ſubmitted to the ſame laws, the Mojos, Chiquitos, and the Avipones. They likewiſe were making progreſſes in the ſouth of Chili, towards the isle of Chiloé; and a few years since, they have opened themselves a road from that province to Peru, passing through the country of the Chiquitos, which is a shorter way than that which was followed till then. In all the countries into which they penetrated, they erected posts, on which they placed their motto; and on the map of their colonies, which they have settled, the latter are placed under the denomination of *Oppida Christianorum.*

It was expected, that in seizing the effects of the Jesuits in this province, very considerable sums of money would be found: however, what was obtained that way, amounted to a mere trifle. Their magazines indeed were furnished with merchandizes of all sorts, both of the products of the country, and of goods imported from Europe. There were even many sorts which could not have a sale in these provinces. The number of their slaves was considerable, and in their house at Cordoua alone, they reckoned three thousand five hundred.

I cannot enter into a detail of all that the public of Buenos Ayres pretends to have found in the papers of the Jesuits; the animosity is yet too recent to enable me to distinguish true imputations from false ones. I will rather do justice to the majority of the members of this society, who were not interested in its temporal affairs. If there were some intriguing men in this body,

body, the far greater number, who were sincerely pious, did not consider any thing in the institution, besides the piety of its founder, and worshipped God, to whom they had consecrated themselves, in spirit and in truth. I have been informed, on my return to France, that the marquis de Bucarelli set out from Buenos Ayres for the missions, the 14th of May, 1768; and that he had not met with any obstacle, or resistance, to the execution of his most catholic majesty's orders. My readers will be able to form an idea of the manner in which this interesting event was terminated, by reading the two following pieces, which contain an account of the first scene. It is a narrative of what happened at the colony of Yapegu, situated upon the Uraguay, and which lay the first in the Spanish general's way; all the others have followed the example of this.

Translation of a letter from a captain of the grenadiers of the regiment of Majorca, commanding one of the detachments of the expedition into Paraguay.

Yapegu, the 19 h July, 1768.

Account of the governor general's entry into the missions.

" YESTERDAY we arrived here very happily; the re-
" ception given to our general has been most magni-
" ficent, and such as could not be expected from so
" simple a people, so little accustomed to shows.
" Here is a college, which has very rich and numerous

" church

" church ornaments; there is likewife a great quantity " of plate. The fettlement is fomewhat lefs than Mon- " tevideo, but more regularly difpofed, and well peopled. " The houfes are fo uniform, that after feeing one, you " have feen them all; and the fame, after you have feen " one man and woman, you have feen them all, there " being not the leaft difference in the manner in which " they are dreffed. There are many muficians, but they " are only middling performers.

" As foon as we arrived near this miffion, the go- " vernor-general gave orders to go and feize the father " provincial of the Jefuits, and fix other fathers, and to " bring them to a place of fafety. They are to embark " in a few days on the river Uraguay. However, we " believe they will ftay at Salto, in order to wait till the " reft of their brethren have undergone the fame fate. " We expected to make a ftay of five or fix days at Ya- " pegu, and then to continue our march to the laft " miffion. We are very well pleafed with our general, " who has procured us all poffible refrefhments. Yefter- " day we had an opera, and fhall have another repre- " fentation of it to-day. The good people do all they " can, and all they know.

" Yefterday we likewife faw the famous Nicolas, " the fame whom people were fo defirous to confine. " He was in a deplorable fituation, and almoft naked.

" He

" He is ſeventy years of age, and ſeems to be a very " ſenſible man. His excellency ſpoke with him a long " time, and ſeemed very much pleaſed with his con- " verſation.

" This is all the news I can inform you of."

Relation publiſhed at Buenos Ayres of the entry of his excellency Don Franciſco Bucarelli y Urſua, in the miſſion of Yapegu, one of thoſe belonging to the Jeſuits, among the nations of Guaranis, on his arrival there the 18th of July, 1768.

" At eight o'clock in the morning, his excellency " went out of the chapel of St. Martin, at one league's " diſtance from Yapegu. He was accompanied by his " guard of grenadiers and dragoons, and had detached " two hours before the companies of grenadiers of " Majorca, in order to take poſſeſſion of, and get ready " every thing at the river of Guavirade, which muſt be " croſſed in canoes and ferries. This rivulet is about " half a league from the colony.

" As ſoon as his excellency had croſſed the rivulet, " he found the caciques and corregidors of the miſſions, " who attended with the Alferes of Yapegu, bearing the " royal ſtandard. His excellency having received all " the honours and compliments uſual on ſuch occa- " ſions, got on horſeback, in order to make his public " entry

"The dragoons began the march; they were follow-"ed by two adjutants, who preceded his excellency af-"ter whom came the two companies of grenadiers of Ma-"jorca, followed by the retinue of the Caciques and "Corregidores, and by a great number of horsemen from "these parts.

"They went to the great place facing the church. His "excellency having alighted, Don Francisco Martinez, "chaplain of the expedition, attended on the steps be-"fore the porch to receive him; he accompanied him "to the *Presbyterium*, and began the *Te Deum*; which was "sung and performed by musicians, entirely consisting "of guaranis. During this ceremony, there was a triple "discharge of the artillery. His excellency went after-"wards to the lodgings, which he had chosen for him-"self, in the college of the fathers; round which the "whole troop encamped, till, by his order, they went "to take their quarters in the *Guatiguasa*, or *la Casa de las* "*recogidas*, house of retirement for women *."

Let

* The Jesuits in Paraguay have been so much the object of private conversation, and of public contest, that it is a wonder the public is still at a loss, in regard to the real situation of their affairs. The account published here by Mr Bougainville, must, no doubt, greatly contribute to throw a light on the transactions in Paraguay, of which so little is known with any degree of certainty. A few remarks taken from the ingenious *Marquis de Pau's Recherches sur les Americains*, will, we hope, not be disagreeable to the readers.

In the year 1731, the Audienca of Chuquisaca, in the province of *las Charcas* found it necessary to empower the *Protector of the Indians*, i. e. the solicitor general for them, and a member of their body, to visit the famous Paraguay missions, and to inquire into the truth of the various unfavourable reports spread about them

Don

Let us now continue the account of our voyage; in which the detail of the revolution that happened in the missions,

Don Joseph de Antequera, a man of abilities, great integrity, and superior courage, was then invested with the dignity of Protector of the Indians. Accompanied only by one *Alguazil Mayor*, called *Joseph de Mena*; and with the deed, impowering him with the visitation of the missions, he went with spirit on his business; and after his arrival at the city of Assumption, he acquainted the Jesuits with the commission. The reverend fathers told him, that he had taken in vain the pains of coming to their missions, where he would never get admittance; and if he should attempt to force his way, he would repent of it. Antequera did neither know the bad character of these people, nor did he fear their threats, and went therefore on his intended journey. But he was soon surrounded by a large detachment of armed Indians, with Jesuits at their head, who fell upon him; and he escaped by a sudden flight only.

The unfortunate Alguazil, being willing to encounter a German Jesuit, was dangerously wounded. The Jesuits, not contented with this inconsiderate step, accused Antequera, as an adventurer, who had attempted to assume the dignity of a king of Paraguay, at the city of Assumption; but that the reverend fathers, as faithful subjects to his Catholic Majesty, had driven him out by main force; and they requested, therefore, to be recompensed for this signal service to their sovereign.

Don Armendariz, Marquis de Castel Fuerte, thirty-third viceroy of Peru, entirely devoted to the Jesuits, sent Don Joseph de Antequera, in consequence of this accusation, immediately to a dungeon. He was examined; and though his counsellors had written five thousand sheets in his defence, he was, however, hanged for the crime of revolting against his sovereign, the fifth of June, together with his assistant Joseph de Mena, who was still very ill from the wound received at Assumption.

Lima and all Peru revolted against their viceroy, on the account of so shocking and tyrannical an action. The troops were sent to quell the riots. The blood of thousands flowed in the streets of Lima, and stained the vallies of Peru. All the men of integrity and honour at Lima, Cusco, Cuença, and Chuquisaca took up mourning for Antequera, the innocent victim of the revenge of the pious fathers, and of the despotism of the arbitrary viceroy, their tool. This transaction ruined the credit of the Jesuits in Peru.

The reverend missionaries found means to settle extensive establishments on the Uraguay, and the interior parts of Paraguay, upon the Pilco Mayo, and other rivers. They collected first, by gentle means, some of the Indian tribes into small settlements, taught them husbandry, and the most necessary arts; and afterwards, music, painting, and sculpture; all were instructed in the use of arms. By the help of these first colonies, they often forced the free rambling tribes of interior America, under the holy yoke of the gospel, and into subjection to these zealous missionaries. The poor wretches were then cloathed with a callico shirt, and got their allowance of meat,

missions, has been one of the most interesting circumstances.

meat, maize, and caamini; but they were in return obliged to drudge for the good fathers, in planting the Paraguay tea, cotton, tobacco, and sugar. Every ounce of cotton and caamini raised by these slaves must be delivered into the society's storehouses, from whence they were transported and sold for the benefit of the missionaries: those who concealed any of the above articles, got twelve lashes, in honour of the twelve apostles, and were confined to fasting during three days in the public work-house. Benedict XIV. the head of the Romish church, a man, whose humanity and extensive learning is so universally known, published two bulls against the Jesuits, wherein he excommunicates them, for the practice of enslaving the poor proselytes, and keeping them no better than animals; (whom men deprive of their liberty, and domesticate them with a view of making use of them in the most laborious employments) and for using religion as a cloak to oppression, despotism, and tyranny; in order to deprive free-born beings and their fellow-creatures of liberty, the first and most precious of all their enjoyments and privileges in this present life. These bulls will be for ever the strongest proofs of the truth of these assertions, and of the specious tyranny of the Jesuits.

The iniquitous practices in regard to the trade of the Paraguay-tea, are so well stated, that whole tribes of Indians were brought to the dilemma either to enlist as bondmen to the Jesuits, or to be starved; the complaints of so many Indian plantations of South-Sea tea destroyed by the Jesuits, were always heard, examined, and reported to the court of Spain; but the influence of the Jesuits prevented the council of the Indies from taking any steps for the punishment of the pious fathers; and they would still remain unknown and unpunished, had not this society been so suddenly involved in their ruin, by the precaution of the court of Spain. F.

CHAP.

CHAP. VIII.

Departure from Montevideo; run to Cape Virgin; entrance into the Straits; interview with the Patagonians; navigation to the isle of St. Elizabeth.

Nimborum in patriam, loca foeta furentibus austris. Virg. Æneid. Lib. 1.

The Etoile comes down from Baragan to Montevideo.

THE repair and loading of the Etoile took us up all October, and cost us a prodigious expence; we were not able to balance our accounts with the provisor-general, and the other Spaniards who had supplied our wants, till the end of this month. I paid them with the money I received, as a reimbursement for the cession of the Malouines, which I thought was preferable to a draught upon the king's treasury. I have continued to do the same in regard to all the expences, at the various places we had occasion to touch at in foreign countries. I have bought what I wanted much cheaper, and obtained it much sooner by this means

Difficulty of this navigation.

1767. November.

The 31st of October, by break of day, I joined the Etoile, some leagues from the Encenada; she having sailed from thence for Montevideo the preceding day. We anchored there on the third of November, at seven in the evening. The necessity of finding out a channel, by constant soundings, between the Ortiz sand-

 bank,

bank, and another little bank to the ſouthward of it, both of which have no beacons on them, makes this navigation ſubject to great difficulties: the low ſituation of the land to the ſouth, which therefore cannot be ſeen with eaſe, increaſes the difficulties. It is true, chance has placed a kind of beacon almoſt at the weſt point of the Ortiz bank. Theſe were the two maſts of a Portugueſe veſſel, which was loſt there, and happily ſtands upright. In the channel you meet with four, four and a half, and five fathoms of water; and the bottom is black ooze; on the extremities of the Ortiz-bank, it is red ſand. In going from Montevideo to the Encenada, as ſoon as you have made the beacon in E. by S. and have five fathoms of water, you have paſſed the banks. We have obſerved 15 deg. 30. min. N. E. variation in the channel.

Loſs of three ſailors.

This ſmall paſſage coſt us three men, who were drowned; the boat getting foul under the ſhip, which was wearing, went to the bottom; all our efforts ſufficed only to ſave two men and the boat, which had not loſt her mooring-rope. I likewiſe was ſorry to ſee, that, notwithſtanding the repairs the Etoile had undergone, ſhe ſtill made water; which made us fear that the fault lay in the caulking of the whole water-line; the ſhip had been free of water till ſhe drew thirteen feet.

Preparations for leaving Rio de la Plata.

We employed some days to stow all the victuals into the Boudeuse, which she could hold, and to caulk her over again; which was an operation, that could not be done sooner, on account of the absence of her caulkers, who had been employed in the Etoile; we likewise repaired the boat of the Etoile; cut grass for the cattle we had on board; and embarked whatever we had on shore. The tenth of November was spent in swaying up our top-masts and lower yards, and setting up our rigging, &c. We could have sailed the same day, if we had not grounded. On the 11th, the tide coming in, the ships floated, and we cast anchor at the head of the road; where vessels are always a-float. The two fol lowing days we could not sail, on account of the high sea; but this delay was not entirely useless. A schooner came from Buenos Ayres, laden with flour, and we took sixty hundred weight of it, which we made shift to stow in our ships. We had now victuals for ten months; though it is true, that the greatest part of the drink consisted of brandy. The crew was in perfect health. The long stay they made in Rio de la Plata, during which a third part of them alternately lay on shore, and the fresh meat they were always fed with, had prepared them for the fatigues and miseries of all kinds, which we were obliged to undergo. I left at

Condition the crews, our sailing from Monvideo.

 Monte-

Montevideo my pilot, my maſter-carpenter, my armourer, and a warrant-officer of my frigate; whom age and incurable infirmities prevented from undertaking the voyage. Notwithſtanding all our care, twelve men, ſoldiers and ſailors, deſerted from the two ſhips. I had, however, taken ſome of the ſailors at the Malouines, who were engaged in the fiſhery there; and likewiſe an engineer, a ſupercargo, and a ſurgeon; by this means my ſhip had as many hands as at her departure from Europe; and it was already a year ſince we had left the river of Nantes.

Departure from Montevideo.

The 14th of November, at half paſt four in the morning, wind due north, a fine breeze, we ſailed from Montevideo. At half paſt eight we were N. and S. off the iſle of Flores; and at noon twelve leagues E. and E. by S. from Montevideo; and from hence I took my point of departure in 34° 54′ 40″ S. lat. and 58° 57′ 30″ W. long. from the meridian of Paris. I have laid down the poſition of Montevideo, ſuch as M. Verron has determined it by his obſervations; which places its longitude 40′ 30″ more W. than Mr. Bellin lays it down in his chart. I had likewiſe profited of my ſtay on ſhore, to try my octant upon the diſtances of known ſtars; this inſtrument always made the altitude of every ſtar too little by two minutes; and I have always ſince attended to this correction. I muſt mention here, that

[It]s poſition [a]ſtronomical[l]y determined.

in

in all the courſe of this Journal, I give the bearings of the coaſts, ſuch as taken by the compaſs; whenever I give them corrected, according to the variations, I ſhall take care to mention it.

Soundings and navigation to the ſtraits of Magalhaens.

On the day of our departure, we ſaw land till ſun-ſet; our ſoundings conſtantly encreaſed, and changed from an oozy to a ſandy bottom; at half paſt ſix of the clock we found thirty-five fathom, and a grey ſand; and the Etoile, to whom I gave a ſignal for ſounding on the fifteenth in the afternoon, found ſixty fathom, and the ſame ground: at noon we had obſerved 36° 1 of latitude. From the 16th to the 21ſt we had contrary winds, a very high ſea, and we kept the moſt advantageous boards in tacking under our courſes and cloſe reefed top-ſails; the Etoile had ſtruck her top-gallant maſts, and we ſailed without having our's up. The 22d it blew a hard gale, accompanied with violent ſqualls and ſhowers, which continued all night; the ſea was very dreadful, and the Etoile made a ſignal of diſtreſs; we waited for her under our fore-ſail and main-ſail, the lee clue-garnet hauled up. This ſtore-ſhip ſeemed to have her fore top-ſail-yard carried away. The wind and ſea being abated the next morning, we made ſail, and the 24th I made the ſignal for the Etoile to come within hail, in order to know what ſhe had ſuffered in the laſt gale. M. de la Giraudais informed

informed me, that besides his fore top-sail yard, four of his chain plates * had likewise been carried away; he added, that all the cattle he had taken in at Montevideo, had been lost, two excepted: this misfortune we had shared with him; but this was no consolation, for we knew not when we should be able to repair this loss. During the remaining part of this month, the winds were variable, from S. W. to N. W; the currents carried us southward with much rapidity, as far as 45° of latitude, where they became insensible. We sounded for several days successively without finding ground, and it was not till the 27th at night, being in the latitude of about 47°, and, according to our reckoning, thirty-five leagues from the coast of Patagonia, that we sounded seventy fathom, oozy bottom, with a fine black and grey sand. From that day till we saw the land, we had soundings in 67, 60, 55, 50, 47, and at last forty fathom, and then we first got sight of Cape Virgins †. The bottom was sometimes oozy, but always of a fine sand, which was grey, or yellow, and sometimes mixed with small red and black gravel.

[H]idden rock [n]ot taken [n]otice of in [th]e charts.

I would not approach too near the coast till I came in latitude of 49°, on account of a sunken rock or vigie,

* Chaines de haubans.

† Cap des Vierges, called Cape Virgin Mary by Lord Anson and Sir John Narborough. F.

which

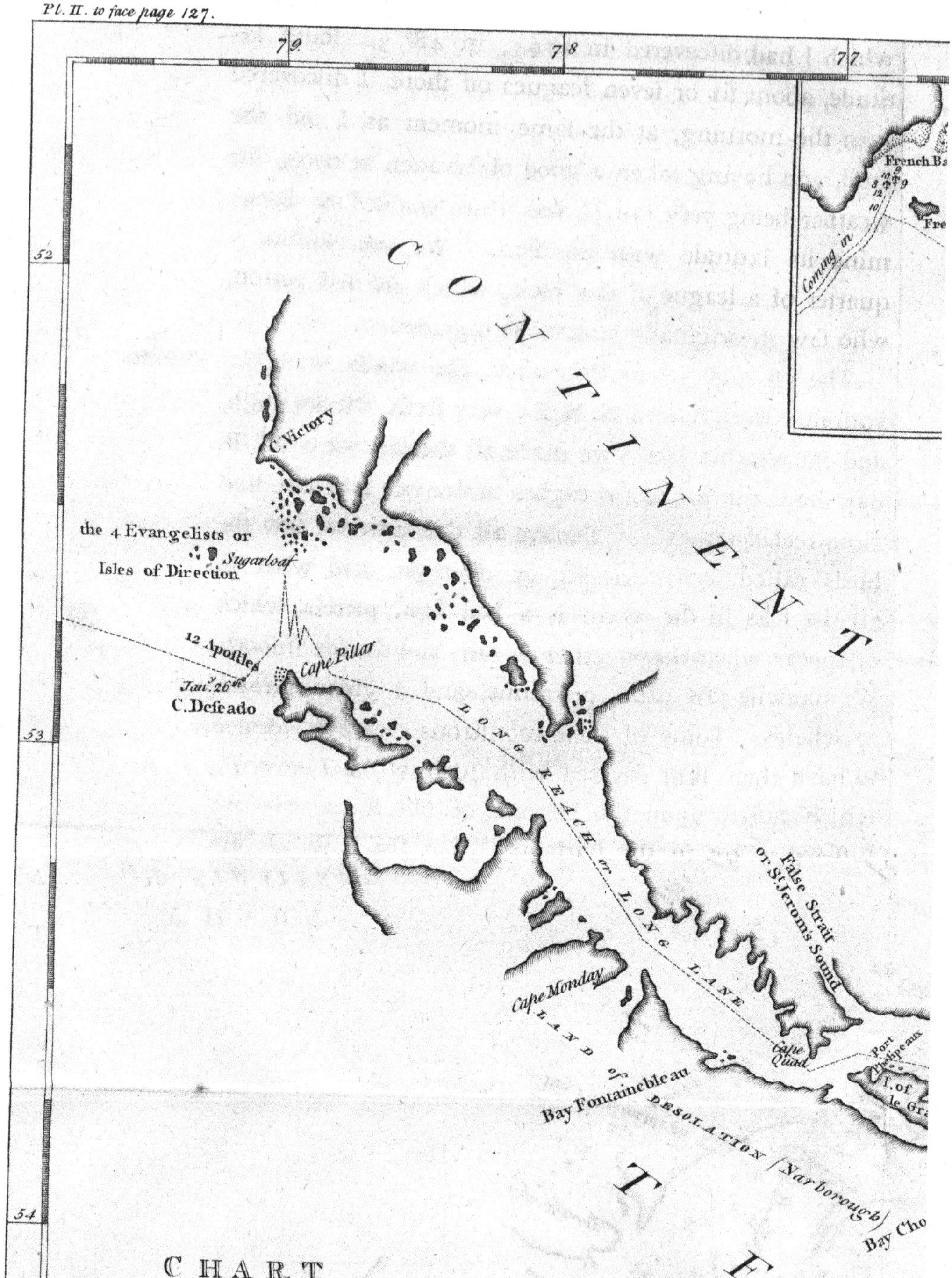

79 78 77

52

CONTINENT

C. Victory

the 4 Evangelists or Isles of Direction

Sugarloaf

12 Apostles

Jan.y 26th

C. Deseado

Cape Pillar

53

LONG REACH or LONG LANE

False Strait or St. Jerom's Sound

Cape Monday

LAND of DESOLATION (Narborough)

Cape Quad

Port Phelipeaux

I. of le Gra

Bay Fontainebleau

Bay Cho

Chann

French Ba

Fre

Coming in

54

TERR

Interspersed with

CHART of the STRAITS of MAGALHAENS or MAGELLAN, with the Track of the Boudeuse & Etoile.

79 78 77

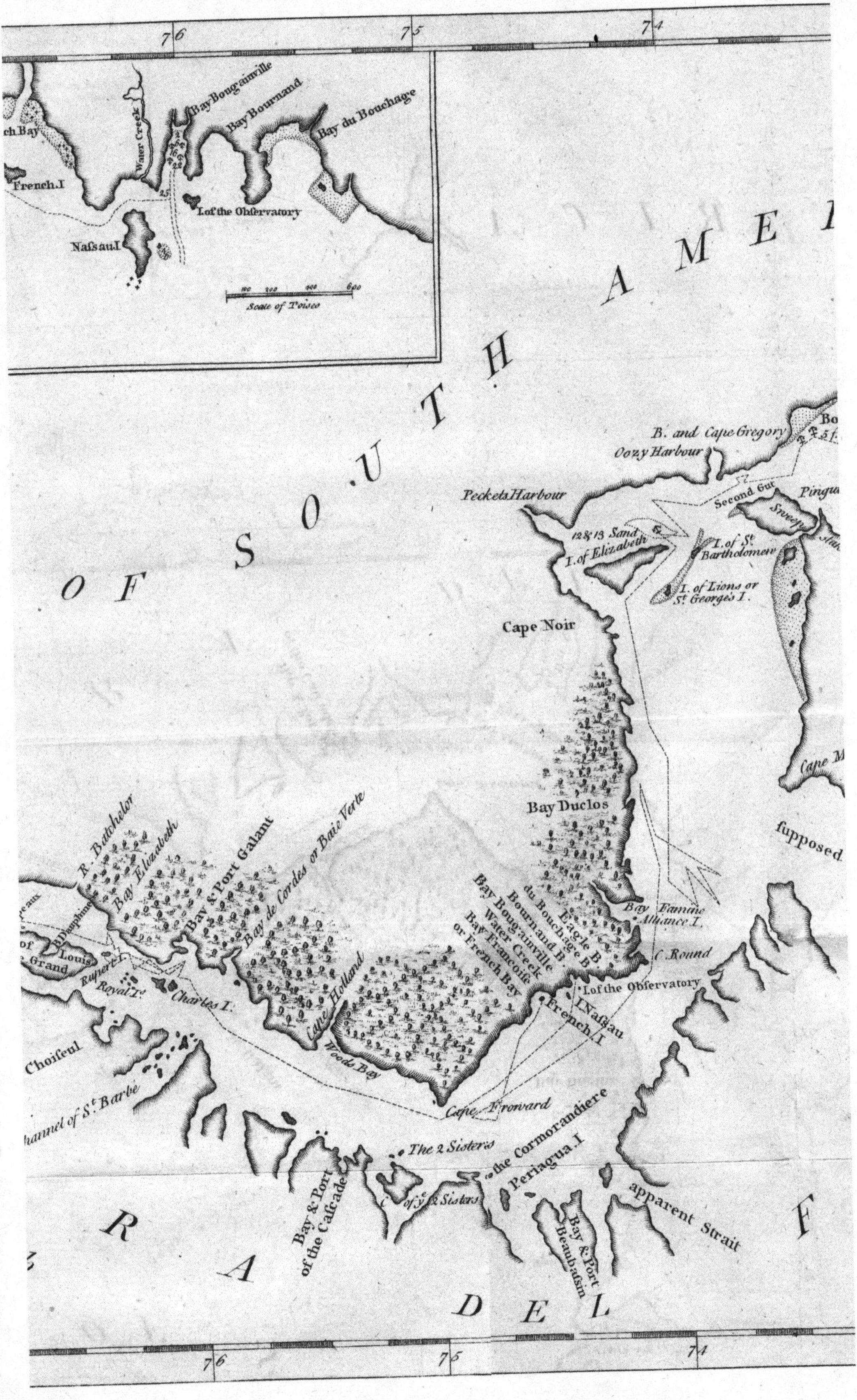

76
75
74
Bay Bougainville
Bay Bournand
Bay du Bouchage
Water Creek
French.I
I. of the Observatory
Nassau.I
Scale of Toises
OF SOUTH AMER
B. and Cape Gregory
Oozy Harbour
Pecketa Harbour
Second Gut
12 & 13 Sand
I. of Elizabeth
I. of St. Bartholomew
I. of Lions or St. George's I.
Cape Noir
Bay Duclos
supposed
R. Batchelor
Bay Elizabeth
Bay & Port Galant
Bay de Cordes or Baie Verte
B. Dauphine
Louis
Grand
Rupert I.
Royal I.
Charles I.
Cape Holland
Woods Bay
Eagle B.
du Bouchage B.
Bournand B.
Bay Bougainville
Water Creek
Bay Françoise or French Bay
Bay Famine
Alliance I.
C. Round
I. of the Observatory
I. Nassau
French I.
Choiseul
Channel of St. Barbe
Cape Froward
The 2 Sisters
the Cormorandiere
Periagua I.
Bay & Port of the Cascade
of ye 2 Sisters
Bay & Port Beaubassin
apparent Strait
R A
DEL
F
76
75
74

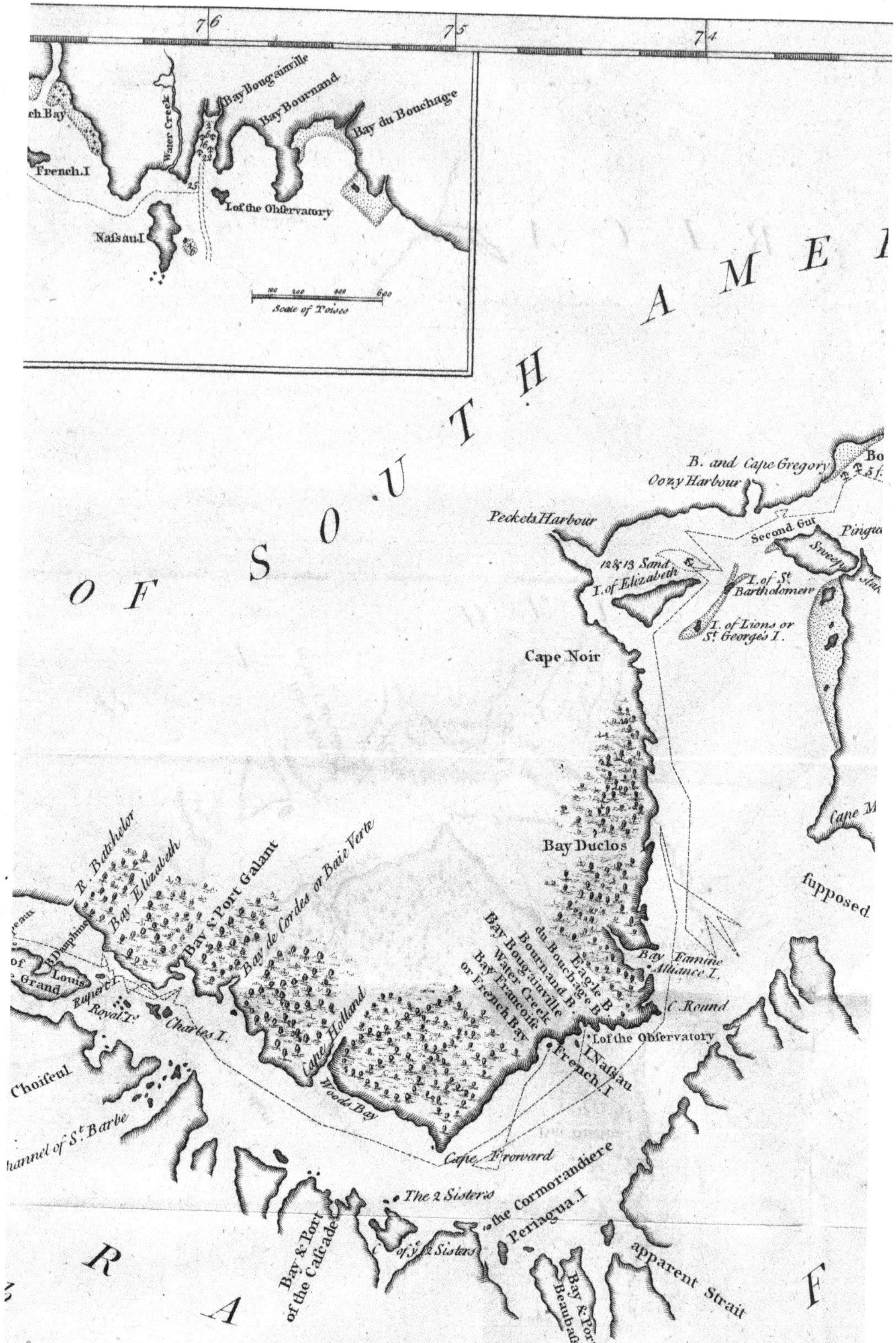

76 75 74

ch Bay
Water Creek
Bay Bougainville
Bay Bournand
Bay du Bouchage
French.I
I.of the Observatory
Nassau.I
Scale of Toises
100 200 400 600

OF SOUTH AMER

B. and Cape Gregory
Oozy Harbour
Peckets Harbour
Second Gut
12&13 Sand
I. of Elizabeth
I. of St. Bartholomew
I. of Lions or St. George's I.
Cape Noir
Bay Duclos
supposed
B. Batchelor
Bay Elizabeth
Bay & Port Galant
Bay de Cordes or Baie Verte
B. Dauphine
of Louis Grand
Rupert I.
Royal I.
Charles I.
Choiseul
Channel of St. Barbe
Cape Holland
Woods Bay
Bay Francoise or French Bay
Water Creek
Bay Bougainville
Bournand B.
du Bouchage B.
Eagle B.
Bay Famine
Alliance I.
C. Round
I. of the Observatory
I. Nassau
French I.
Cape Frovard
The 2 Sisters
the Cormorandiere
Periagua I.
C. of ye 2 Sisters
Bay & Port of the Cascade
Bay & Port Beaubassin
apparent Strait
R A D E L F

76 75 74

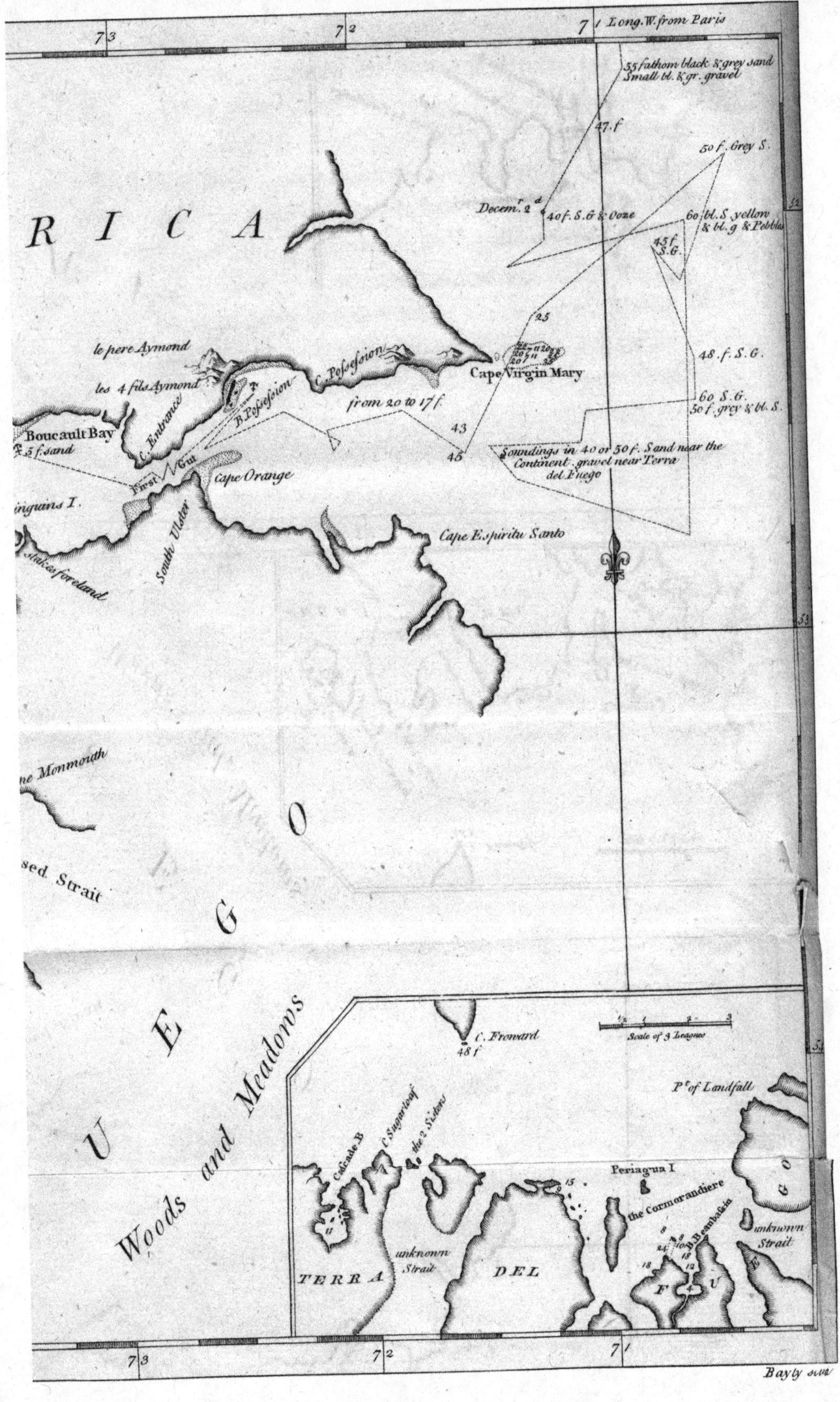

73
72
71 Long. W. from Paris
55 fathom black & grey sand
Small bl. & gr. gravel
47 f
50 f. Grey S.
R I C A
Decem.r 2d
40 f. S.G & Ooze
60 bl. S. yellow
& bl. g & Pebbles
45 f.
S.G.
25
le pere Aymond
48. f. S. G
Cape Virgin Mary
les 4 fils Aymond
C. Possession
60 S.G.
50 f. grey & bl. S.
C. Entrance
B. Possession
from 20 to 17 f.
Boucault Bay
43
5 f. sand
45
Soundings in 40 or 50 f. Sand near the
Continent gravel near Terra
del Fuego
First Gut
Cape Orange
South Ulster
Cape Espiritu Santo
Monmouth
sed Strait
O
G
E
U
Woods and Meadows
C. Froward
48 f
Scale of 3 Leagues
Pt. of Landfall
Cascade B
C. Sugarloaf
the 2 Sisters
Periagua I
the Cormorandiere
unknown Strait
B. Beambasin
unknown Strait
TERRA
DEL
F
U
73
72
71
Bayly sc.

which I had discovered in 1765, in 48° 30′ south latitude, about six or seven leagues off shore. I discovered it in the morning, at the same moment as I did the land, and having taken a good observation at noon, the weather being very fair, I was thus enabled to determine its latitude with precision. We ran within a quarter of a league of this rock, which the first person who saw it, originally took to be a *grampus*.

1767. December.

The 1st and 2d of December, the winds were favourable from N. and N. N. E; very fresh, the sea high, and the weather hazy; we made all the sail we could in day time, and passed the nights under our fore-sail, and close-reefed top-sails. During all this time we saw the birds called *Quebrantahuessos*, or *Albatrosses*, and what in all the seas in the world is a bad sign, petrels, which disappear when the weather is fair, and the sea smooth. We likewise saw seals, penguins, and a great number of whales. Some of these monstrous creatures seemed to have their skin covered with such white vermiculi, which fasten upon the bottoms of old ships that are suffered to rot in the harbours. On the 30th of November, two white birds, like great pigeons, perched on our yards. I had already seen a flight of these birds cross the bay of the Malouines.

Sight of C Virgins.

On the 2d of December in the afternoon, we discovered Cape Virgins, and we found it bore S. about seven

ſeven leagues diſtant. At noon I had obſerved 52° S. lat. and I was now in 52° 3′ 30″ of latitude, and in 71° 12′ 20″ of longitude weſt from Paris. This poſition of the ſhip, together with the bearing, places Cape Virgins in 52° 23′ of latitude, and in 71° 25′ 20″ of longitude weſt from Paris. As Cape Virgins is an intereſting point in geography, I muſt give an account of the reaſons which induced me to believe that the poſition I give is nearly exact.

Its poſition.

Diſcuſſion upon the poſition given to Cape Virgin.

The 27th of November in the afternoon, the chevalier du Bouchage had obſerved eight diſtances of the moon from the ſun, of which the mean reſult had given him the weſt longitude of the ſhip, in 65° 0′ 30″ for one hour, 43 min. 26 ſec. of true time: M. Verron, on his part, had obſerved five diſtances, the reſult of which gave for our longitude, at the ſame inſtant, 64° 57′. The weather was fair, and extremely favourable for obſervations. The 29th at 3 hours 57 min. 35 ſec. true time, M. Verron, by five obſervations of the diſtance of the moon from the ſun, determined the ſhip's weſt longitude, at 67° 49′ 30″.

Now, by following the longitude determined the 27th of November, taking the medium between the reſult of the obſervations of the chevalier du Bouchage and thoſe of M. Verron, in order to fix the longitude of the ſhip, when we got ſight of Cape Virgins, the longitude

gitude of that Cape will be 71° 29′ 42″ weſt from Paris. The obſervations made the 29th afternoon, likewiſe referred to the place of the ſhip, when we made the Cape, would give a reſult of 38′ 47″ more weſtward. But it ſeems to me that thoſe of the 27th ought rather to be followed, though two days more remote, becauſe they were made in a greater number by two obſervers, who did not communicate their obſervations to each other, and however did not differ more than 3′ 30″. They carry an appearance of probability which cannot well be objected to. Upon the whole, if a medium is to be taken between the obſervations of both days, the longitude of Cape Virgins will be 71° 49′ 5″, which differs only four leagues from the firſt determination, which anſwers within a league to that which the reckoning of my courſe gave me, and which I follow for this reaſon.

This longitude of Cape Virgin is more weſterly by 42′ 20″ than that which M. Bellin places it in, and this is the ſame difference which appears in his poſition of Montevideo, of which we have given an account in the beginning of this chapter. Lord Anſon's chart aſſigns for the longitude of Cape Virgins, 72°. weſt from London, which is near 75° weſt from Paris *; a much

* 74° 25′; Paris being 2° 25′ E. from London: vide Ferguſon's Tables. F.

more considerable error, which he likewise commits at the mouth of the river Plata, and generally along the whole coast of Patagonia.

Digression upon the instruments proper for observing the longitude at sea.

The observations which we have now mentioned, have been made with the English octant. This method of determining the longitude, by means of the distances of the moon from the sun, or from the stars in the zodiac, has been known for several years. Mess. de la Caille and Daprès have particularly made use of it at sea, likewise employing Hadley's octant. But as the degree of accuracy obtained by this method depends in a great measure upon the accuracy of the instrument with which you observe, it follows that M. Bouguer's heliometer, if one could measure great angles with it, would be very fit for rectifying these observations of distances. The Abbé de la Caille probably has thought of that, because he got one made, which would measure arcs of six or seven degrees; and if in his works he does not speak of it as an instrument fit for observing at sea, it is because he foresaw the difficulty of using it on board a ship.

M. Verron brought on board with him an instrument called a megameter, which he has employed in the other voyages he made with M. de Charnieres, and which he has likewise made use of on this. This instrument appeared to be very little different from the heliometer

of

of M. Bouguer, except that the ſcrew by which the objectives move, being longer, it places them at a greater diſtance aſunder, and by that means makes the inſtrument capable of meaſuring angles of ten degrees, which was the limit of M. Verron's megameter. It is to be wiſhed, that by lengthening the ſcrew, we were able to augment its extenſion ſtill more, it being confined in too narrow bounds to allow a frequent repetition, and even to make the obſervations exact; but the laws of dioptrics limit the removing of the objectives. It is likewiſe neceſſary to remedy the difficulty which the Abbé de la Caille foreſaw, I mean, that which ariſes from the element on which the obſervation muſt be made. In general, it ſeems that the reflecting quadrant of Hadley would be preferable, if it were equally accurate.

Difficulties on entering the ſtraits.

From the 2d of December in the afternoon, when we got ſight of Cape Virgins, and ſoon after of Terra del Fuego, the contrary wind and the ſtormy weather oppoſed us for ſeveral days together. We plyed to windward the 3d till ſix in the evening, when the winds becoming more favourable, permitted our bearing away for the entrance of Magalhaens' Straits: this laſted but a ſhort time; at half paſt ſeven it became quite calm, and the coaſts covered with fogs; at ten it blew freſh again, and we paſſed the night by plying to windward. The 4th, at three o'clock in the morning, we made for the

the land with a good northern breeze; but the weather which was rainy and hazy intercepting our ſight of it, we were obliged to ſtand off to ſea again. At five in the morning, in a clear ſpot, we perceived Cape Virgins, and bore away in order to enter the ſtraits; almoſt immediately the wind changed to S. W. whence it ſoon blew with violence, the fog became thicker and we were obliged to lay-to between the two ſhores of Terra del Fuego and the continent.

Our fore-ſail was ſplit the fourth in the afternoon; and we having ſounded, almoſt at the ſame moment, only twenty fathom, the fear of the breakers, which extend S. S. E. off Cape Virgins, made me reſolve to ſcud under our bare poles; eſpecially as this manœuvre facilitated the operation of bending another fore-ſail to the yard. Theſe ſoundings, however, which made me bear away, were not alarming; they were thoſe in the channel, as I have ſince learnt, by ſounding with a clear view of the land. I ſhall add, for the uſe of thoſe who may be plying here in thick weather, that a gravelly bottom ſhews that they are nearer the coaſt of Terra del Fuego than to the continent; where they will find a fine ſand, and ſometimes oozy bottom.

Obſervation on the nature of the ground at the entrance of the ſtraits.

At five o'clock in the evening we brought to again, under the main and mizen ſtay-ſails; at half paſt ſeven

of

of the clock the wind abated, the ſky cleared up, and we made ſail; but with diſadvantageous tacks, which brought us further from the coaſt; and, indeed, though on the 5th the weather was very fair and the wind favourable, we did not ſee the land till two in the afternoon; when it extended from S. by W. to S. W by W. about ten leagues off. At four o'clock we again diſcovered Cape Virgins; and we made ſail in order to double it, at the diſtance of about a league and a half, or two leagues It is not adviſeable to come nearer, on account of a bank, which lies off the Cape, at about that diſtance. I am even inclined to believe, that we paſſed over the tail of that ſand; for as we ſounded very frequently, between two ſoundings, one of twenty five and the other of ſeventeen fathom, the Etoile, which ſailed in our wake, made ſignal of eight fathom; but the moment after ſhe deepened her water.

Nautical re-marks upon the entrance of the ſtraits.

Cape Virgins is a table-land, of a middling height; it is perpendicular at its extremity; the view of it given by lord Anſon, is moſt exactly true. At half paſt nine in the evening, we had brought the north point of the entrance to the ſtraits to bear W. from which a ledge of rocks extends a league into the ſea. We ran under our cloſe-reefed fore-top-ſail and lower ſails hauled up, till eleven o'clock at night, when Cape Virgins bore N. of us. It blew very freſh; and the gloomineſs of the weather

weather ſeeming to threaten a ſtorm, determined me to paſs the night ſtanding off and on.

The 6th, at break of day, I ordered all the reefs out of the top-ſails, and run to W. N. W. We did not ſee land, till half paſt four o'clock, when it appeared to us that the tides had carried us to the S. S. W. At half after five, being about two leagues from the continent, we diſcovered Cape Poſſeſſion, being W. by N. and W. N. W. This Cape is very eaſily known; it is the firſt head-land from the north point, at the entrance of theſe ſtraits. It is more ſoutherly than the reſt of the coaſt, which afterwards forms a great gulf, called *Poſſeſſion Bay*, between this Cape and the next narrow gut. We had likewiſe ſight of Terra del Fuego. The winds ſoon changed to the ordinary points of W. and N. W. and we ran the moſt advantageous tacks for entering the ſtrait, endeavouring to come cloſe to the coaſt of Patagonia, and taking advantage of the tide, which then ſet to the weſtward.

At noon we had an obſervation; and the bearings taken at the ſame time gave me the ſame latitude, within a minute, for Cape Virgins, as that which I had concluded from my obſervations of the third of this month. We likewiſe made uſe of this obſervation, to aſcertain the latitude of Cape Poſſeſſion, and of Cape Eſpiritù Santo, on Terra del Fuego.

We continued to ply to windward, under our courses and top-sails, all the sixth; and the next night, which was very clear, often sounding, and never going further than three leagues from the coast of the continent. We got forward very little, by this disagreeable manœuvre; losing as much by the tides as we gained by them; and the 7th, at noon, we were still at Cape Possession. Cape Orange bore S. W. about six leagues distant. This cape is remarkable by a pretty high hillock; steep towards the sea-side, and forms to the southward the first gut, or narrow pass, in the straits *. Its point is dangerous, on account of a ledge which extends to the N. E. of the cape, at least three leagues into the sea. I have very plainly seen the sea break over it. At one o'clock, after noon, the wind having shifted to N. N. W. we made advantage of it to continue our voyage. At half past two we were come to the entrance of the gut; another obstacle attended us there; we were not able, with a fine fresh breeze, and all our sails set, to stem the tide. At four o'clock it ran six knots a-long side of us, and we went a-stern. We persisted in vain to

Description of Cape Orange.

Its rocks.

* From Cape Virgin, till to the entrance of the first *goulet*, we may reckon 14 or 15 leagues; and the straits are in every part of this interval, between five and seven leagues wide. The north coast, as far as Cape Possession, is uniform, but little elevated, and very healthy. From this cape onward, one must be careful to a void the rocks, which are situated in a part of the bay of the same name. When the hillocks, which I have named the *Quatre fils Aymond*, † only offer two to sight, in form of a gate, you are then opposite the said rocks.

† These rocks are called *Ass's Ears*, by Sir John Narborough. F.

strive

ſtrive againſt it. The wind was leſs conſtant than we were, and obliged us to return. It was to be feared, that we might be becalmed in the gut; expoſed to the current of the tide; which might carry us on the ledges off the capes which form its entrance at E. and W.

Anchoring in Poſſeſſion-bay.

We ſteered N. by E. in ſearch of a good anchoring-ground, in the bottom of Poſſeſſion-bay; when the Etoile, which was nearer the coaſt than we were, having paſſed all at once from twenty fathom to five, we bore away, and ſtood eaſt, in order to avoid a ledge of rocks, which ſeemed to lie in the bottom, and in the whole circuit of the bay. During ſome time we found a bottom of nothing but rocks and pebbles; and it was ſeven at night, being in twenty fathom, the ground mud and ſand, with black and white gravel, when we anchored about two leagues from the land. Poſſeſſion-bay is open to all winds, and has but very bad anchoring-ground. In the bottom of this bay ariſe five hills; one of which is a very conſiderable one; the other four are little and pointed. We have called them *le Pere et les quatre fils Aymond*; they ſerve as a conſpicuous mark for this part of the ſtraits. At night we ſounded at the ſeveral times of the tide, without finding any ſenſible difference in the depth. At half an hour paſt eight it ſet to the weſt; and at three in the morning to the eaſtward.

The

Passing the first *goulet*, or gut.

The eighth in the morning we set sail under courses, and double-reefed top sails; the tide was contrary to us, but we stemmed it with a fine N. W. breeze*. At eight o'clock the wind headed us, and we were obliged to ply to windward; now and then receiving violent squalls of wind. At ten o'clock, the tide beginning to set in westward with sufficient force, we lay to, under our top-sails, at the entrance of the first gut, driving with the current, which carried us to windward; and tacking about whenever we found ourselves too near either coast. Thus we passed the first narrow entrance or gut † in two hours; notwithstanding the wind was right against us, and blew very hard.

Sight of the Patagonians.

This morning the Patagonians, who had kept up fires all night, at the bottom of Possession-bay, hoisted a white flag on an eminence; and we answered it by hoisting that of our ships. These Patagonians certainly are the same which the Etoile saw in June 1766, in Boucault's-bay, and with whom she left this flag, as a

* When one intends to enter the first gut, or narrow passage in the straits, it is proper to coast within a league of Cape Possession; then to steer S. by W. taking care not to fall off too much to the south, on account of the rocks which extend N. N. E. and S. S. W. from Cape Orange, more than three leagues.

† The first gut lies N. N. E. and S. S. W. and is not above three leagues long. Its breadth varies from a league, to a league and a half. I have already given notice of the ledge of rocks at Cape Orange. At coming out of the first gut, you meet with two lesser rocks, extended on each of its extremities. They both project to S. W. There is a great depth of water in the gut.

ſign of alliance. The care they have taken to preſerve it; ſhews that good-nature, a due regard of their word, or, at leaſt, gratitude for preſents received, are the characteriſtics of theſe men.

We likewiſe ſaw, very diſtinctly, when we were in the gut, about twenty men on Terra del Fuego. They were dreſſed in ſkins, and ran as faſt as poſſible along the coaſt, parallel to our courſe. They ſeemed likewiſe from time to time to make ſigns to us with their hands, as if they wanted us to come to them. According to the report of the Spaniards, the nation which inhabits this part of Terra del Fuego, practiſes none of the cruel cuſtoms of moſt other ſavages. They behaved with great humanity to the crew of the ſhip la Conception, which was loſt on their coaſt in 1765. They aſſiſted them in ſaving part of her cargo; and in erecting ſheds, to ſhelter them againſt bad weather. The Spaniards built a bark there of the wreck of their ſhips, in which they went to Buenos Ayres. The xebeck el Andaluz was going to bring miſſionaries to theſe Indians, when we left Rio de la Plata. Lumps of wax, being part of the cargo of the above ſhip, have been carried by the force of currents to the coaſt of the Malouines, where they were found in 1766.

Americans of Terra del Fuego.

I have already obſerved, that we were gone through the firſt gut at noon; after that we made ſail. The wind

[W]e anchor [in] Boucault [Ba]y.

wind was veered to S. and the tide continued to carry us to the weſtward. At three o'clock they both failed us; and we anchored in Boucault's-bay, in eighteen fathom, oozy bottom.

Interview with the Patagonians.

As ſoon as we were at anchor, I hoiſted out one of my boats, and one belonging to the Etoile. We embarked in them, being about ten officers, each armed with our muſkets; and we landed at the bottom of the bay, with the precaution of ordering our boats to be kept a float, and the crew to remain in them. We had hardly ſet foot on ſhore, but we ſaw ſix Americans come to us on horſeback, in full gallop. They alighted about fifty yards from us; and immediately ran towards us, crying, *ſhawa.* When they had joined us, they ſtretched out their arms towards us, and laid them upon ours. They then embraced us, and ſhook hands with us, crying continually, *Shawa, ſhawa,* which we repeated with them. Theſe good people ſeemed very much rejoiced at our arrival. Two of them, who trembled as they came towards us, had their fears very ſoon removed. After many reciprocal careſſes, we ſent for ſome cakes and ſome bread from our boats; which we diſtributed amongſt them, and which they devoured with avidity. Their numbers encreaſed every moment; they were ſoon come to thirty, among whom were ſome young people, and a child of eight or ten years old. They all came to us with entire confidence; and careſſed us all, as the firſt had done. They did not ſeem ſurpriſed to ſee us; and by imitating the report of muſkets with their voice, they ſhewed that they were acquainted with theſe arms. They appeared attentive to do what might give us pleaſure. M. de Commerçon, and ſome of our gentlemen, were buſy in picking up plants: ſeveral Patagonians immediately began to ſearch for them too, and brought what ſpecies they ſaw us take up. One of them ſeeing the chevalier du Bouchage occupied in this manner, came to ſhew him his eye, which was very viſibly affected; and aſked him by ſigns, to point out to him ſome ſimple, by which he could be cured. This ſhews that they have an idea, and make uſe of that ſort of medicine, which requires the knowledge of ſimples, and applies them for the cure of mankind. This was the medicine of Machaon, who was phyſician to the gods; and, I believe, that many Machaons might be found among the Indians in Canada.

We exchanged ſome trifles, valuable in their eyes, againſt ſkins of *guanacoes* and *vicunnas*. They aſked us by ſigns for tobacco; and they were likewiſe very fond of any thing red: as ſoon as they ſaw ſomething of that colour upon us, they came to ſtroke it with their hands, and ſeemed very deſirous of it. At every preſent which we gave them, and at every mark of fondneſs, they re-

4 peated

peated their *ſhawa*, and cried ſo that it almoſt ſtunned us. We gave them ſome brandy; giving each of them only a ſmall draught: as ſoon as they had ſwallowed it, they beat with their hands on their throat, and by blowing with their mouths, uttered a tremulous inarticulate ſound, which terminated in a quick motion of the lips. They all made the ſame droll ceremony, which was a very ſtrange ſight to us.

However, it grew late, and was time to return on board. As ſoon as they ſaw that we were preparing for that purpoſe, they ſeemed ſorry; they made ſigns for us to wait, becauſe ſome more of their people were coming. We made ſigns that we would return the next day, and that we would bring them what they deſired: they ſeemed as if they would have liked our paſſing the night on ſhore much better. When they ſaw that we were going, they accompanied us to the ſea ſhore; a Patagonian ſung during this march. Some of them went into the water up to their knees, in order to follow us further. When we were come to our boats, we were obliged to look after every thing; for they got hold of all that was within their reach. One of them had taken a ſickle, but on its being perceived, he returned it without reſiſtance. Before we were got to any diſtance, we perceived their troops encreaſe, by the arrival of others, who came in full gallop. We did not fail, as as we left them, to ſhout *ſhawa* ſo loud that the whole coaſt reſounded with it.

Deſcription of theſe Americans.

Theſe Americans are the ſame with thoſe ſeen by the Etoile in 1765. One of our ſailors, who was then on board that veſſel, now knew one of theſe Americans again, having ſeen him in the firſt voyage. They have a fine ſhape; among thoſe whom we ſaw, none was below five feet five or ſix inches, and none above five feet nine or ten inches *; the crew of the Etoile had even ſeen ſeveral in the preceding voyage, ſix feet (or ſix feet, 4 728 inches Engliſh) high. What makes them appear gigantic, are their prodigious broad ſhoulders, the ſize of their heads, and the thickneſs of all their limbs. They are robuſt and well fed: their nerves are braced, and their muſcles are ſtrong and ſufficiently hard; they are men left entirely to nature, and ſupplied with food abounding in nutritive juices, by which means they are come to the full growth they are capable of: their figure is not coarſe or diſagreeable; on the contrary, many of them are handſome: their face is round, and ſomewhat flattiſh; their eyes very fiery; their teeth vaſtly white, and would only be ſomewhat too great at Paris; they have long black hair tied up on the top of

* This is to be underſtood in French meaſure, in which the French foot exceeds the Engliſh by ,788 of an inch; accordingly in French meaſure 5 feet 6 inches = 5 feet, 10,334, inches Engliſh; and French 5 feet 10 inches are = 6 feet, 2,5704, inches Engliſh. F.

their

their heads; I have ſeen ſome of them with long but thin whiſkers. Their colour is bronzed, as it is in all the Americans, without exception, both in thoſe who inhabit the torrid zone, and thoſe who are born in the temperate and in the frigid ones. Some of them had their cheeks painted red: their language ſeemed very delicate, and nothing gave us reaſon to fear any ferocity in them. We have not ſeen their women; perhaps they were about to come to us; for the men always deſired that we ſhould ſtay, and they had ſent one of their people towards a great fire, near which their camp ſeemed to be, about a league from us; and they ſhewed us that ſomebody would come from thence.

The dreſs of theſe Patagonians is very nearly the ſame with that of the Indians of Rio de la Plata; they have merely a piece of leather which covers their natural parts, and a great cloak of *guanaco* or *ſorillos* ſkins, which is faſtened round the body with a girdle; this cloak hangs down to their heels, and they generally ſuffer that part which is intended to cover the ſhoulders to fall back, ſo that, notwithſtanding the rigour of the climate, they are almoſt always naked from the girdle upwards. Habit has certainly made them inſenſible to cold; for though we were here in ſummer, Reaumur's thermometer was only one day riſen to ten degrees

degrees above the freezing point. These men have a kind of half boots, of horse-leather, open behind, and two or three of them had on the thigh a copper ring, about two inches broad. Some of my officers likewise observed, that two of the youngest among them had such beads as are employed for making necklaces.

The only arms which we observed among them, are, two round pebbles, fastened to the two ends of a twisted gut, like those which are made use of in all this part of America, and which we have described above. They had likewise little iron knives, of which the blade was between an inch and an inch and a half broad. These knives, which were of an English manufactory, were certainly given them by Mr. Byron. Their horses, which are little and very lean, were bridled and saddled in the same manner as those belonging to the inhabitants of Rio de la Plata. One of the Patagonians had at his saddle, gilt nails; wooden stirrups, covered with plates of copper; a bridle of twisted leather, and a whole Spanish harness. The principal food of the Patagonians seems to be the marrow and flesh of *guanacoes* and *vicunnas*; many of them had quarters of this flesh fastened on their horses, and we have seen them eat pieces of it quite raw. They had likewise little nasty dogs with them, which, like their horses, drink sea-water, it being a very scarce thing to get fresh water on this coast, and even in the country.

None of them had any apparent superiority over the rest; nor did they shew any kind of esteem for two or three old men who were in their troop. It is remarkable that several of them pronounced the Spanish words *manana, muchacha, bueno, chico, capitan.* I believe this nation leads the life of Tartars. Besides rambling through the immense plains of South America, men, women and children being constantly on horseback, pursuing the game, or the wild beasts, with which those plains abound, dressing and covering themselves with skins, they bear probably yet this resemblance with the Tartars, that they pillage the caravans of travellers. I shall conclude this article by adding, that we have since found a nation in the South Pacific Ocean which is taller than the Patagonians.

The soil in the place we landed at is very dry, and in that particular bears great resemblance with that of the Malouines; the botanists have likewise found almost all the same plants in both places. The sea shore was surrounded with the same sea-weeds, and covered with the same shells. Here are no woods, but only some shrubs. When we had anchored in Boucault's bay, the tide was going to set in against us, and whilst we were on shore, we observed that the water rose, and accordingly the flood sets in to eastward. This observation we have been able to make with certainty several times

Quality of the soil in this part o[f] America.

Remarks [on] the tides [in] these par[ts]

during this navigation, and it had ſtruck me already in my firſt voyage. At half paſt nine in the evening, the ebb ſet to weſtward. We ſounded at high water *, and found the depth was encreaſed to twenty-one fathoms, from eighteen, which we had when we caſt anchor.

Second time of anchoring in Boucault bay.

On the 9th, at half an hour paſt four in the morning, the wind being N W. we ſet all our ſails in order to ſtem the tide, ſteering S W. by W. we advanced only one league; the wind veering to S. W. and blowing very freſh, we anchored again in nineteen fathom, bottom of ſand, ooze, and rotten ſhells. The bad weather continued throughout this day and the next The ſhort diſtance we were advanced had brought us further from the ſhore, and during theſe two days, there was not one favourable inſtant for ſending out a boat, for which, the Patagonians were certainly as ſorry as ourſelves. We ſaw the whole troop of them collected at the place where we landed before, and we thought we perceived with our perſpective glaſſes, that they had erected ſome huts there. However, I apprehend that their head quarters were more diſtant, for men on horſeback were conſtantly going and coming. We were very ſorry that we could not bring them what we had promiſed; they might be ſatisfied at a ſmall expence

* A mer étale

The

The difference of the depth at the different times of tide, was only one fathom here. On the 10th, from an obſervation of the moon's diſtance from Regulus, M. Verron calculated our weſt longitude in this anchoring place, at 73° 26′ 15″, and that of the eaſterly entrance of the ſecond gut, at 73° 34′ 30″. Reaumur's thermometer fell from 9° to 8° and 7°.

Loſs of an anchor.

The 11th, at half an hour after midnight, the wind veering to N. E. and the tide ſetting to weſtward an hour before, I made ſignal for weighing. Our efforts to that purpoſe were fruitleſs, though we had got the winding-tackle upon the cable. At two in the morning, the cable parted between the bits and the hawſe, and ſo we loſt our anchor. We ſet all our ſails, and ſoon had the tide againſt us, which we were hardly able to ſtem with a light breeze at N. W. though the tide in the ſecond gut is not near ſo ſtrong as in the firſt. At noon the ebb came to our aſſiſtance, and we paſſed the ſecond gut *, the wind having been variable till three in the afternoon, when it blew very freſh from S. S. W. and S. S. E. with rain and violent ſqualls †.

Paſſing the ſecond gut

* The diſtance between the W. point or end of the firſt gut, and the entrance of the ſecond, is about ſix or ſeven leagues, and the breadth of the ſtraits there is likewiſe about ſeven leagues. The ſecond gut lies N. E. by E. and S. W. by W. it is about a league and a half broad, and three or four long.

† In paſſing the ſecond gut, it is neceſſary to keep along the coaſt of Patagonia, becauſe, when you come out of the gut, the tides run ſouthward, and you muſt be careful to avoid a low point, projecting below the head-land of St George's iſle, and though this apparent cape is high and ſteep, the low land advances far to W. N. W.

In two boards we came to the anchoring-place, to the northward of the iſle of Elizabeth, where we anchored, two miles off ſhore, in ſeven fathom, grey ſand with gravel and rotten ſhells. The Etoile anchored a quarter of a league more to the S. E. than we did, and had ſeventeen fathom of water.

We anchor near the iſle of Elizabeth.

We were obliged to ſtay here the 11th and 12th, on account of the contrary wind, which was attended with violent ſqualls, rain, and hail On the 12th in the afternoon we hoiſted out a boat, in order to go on ſhore on the iſle of Elizabeth * We landed in the N E part of the iſland. Its coaſts are high and ſteep, except at the S. W. and S. E. points, where the ſhore is low. However, one may land in every part of it, as there is always a ſmall ſlip of flat land under the high perpendicular ſhores. The ſoil of the iſle is very dry; we found no other water than that of a little pool in the S. W. part of the iſle, but it was very brackiſh. We likewiſe ſaw ſeveral dried marſhes, where the earth is in ſome places covered with a thin cruſt of ſalt. We found ſome buſtards, but

Deſcription of this iſle.

* The iſle of Elizabeth † lies N. N. E. and S S. W. with the weſt point of the ſecond gut, on the Patagonian ſide. The iſles of St. Barthelemi (St. Bartholomew) and of Lions likewiſe, lie N. N. E. and S. S. W. between them and the weſt point of the ſecond gut on St. George's iſland.

† The French call it Sainte Elizabeth. F.

2 they

they were in ſmall number, and ſo very ſhy, that we were never able to come near enough to ſhoot them: they were however ſitting on their eggs. It appears that the ſavages come upon this iſland. We found a dead dog, ſome marks of fire places, and the remnants of ſhells, the fiſh of which had been feaſted upon. There is no wood on it, and a ſmall ſort of heath is the only thing that may be uſed as fuel. We had already collected a quantity of it, fearing to be obliged to paſs the night on this iſle, where the bad weather kept us till nine of the clock in the evening: we ſhould have been both ill lodged and ill fed on it

CHAP.

C H A P. IX.

The run from the isle of Elizabeth, through the straits of Magalhaens. Nautical details on this navigation.

Difficulties of the navigation along the isle of Elizabeth.

WE were now going to enter the woody part of the the straits of Magalhaens; and the first difficult steps were already made.

It was not till the 13th in the afternoon, the wind being N. W. that we weighed, notwithstanding the force with which it blew, and made sail in the channel, which separates the isle of Elizabeth from the isles of St. Barthelemi and of Lions *. We were forced to carry sail; though there were almost continually very violent squalls coming off the high land of Elizabeth island; along which we were obliged to sail, in order to avoid the breakers, which extend around the other two isles †.

The

* The isles of St. Barthelemi and of Lions, are connected together by a shoal. There are likewise two shoals; one S. S. W. of the isle of Lions, and the other W. N. W. of St. Barthelemi, one or two leagues distant; so that these three shoals, and the two isles form a chain; between which, to E. S. E. and the isle of St. Elizabeth to W. N. W. is the channel through which you advance into the straits. This channel runs N. N. E. and S S. W.

I do not think it practicable to sail on the south side of the isles of St. Barthelemi and of Lions, nor between the isle of Elizabeth and the main land.

† From the end of the second gut, to the N. E. point of the isle of Elizabeth, the distance is about four leagues. Elizabeth island extends S. S. W. and N. N. E.

for

The tide in this channel ſets to the ſouthward, and ſeemed very ſtrong to us. We came near the ſhore of the main-land, below Cape Noir; here the coaſt begins to be covered with woods; and its appearance from hence is very pleaſant. It runs ſouthward; and the tides here are not ſo ſtrong as in the above place.

It blew very freſh and ſqually, till ſix o'clock in the evening; when it became calm and moderate. We ſailed along the coaſt, at about a league's diſtance, the weather being clear and ſerene; flattering ourſelves to be able to double Cape Round during night; and then to have, in caſe of bad weather, Port Famine to leeward. But theſe projects were fruſtrated; for, at half an hour after mid-night, the wind ſhifted all at once to S. W. the coaſt became foggy; the continual and violent ſqualls brought rain and hail with them; and, in ſhort, the weather ſoon became as foul, as it had been fair the moment before. Such is the nature of this climate; the changes of weather are ſo ſudden and frequent, that it is impoſſible to foreſee their quick and dangerous revolutions.

Bad weathe and diſagre able night.

Our main-ſail having been ſplit, when in the brails, we were forced to ply to windward, under our fore-ſail,

for the length of about three leagues and a half. It is neceſſary to keep this ſhore on board, in paſſing through the above channel.

From the S. W. point of Elizabeth iſland, to Cape Noir, the diſtance is not above a league ‡.

‡ This Cape Noir is not mentioned in M. de B's. map; but ſhould be carefully diſtinguiſhed from *Cape Noir*, or *Cabo Negro*, ſeen by lord Anſon upon Terra del Fuego, in about 54° S. lat. F

6 main-

main-ſtay-ſail, and cloſe-reefed top-ſails, endeavouring to double Point St. Anne, and to take ſhelter in Port Famine. This required our gaining a league to windward; which we could never effect. As our tacks were ſhort, and being obliged to wear, a ſtrong current was carrying us into a great inlet in Terra del Fuego; we loſt three leagues in nine hours on this manœuvre, and were obliged to go along the coaſt in ſearch of anchorage to leeward. We ranged along it, and kept ſounding continually; and, about eleven o'clock in the morning, we anchored a mile off ſhore, in eight fathom and a half, oozy ſand, in a bay, which I named Bay Duclos *; from the name of M. Duclos Guyot, a captain of a fire-ſhip, who was the next in command after me on this voyage; and whoſe knowledge and experience have been of very great uſe to me.

We anchor in Bay Duclos.

Deſcription of this bay.

This bay is open to the eaſtward, and its depth is very inconſiderable. Its northern point projects more into the ſea, than the ſouthern one; and they are about a league diſtant from each other. The bottom is very good in the whole bay; and there is every where ſix or eight fathom of water, within a cable's length from the ſhore. This is an excellent anchorage; becauſe the

* From Cape Noir the coaſt runs S S. E. to the northern point of Bay Duclos; which is about ſeven leagues diſtant from it.

Oppoſite Bay Duclos, there is a prodigious inlet in Terra del Fuego; which I ſuſpect to be a channel, diſemboguing eaſtward of Cape Horn. Cape Monmouth forms the north point of it.

weſterly

westerly winds, which prevail here, blow over the coast, which is very high in this part. Two little rivers discharge themselves into the bay; the water is brackish at their mouth, but very good five hundred yards above it. A kind of meadow lies along the landing-place, which is sandy. The woods rise behind it in form of an amphitheatre; but the whole country seems entirely without animals. We have gone through a great track of it, without finding more than two or three snipes, some teals, ducks, and bustards in very small number: we have likewise perceived some *perrokeets* *; the latter are not afraid of the cold weather.

At the mouth of the most southerly river, we found seven huts, made of branches of trees, twisted together, in form of an oven; they appeared to have been lately built, and were full of calcined shells, muscles, and limpets. We went up a considerable way in this river, and saw some marks of men. Whilst we were on shore, the tide rose one foot, and the flood accordingly came from east, contrary to the observations we had made after doubling Cape Virgin; having ever since seen the water rise when the tide went out of the straits. But it seems to me, after several observations, that having passed the guts, or narrows, the tides cease to be regular in all that part of the straits, which runs north and south.

New observations on tides.

* *Perruches*, probably *sea-parrots*, or auks. F.

The number of channels, which divide Terra del Fuego in this part, ſeem neceſſarily to cauſe a great irregularity in the motion of the water. During the two days which we paſſed in this anchoring-place, the thermometer varied from eight to five degrees. On the 15th, at noon, we obſerved 53° 20′ of latitude there; and that day we employed our people in cutting wood; the calm not permitting us then to ſet ſail.

Nautical obſervations.

Towards night the clouds ſeemed to go to weſtward, and announced us a favourable wind. We hove a-peek upon our anchor; and, actually, on the 16th, at four o'clock in the morning, the breeze blowing from the point whence we expected it, we ſet ſail. The ſky, indeed, was cloudy; and, as is uſual in theſe parts, the eaſt and north-eaſt winds, accompanied with fog and rain. We paſſed Point St. Anne * and Cape Round † The former is a table-land, of a middling height; and covers a deep bay, which is both ſafe and convenient for anchoring. It is that bay, which, on account of the unhappy fate of the colony of Philippeville, eſtabliſhed by the preſumptuous Sarmiento, has got the name of *Port Famine* Cape Round is a high land, remarkable

* The diſtance from Bay Duclos to Point St. Anne, is about five leagues; and the bearing being S. E. by S. there is nearly the ſame diſtance from Point St. Anne to Cape Round, which bear reſpectively N. N. E. and S. S. W.

† From the ſecond gut to Cape Round, the breadth of the ſtraits varies from ſeven to five leagues; they grow narrow at Cape Round, where their breadth does not exceed three leagues

on

on account of the figure which its name expresses; the shores, in all this tract, are woody and steep; those of Terra del Fuego appear cut through by several straits. Their aspect is horrible; the mountains there are covered with a blueish snow, as old as the creation. Between Cape Round and Cape Forward there are four bays, in which a vessel may anchor.

Description of a singular cape.

Two of these are separated from each other by a cape; the singularity of which fixed our attention, and deserves a particular description. This cape rises upwards of a hundred and fifty feet above the level of the sea; and consists entirely of horizontal strata, of petrified shells. I have been in a boat to take the soundings at the foot of this monument, which marks the great changes our globe has undergone; and I have not been able to reach the bottom, with a line of a hundred fathom.

Description of Cape Forward.

The wind brought us to within a league and a half of Cape Forward; we were then becalmed for two hours together. I profited of this time to go in my pinnance, near Cape Forward, to take soundings and bearings. This cape is the most southerly point of America, and of all the known continents. From good observations we have determined its south lat. to be 54° 5′ 45″. It shews a surface with two hillocks, extending about three quarters of a league; the eastern hillock being higher

higher than the weſtern one. The ſea is almoſt unfathomable below the cape; however, between the two hillocks or heads, one might anchor in a little bay provided with a pretty conſiderable rivulet, in 15 fathom, ſand and gravel; but this anchorage being dangerous in a ſoutherly wind, ought only to ſerve in a caſe of neceſſity. The whole cape is a perpendicular rock, whoſe elevated ſummit is covered with ſnow. However, ſome trees grow on it; the roots of which are fixed in the crevices, and are ſupplied with perpetual humidity. We landed below the cape at a little rock, where we found it difficult to get room for four perſons to ſtand on. On this point, which terminates or begins a vaſt continent, we hoiſted the colours of our boat; and theſe wild rocks reſounded, for the firſt time, with the repeated ſhouts of *vive le Roi*. From hence we ſet out for Cape Holland, bearing W. 4° N. and accordingly the coaſt begins here to run northward again.

Anchoring in Bay Françoiſe.

We returned on board at ſix o'clock in the evening; and ſoon after the wind veering to S. W. I went in ſearch of the harbour, which M. de Gennes named the French Bay *(Baie Françoiſe)*. At half an hour paſt eight o'clock we anchored there in ten fathom, ſandy and gravelly bottom; between the two points of the bay, of which the one bore N. E. ½ E. and the other S. ½ W. and the little iſland in the middle, N. E. As we wanted to take in water and wood for our courſe acroſs the Pacific Ocean, and the remaining part of the ſtraits was unknown to me; being in my firſt voyage, come no further than near Bay Francoiſe, I reſolved to take in thoſe neceſſaries here; eſpecially as M. de Gennes repreſents it very ſafe and convenient for this purpoſe: accordingly that very evening we hoiſted all our boats out. During night the wind veered all round the compaſs; blowing in very violent ſqualls; the ſea grew high, and broke round us upon a ſand, which ſeemed to ly all round the bottom of the bay. The frequent turns, which the changes of the wind cauſed our ſhip to make round her anchor, gave us room to fear that the cable might be foul of it; and we paſſed the night under continual apprehenſions.

Advice with regard to th[is] harbour.

The Etoile lying more towards the offing than we did, was not ſo much moleſted. At half paſt two in the morning, I ſent the little boat to ſound the mouth of the river, to which M. de Gennes has given his name. It was low water; and the boat did not get into the river, without running a ground upon a ſand at its mouth; at the ſame time they found, that our large boats could only get up at high-water; and thus could hardly make above one trip a day. This difficulty of watering, together with the anchorage not appearing ſafe to me, made me reſolve to bring the ſhips into a little

bay,

bay, a league to the eastward of this. I had there, without difficulty, in 1765, taken a loading of wood for the Malouines, and the crew of the ship had given it my name. I wanted previously to go and be sure, whether the crews of both ships could conveniently water there. I found, that besides the rivulet, which falls into the bottom of the bay itself; and which might be adapted for the daily use, and for washing, the two adjoining bays had each a rivulet proper to furnish us easily with as much water as we wanted; and without having above half a mile to fetch it.

In consequence of this, we sailed on the 17th, at two o'clock in the afternoon, with our fore and mizen-top-sails. We passed without the little isle, in Bay Françoise; and, afterwards, we entered into a very narrow pass, in which there is deep water, between the north point of this bay and a high island, about half a quarter of a league long. This pass leads to the entrance of Bougainville's bay; which is, moreover, covered by two other little isles; the most considerable of which, has deserved the name of Isle of the Observatory, *(Islot de l'Observatoire)**.

The bay is two hundred toises † long, and fifty deep; high mountains surround it, and secure it against all

* From Cape Round, to the Isle of the Observatory, the distance is about four leagues; and the coast runs W S. W. In this distance there are three good anchoring-places.

† A French *toise* is six feet Paris measure F.

winds;

winds; and the sea there is always as smooth as in a bason.

We anchor in Bay Bougainville.

We anchored at three o'clock in the entrance of the bay, in twenty-eight fathom of water; and we immediately sent our tow-lines on shore, in order to warp into the bottom of the bay. The Etoile having let go her off anchor in too great a depth of water, drove upon the Isle of the Observatory; and before she could haul-tight the warps which she had sent a-shore, to steady her, her stern came within a few feet of this little isle, though she had still thirty fathom of water. The N. E. side of this isle is not so steep. We spent the rest of the day in mooring, with the head towards the offing, having one anchor a-head in twenty-three fathom oozy sand; a kedge-anchor a-stern, almost close to the shore; and two hawsers fastened to the trees on the larboard-side; and two on board the Etoile, which was moored as we were. Near the rivulet we found two huts, made of branches, which seemed to have been abandoned long ago. In 1765 I got one of bark constructed there, in which I left some presents for the Indians, which chance might conduct thither; and at the top of it I placed a white flag: we found the hut destroyed; the flag, with the presents, being carried off.

On the 18th, in the morning, I established a camp on shore, in order to guard the workmen, and the various

rious effects which we landed; we likewise sent all our casks on shore, to refit them and prepare them with sulphur; we made pools of water for the use of those who were employed in washing, and hauled our long boat a-shore, because she wanted a repair. We passed the remainder of December in this bay, where we provided ourselves with wood; and even with planks at our ease. Every thing facilitated this work: the roads were ready made through the woods; and there were more trees cut down than we wanted, which was the work of the Eagle's crew in 1765. Here we likewise heeled ship, boot-topped and mounted eighteen guns. The Etoile had the good fortune to stop her leak; which, since her departure from Montevideo, was grown as considerable as before her repair at the Encenada. By bringing her by the stern, and taking off part of the sheathing forward, it appeared that the water entered at the scarfing of her stem. This was remedied; and it was during the whole voyage, a great comfort to the crew of that vessel, who were almost worn out by the continual exercise of pumping.

Observations astronomical and meteorological.

M. Verron, in the first days, brought his instruments upon the Isle of the Observatory; but past most of his nights there in vain. The sky of this country, which is very bad for astronomers, prevented his making any observation for the longitude; he could only determine

mine by three obſervations with the quadrant, that the ſouth latitude of the little iſle is 53° 50′ 25″. He has likewiſe determined the flowing of the tide in the entrance to the bay, at 00^h 59′. The water never roſe here above ten feet. During our ſtay here the thermometer was generally between 8° and 9°, it fell once to 5°, and the higheſt it ever roſe to was 12 ½°. The ſun then appeared without clouds, and its rays, which are but little known here, melted part of the ſnow that lay on the mountains of the continent. M de Commerçon, accompanied by the prince of Naſſau, profited of ſuch days for botanizing. He had obſtacles of every kind to ſurmount, yet this wild ſoil had the merit of being new to him, and the ſtraits of Magalhaens have filled his herbals with a great number of unknown and intereſting plants. We were not ſo ſucceſsful in hunting and fiſhing, by which we never got any thing, and the only quadruped we ſaw here, is a fox, almoſt like an European one, which was killed amidſt the workmen.

Deſcription of this part of the ſtrai[ts]

We likewiſe made ſeveral attempts to ſurvey the neighbouring coaſts of the continent, and of Terra del Fuego; the firſt was fruitleſs. I ſet out on the 22d at three o'clock in the morning with Meſſ. de Bournand and du Bouchage, intending to go as far as Cape Holland, and to viſit the harbours that might be found on that part of the coaſt. When we ſet out it was calm and very fine weather. An hour afterwards, a light breeze at N. W. ſprung up, but immediately after, the wind ſhifted to S. W. and blew very freſh. We ſtrove againſt it for three hours together, under the lee of the ſhore, and with ſome difficulty got into the mouth of a little river, which falls into a ſandy creek, covered by the eaſtern head of Cape Forward. We put in here, hoping that the foul weather would not laſt long. This hope ſerved only to wet us thoroughly by the rain, and to make us quite chilled with cold. We made us a hut of branches of trees in the woods, in order to paſs the night there a little more under ſhelter. Theſe huts ſerve as palaces to the natives of theſe climates; but we had not yet learnt their cuſtom of living in them. The cold and wet drove us from our lodging, and we were obliged to have recourſe to a great fire, which we took care to keep up, endeavouring to ſhelter us againſt the rain, by ſpreading the ſail over us which belonged to our little boat. The night was dreadful, wind and rain encreaſed, and we could do nothing elſe but return at break of day. We arrived on board our frigate at eight of the clock in the morning, happy to have been able to take ſhelter there; for the weather became ſo much worſe ſoon after, that we could not have thought of coming back again. During two days there was a real tempeſt, and the mountains were all covered with ſnow

ſnow again. However, this was the very middle of ſummer, and the ſun was near eighteen hours above the horizon.

Diſcovery of ſeveral ports on Teira del Fuego.

Some days after I undertook a new courſe, more ſucceſsfully, for viſiting part of Terra del Fuego, and to look for a port there, oppoſite Cape Forward; I then intended to croſs the ſtraits to Cape Holland, and to view the coaſts from thence till we came to Bay Françoiſe, which was what we could not do on our firſt attempt. I armed the long boat of the Boudeuſe, and the Etoile's barge, with ſwivel guns and muſkets, and on the 27th, at four o'clock in the morning, I went from on board with Meſſrs. de Bournand, d'Oraiſon, and the prince of Naſſau. We ſet ſail at the weſt point of Bay Francoiſe, in order to croſs the ſtraits to Terra del Fuego, where we landed about ten o'clock, at the mouth of a hide river, in a ſandy creek, which is inconvenient even for boats. However, in a caſe of neceſſity, the boats might go up the river at high water, where they would find ſhelter. We dined on its banks, in a pleaſant wood, under the ſhade of which were ſeveral huts of the ſavages. From this ſtation, the weſtern point of Bay Françoiſe bore N. W. by W. ½ W. and we reckoned ourſelves five leagues diſtant from it.

After dinner we proceeded by rowing along the coaſt of Terra del Fuego; it did not blow much from the weſt-

weſtward, but there was a hollow ſea We croſſed a great inlet, of which we could not ſee the end. Its entrance, which is about two leagues wide, is barred in the middle by a very high iſland. The great number of whales which we ſaw in this part, and the great rolling ſea, inclined us to imagine that this might well be a ſtrait leading into the ſea pretty near Cape Horn. Being almoſt come to the other ſide, we ſaw ſeveral fires appear, and become extinct; afterwards they remained lighted, and we diſtinguiſhed ſome ſavages upon the low point of a bay, where I intended to touch. We went immediately to their fires, and I knew again the ſame troop of ſavages which I had already ſeen on my firſt voyage in the ſtraits. We then called them *Pécherais*, becauſe that was the firſt word which they pronounced when they came to us, and which they repeated to us inceſſantly, as the Patagonians did their *ſhawa*. For this reaſon we gave them that name again this time. I ſhall hereafter have an opportunity to deſcribe theſe inhabitants of the wooded parts of the ſtrait. The day being upon the decline, we could not now ſtay long with them. They were in number about forty, men, women, and children; and they had ten or a dozen canoes in a neighbouring creek. We left them in order to croſs the bay, and enter into an inlet, which, the night coming on, prevented us from executing. We

Meeting with ſavages.

paſſed

paſſed the night on the banks of a pretty conſiderable river, where we made a great fire, and where the ſails of our boats, which were pretty large, ſerved us as tents; the weather was very fine, although a little cold.

The next morning we ſaw that this inlet was actually a port, and we took the ſoundings of it, and of the bay. The anchorage is very good in the bay, from forty to twelve fathoms, bottom of ſand, ſmall gravel and ſhells. It ſhelters you againſt all dangerous winds. Its eaſterly point may be known by a very large cape, which we called the *Dome*. To the weſtward is a little iſle, between which and the ſhore, no ſhip can go out of the bay; you come into the port by a very narrow paſs, and in it you find ten, eight, ſix, five, and four fathoms, oozy bottom; you muſt keep in the middle, or rather come nearer the eaſt ſide, where the greateſt depth is. The beauty of this anchoring place determined us to give it the name of bay and port of *Beaubaſſin*. If a ſhip waits for a fair wind, ſhe need anchor only in the bay. If ſhe wants to wood and water, or even careen, no properer place for theſe operations can be thought of than the port of *Beaubaſſin*.

Bay and port of Beaubaſſin

Its deſcription.

I left here the chevalier de Bournand, who commanded the long boat, in order to take down as minutely as poſſible all the information relative to this important

portant place, and then to return to the ſhips. For my part, I went on board the Etoile's barge with Mr. Landais, one of the officers of that ſtore-ſhip, who commanded her, and I continued my ſurvey. We proceeded to the weſtward, and firſt viewed an iſland, round which we went, and found that a ſhip may anchor all round it, in twenty-five, twenty-one, and eighteen fathoms, ſand and ſmall gravel. On this iſle there were ſome ſavages fiſhing. As we went along the coaſt, we reached a bay before ſun-ſet, which affords excellent anchorage for three or four ſhips. I named it bay *de la Cormorandiere*, on account of an apparent rock, which is about a mile to E. S. E. of it. At the entrance of the bay we had fifteen fathoms of water, and in the anchoring place eight or nine; here we paſſed the night.

Bay de la Cormorandiere.

On the 29th at day break we left bay *de la Cormorandiere*, and went to the weſtward by the aſſiſtance of a very ſtrong tide. We paſſed between two iſles of unequal ſize, which I named the two Siſters (*les deux Soeurs*). They bear N. N. E. and S. S. W. with the middle of Cape Forward, from which they are about three leagues diſtant. A little farther we gave the name of Sugar-loaf *(Pain de ſucre)* to a mountain of this ſhape, which is very eaſy to be diſtinguiſhed, and bears N. N. E. and S. S. W. with the ſouthern point of the ſame cape; and about five leagues from the *Cormorandiere* we diſ-

covered

covered a fine bay, with an amazing fine port at the bottom of it; a remarkable water-fall in the interior part of the port, determined me to call them *Bay and Port of the Cascade*. The middle of this bay bears N. E. and S. W with Cape Forward. The safe and convenient anchorage, and the facility of taking in wood and water, shew that there is nothing wanting in it.

Bay and Port of the Cascade.

Description of the country.

The cascade is formed by the waters of a little river, which runs between several high mountains; and its fall measures about fifty or sixty toises, (*i. e.* 300 or 360 feet French measure): I have gone to the top of it. The land is here and there covered with thickets, and has some little plains of a short spungy moss; I have here been in search of vestiges of men, but found none; for the savages of this part seldom or never quit the sea-shores, where they get their subsistence. Upon the whole, all that part of Terra del Fuego, reckoning from opposite Elizabeth island, seems to me, to be a mere cluster of great, unequal, high and mountainous islands, whose tops are covered with eternal snow. I make no doubt but there are many channels between them into the sea. The trees and the plants are the same here as on the coast of Patagonia; and, the trees excepted, the country much resembles the Malouines.

Usefulness of the three ports before described.

I here add a particular chart which I have made of this interesting part of the coast of Terra del Fuego.

Till

Till now, no anchoring place was known on it, and ſhips were careful to avoid it. The diſcovery of the three ports which I have juſt deſcribed on it, will facilitate the navigation of this part of the ſtraits of Magalhaens. Cape Forward has always been a point very much dreaded by navigators. It happens but too frequently, that a contrary and boiſterous wind prevents the doubling of it, and has obliged many to put back to Bay Famine. Now, even the prevailing winds may be turned to account, by keeping the ſhore of Terra del Fuego on board, and putting into one of the abovementioned anchoring places, which can be done almoſt at any time, by plying in a channel where there is never a high ſea for ſhips. From thence all the boards are advantageous, and if one takes care to make the beſt of the tides, which here begin to have more effect again, it will no longer be difficult to get to Port Galant.

We paſſed a very diſagreeable night in Port Caſcade. It was very cold, and rained without intermiſſion. The rain continued throughout almoſt the whole 30th day of December. At five o'clock in the morning we went out of the port, and ſailed acroſs the ſtrait with a high wind and a great ſea, conſidering the little veſſel we were in. We approached the coaſt nearly at an equal diſtance between Cape Holland and Cape Forward. It was not now

in

in question to view the coast, being happy enough to run along it before the wind, and being very attentive to the violent squalls, which forced us to have the haliards and sheets always in hand. A false movement of the helm was even very near oversetting the boat, as we were crossing Bay Francoise. At last I arrived on board the frigate, about ten o'clock in the morning. During my absence, M. Duclos Guyot had taken on board what we had on shore, and made every thing ready for weighing; accordingly, we began to unmoor in the afternoon.

Departure from Bougainville Bay.

The 31st of December at four of the clock in the morning we weighed, and at six o'clock we left the bay, being towed by our boats. It was calm; at seven a light breeze sprung up at N. E. which became more fresh in the day; the weather was clear till noon, when it became foggy and rainy. At half an hour past eleven, being in the middle of the strait *, we discovered, and set the Cascade bearing S. E. the Sugar Loaf S. E. by E. ½ E. Cape Forward † E. by N. Cape Holland ‡ W. N. W. ½ W. From noon till six in the even-

* A mi-canal.

† From the isle of the Observatory, Cape Forward is about six leagues distant, and the coast runs nearly W. S. W. The strait is there between three and four leagues broad.

‡ In the space of about five leagues, which are between Cape Forward and Cape Holland, there are two other capes, and three creeks, of little depth. I know of no anchorage there. The breadth of the straits varies from three to four leagues.

ing we doubled Cape Holland. It blew a light breeze, which abating in the evening, and the ſky being covered, I reſolved to anchor in the road of Port Galant, where we anchored in ſixteen fathoms, coarſe gravel, ſand and ſmall coral; Cape Galant bearing S. W. 3° W * We had ſoon reaſon to congratulate ourſelves on being in ſafety; for, during the night, it rained continually, and blew hard at S. W.

Anchorage in Forteſcue Bay.

1768. January.

We began the year 1768 in this bay, called Bay Forteſcue, at the bottom of which is Port Galant †. The plan of the bay and port is very exact in M. de Gennes. We have had too much leiſure to confirm it, having been confined there for three weeks together, by ſuch weather as one cannot form any idea of, from the worſt winter at Paris. It is but juſt to let the reader partake in ſome meaſure of the diſagreeable circumſtances on theſe unlucky days, by giving the ſketch of our ſtay in this place.

Account of the obſtacles we met with.

* Cape Holland and Cape Galant bear among themſelves E. 2° S. and W. 2° N. and the diſtance is about eight leagues. Between theſe two capes there is one, leſs projecting, called Cape Coventry. They likewiſe place ſeveral bays there, of which we have only ſeen Bay Verte, or Green Bay, or Bay De Cordes, which has been viſited by land. It is great and deep, but there ſeem to be ſeveral ſhallows in it.

† Bay Forteſcue is about two miles broad from one point to the other, and not quite ſo deep, from its entrance, till to a peninſula, which, coming from the weſt-ſide of the bay, extends E. S. E. and covers a port, well ſheltered from all the winds. This is Port Galant, which is a mile deep towards the W. N. W. Its breadth is from four hundred to five hundred yards. There is a river at the bottom of the port, and two more on the N. E. ſide. In the middle of the port there is four or five fathoms of water, bottom of ooze and ſhells.

My

My first care was to send out people to view the coast as far as Bay Elizabeth, and the isles with which the straits of Magalhaens are full in this part. From our anchoring-place we perceived two of these isles, which Narborough * calls Charles and Monmouth. Those which are farther off he calls the Royal Isles, and the westermost of all, he names Rupert Island. The west winds preventing us from making sail, we moored with a stream-anchor. The rain did not keep our people from going on shore, where they found vestiges of the passage and touching of English ships; viz. some wood, lately sawed and cut down; some spice-laurel trees †, lately stripped of their bark; a label of wood, such as in marine arsenals, are generally put upon pieces of cloth, &c. on which we very distinctly read the words, *Chatham, March*, 1766; they likewise found upon several trees, initial letters and names, with the date of 1767.

Vestiges we found of the passage of English ships.

M. Verron, who had got all his instruments carried upon the peninsula that forms the harbour, made an observation there at noon, with a quadrant; and found 53° 40′ 41″ S. lat. This observation, and the bearings

Astronomical and nautical observations.

* Sir John Narborough. F.

† *Laurier-epice*, spice-laurel is probably the famous *Winters-bark*, mentioned by Sir John Narborough, and afterwards well drawn and described by Sir Hans Sloane, in his History of Jamaica, vol. ii. p. 87. t. 19. f. 2. and Plukenet. Almagest. 89. t. 81. f. 1. and t. 160. f. 7. F.

 of

of Cape Holland, taken from hence; and those of the same cape, taken the 16th of December, upon the point from Cape Forward, determine the distance of Port Galant to Cape Forward, to twelve leagues. Here he likewise observed, by the azimuth-compass, the declination of the needle 22° 30′ 32″ N. E. and its inclination from the elevation of the pole 11° 11′. These are the only observations he was able to make, during almost a whole month; the nights being as gloomy as the days. On the third of January, there was a fine opportunity of determining the longitude of this bay; by means of an eclipse of the moon, which began here at 10 hours, 30′ in the evening; but the rain, which had been continual in the day-time, lasted likewise through the whole night.

The 4th and 5th the weather was intolerable; we had rain, snow, a sharp cold air, and a storm; it was such weather as the Psalmist describes, saying, *Nix, grando, glacies, spiritus procellarum.* On the third I had sent out a boat on purpose, to endeavour to find out an anchorage on the coast of Terra del Fuego; and they found a very good one S. W. of the isles Charles and Monmouth. I likewise gave them orders to observe the direction which the tide took in that channel. With their assistance, and the knowledge of anchoring-places, both to the northward and southward, I would have

made

made sail, even though the wind should be contrary; but it was never moderate enough for me to do it. Upon the whole, during our stay in this part of the straits, we observed constantly, that the tides set in as in the part of the narrows or guts; i. e. that the flood sets to the eastward, and the ebb to the westward.

Interview with, and d scription of the Pecherai

On the 6th, in the afternoon, we had some fair moments; and the wind too seemed to blow from S. E. we had already unmoored; but the moment we were setting sail, the wind came back to W. N. W. in squalls, which obliged us to moor again immediately. That day some savages came to visit us. Four periaguas appeared in the morning, at the point of Cape Galant; and, after stopping there for some time, three advanced into the bottom of the bay, whilst one made towards our frigate. After hesitating for about half an hour, they at last brought her along-side of us, with repeated shouts of *Pecherais*. In this boat were a man, a woman, and two children. The woman remained to take care of the periagua; and the man alone came on board, with much confidence, and with an air of gaiety. Two other pariaguas followed the example of the first; and the men came on board the frigate with their children. Here they were soon very happy and content. We made them sing, and dance, let them hear music; and, above all, gave them to eat, which they did with much appetite.

tite. They found every thing good; whether bread, ſalt meat, or fat, they devoured what was offered to them. We found it rather difficult to get rid of theſe troubleſome and diſguſting gueſts; and we could not determine them to return to their periaguas, till we ſent pieces of ſalt fleſh down into them, before their faces. They ſhewed no ſurpriſe; neither at the ſight of the ſhips, nor at the appearance of various objects, that offered themſelves to their eyes; this certainly ſhews, that in order to be capable of being ſurpriſed at the work of art, one muſt have ſome fundamental ideas of it. Theſe unpoliſhed men, conſidered the maſter-pieces of human induſtry, in the ſame light as the laws of nature and its phenomena. We ſaw them often on board, and on ſhore, during ſeveral days which they ſtayed in Port Galant.

Theſe ſavages are ſhort, ugly, meagre, and have an inſupportable ſtench about them. They are almoſt naked; having no other dreſs than wretched ſeal-ſkins, too little for them to wrap themſelves in; theſe ſkins ſerve them equally as roofs to their huts, and as ſails to their periaguas. They have likewiſe ſome guanaco-ſkins; but they are in ſmall number. Their women are hideous, and ſeemed little regarded by the men. They are obliged to ſteer their periaguas, and to keep them in repair; often ſwimming to them, notwithſtanding the cold, through the ſea-weeds, which ſerve as a harbour to theſe periaguas, at a pretty diſtance from the ſhore, and ſcooping out the water that may have got into them. On the ſhore they gather wood and ſhells, without the men partaking in any thing of their labour; nor are thoſe women, who have children at their breaſt, exempted from their taſk. They carry their children on their backs, folded in the ſkins, which ſerve them as dreſſes.

Their periaguas are made of bark, ill connected with ruſhes, and caulked with moſs in the ſeams. In the middle of each is a little hearth of ſand, where they always keep up ſome fire. Their arms are bows and arrows, made of the wood of a holly-leaved berberry-buſh, which is common in the ſtraits; the bow-ſtring is made of a gut, and the arrows are armed with points of ſtone, cut with ſufficient ſkill; but theſe weapons are made uſe of, rather againſt game, than againſt enemies for they are as weak as the arms, which are deſtined to manage them. We likewiſe ſaw amongſt them, ſome bones of fiſh, about a foot long, ſharp at the end, and toothed along one ſide. This is, perhaps, a dagger; or rather, as I think, an inſtrument for fiſhing: they fix it to a long pole, and uſe it as a harpoon. Theſe Indians, men, women, and children, live promiſcuouſly in their huts, in the middle of which they light a fire.

They

They live chiefly on ſhell-fiſh; however, they have likewiſe dogs, and nooſes, or ſpringes, made of whalebone. I have obſerved, that they had all of them bad teeth; and, I believe, we muſt attribute that to their cuſtom of eating the ſhell-fiſh boiling hot, though half raw.

Upon the whole, they ſeem to be good people; but they are ſo weak, that one is almoſt tempted to think the worſe of them on that account. We thought we obſerved that they were ſuperſtitious and believed in evil genii; and among them, the ſame perſons, who conciliate the influence of thoſe ſpirits, are their phyſicians and prieſts. Of all the ſavages I ever ſaw, the Pecherais are thoſe who are moſt deprived of every convenience; they are exactly, in what may be called, a ſtate of nature; and, indeed, if any pity is due to the fate of a man, who is his own maſter, has no duties or buſineſs to attend, is content with what he has, becauſe he knows no better, I ſhould pity theſe men; who, beſides being deprived of what renders life convenient, muſt ſuffer the extreme roughneſs of the moſt dreadful climate in the world. Theſe Pecherais, likewiſe, are the leaſt numerous ſociety of men I have met with in any part of the world; however, as will appear in the ſequel, there are quacks among them: but as ſoon as more than one family is together, (by family, I underſtand father, mother,

ther, and children) their interefts become complicated, and the individuals want to govern, either by force or by impofture. The name of family then changes into that of fociety; and though it were eftablifhed amidft the woods, and compofed only of coufins-german, a fkilful obferver would there difcover the origin of all the vices, to which men, collected into whole nations, have, by growing more civilized, given names; vices that caufed the origin, progrefs, and ruin of the greateft empires. Hence it follows, by the fame principle, that in civilized focieties, fome virtues fpring up, of which thofe who border on a ftate of nature are not fufceptible.

The 7th and 8th the weather was fo bad, that we could not by any means go from on board; in the night we drove, and were obliged to let go our fheet-anchor. At fome intervals the fnow lay four inches deep on the deck; and, at day-break, we faw that all the ground was covered with it, except the flat lands, the wetnefs of which melted the fnow. The thermometer was about 5° and 4°; but fell to two degrees below the freezing-point. The weather was bad on the ninth in the afternoon. The Pecherais fet out in order to come on board us. They had even fpent much time at their toilet; I mean, they had painted their bodies all over, with red and white fpots: but feeing our boats go from the fhips, towards their

 huts,

huts, they followed them; but one periagua came on board the Etoile. She ſtayed but a ſhort time there, and joined the others; who were very much the friends of our people. The women were, however, all retired into one hut; and the ſavages ſeemed uneaſy, whenever one of our men attempted to go in. They invited them rather to come into the other huts, where they preſented our gentlemen with muſcles, which they ſucked before they gave them away. They got ſome little preſents, which they gladly accepted. They ſung, danced, and appeared more gay, than one might expect from ſavages, whoſe outward behaviour is commonly ſerious.

Unlucky accident, which befalls one of them.

Their joy was but of very ſhort duration. One of their children, about twelve years old, the only one in the whole troop whoſe figure engaged our attention, was all at once ſeized with ſpitting of blood, and violent convulſions. The poor creature had been on board the Etoile, where the people had given him bits of glaſs, not foreſeeing the unhappy effect, which this preſent might have. Theſe ſavages have a cuſtom of putting pieces of talc into their throat and noſtrils. Perhaps their ſuperſtition combines ſome powers with this kind of taliſman; or, perhaps, they look on it as a preſervative againſt ſome ſickneſs they are ſubject to. The child, probably, had made the ſame uſe of this glaſs. His lips, gums, and palate, were cut in ſeveral places, and he bled continually.

This accident ſpread conſternation and miſtruſt amongſt them, They certainly ſuſpected us of ſome bad action; for the firſt thing their juggler did, was to ſtrip the child immediately of a linen jacket, which had been given him. He wanted to return it to the French; and upon their refuſing it, he threw it at their feet. However, another ſavage, who, doubtleſs, loved clothes more than he feared enchantments, took it up immediately.

The juggler firſt laid the child down upon his back, in one of the huts; and, kneeling down between his legs, he bent himſelf upon him, and with his head and hands preſſed the child's belly as much as he could, crying out continually, without our being able to diſtinguiſh any articulate ſounds in his cries. From time to time he got up, and ſeeming to hold the diſeaſe in his joined hands, he opened them all at once into the air, blowing as if he wanted to drive away ſome evil ſpirit. During this ceremony, an old woman in tears, howled in the ſick child's ears, enough to make him deaf. This poor wretch ſeemed to ſuffer as much from the remedy, as from the hurt he had received. The juggler gave him ſome reſpite, and went to fetch his habit of ceremony; after which, having his hair powdered, and his head adorned with two white wings, like thoſe on Mercury's cap, he began his rites again, with more con-

 ſidence,

fidence, but with no better ſucceſs. The child then appearing to be worſe, our chaplain adminiſtred baptiſm to him by ſtealth.

The officers returned on board, and told me what had happened on ſhore. I went thither immediately with M. de la Porte, our ſurgeon, who brought ſome milk and gruel with him. When we arrived, the patient was out of the hut; the juggler, who had now got a companion in the ſame dreſs, had begun again with his his operation on the belly, thighs, and back of the child. It was a pity to ſee them torment the poor creature, who ſuffered without complaining. His body was already bruiſed all over; and the doctors ſtill continued to apply their barbarous remedy, with abundance of conjurations. The grief of the parents, their tears, the part which the whole troop took in this accident, and which broke out in the moſt expreſſive ſigns, afforded us a moſt affecting ſcene. The ſavages certainly perceived that we partook of their diſtreſs; at leaſt they ſeemed to be leſs miſtruſtful. They ſuffered us to come near the patient; and our ſurgeon examined his bloody mouth, which his father and another Pecherais ſucked alternately. We had much trouble to perſuade them to uſe milk; we were obliged to taſte it before them ſeveral times; and, notwithſtanding the invincible objection of their jugglers, the father at laſt reſolved to let his

ſon

ſon drink it; he even accepted a pot-full of gruel. The jugglers were jealous of our ſurgeon; whom, however, they ſeemed at laſt to acknowledge as an able juggler. They even opened for him a leather bag, which they always wear hanging by their ſide; and which contains their feathered cap, ſome white powder, ſome talc, and other inſtruments of their art; but he had hardly looked into it, when they ſhut it again. We likewiſe obſerved, that whilſt one of the jugglers was conjuring the diſtemper of the patient, the other ſeemed to be buſied ſolely in preventing, by his enchantments, the effect of the bad luck, which they ſuſpected we had brought upon them.

We returned on board, towards night, and the child ſeemed to ſuffer leſs; however, he was plagued with almoſt continual puking, which gave us room to fear that ſome glaſs was got down into his ſtomach. We had afterwards ſufficient reaſon to believe our conjectures had been true; for about two o'clock in the morning, we on board heard repeated howls; and, at break of day, though the weather was very dreadful, the ſavages went off. They, doubtleſs, fled from a place defiled by death, and by unlucky ſtrangers, who they thought were come merely to deſtroy them. They were not able to double the weſtermoſt point of the bay: in a more moderate interval they ſet ſail again; a violent ſquall carried

carried them out into the offing, and dispersed their feeble vessels. How desirous they were of getting away from us! They left one of their periaguas, which wanted a repair on the shore, *Satis est gentem effugisse nefandam.* They are gone away, considering us as mischievous beings: but who would not pardon their resentment on this occasion? and, indeed, how great is the loss of a youth, who has escaped from all the dangers of childhood, to a body of men so very inconsiderable in number!

Continuation of bad weather.

The wind blew east with great violence, and almost without intermission, till the 13th, when the weather was mild enough in day-time; and we had even conceived hopes of weighing in the afternoon. The night between the 13th and 14th was calm. At half an hour past two in the morning we had unmoored, and hove a-peak. At six o'clock we were obliged to moor again, and the day was dreadful. The 15th, the sun shone almost the whole day; but the wind was too strong for us to leave the harbour.

Danger which the frigate is exposed to.

The 16th, in the morning, it was almost a calm; then came a breeze from the north, and we weighed, with the tide in our favour: it was then ebbing, and set to the westward. The winds soon shifted to W. and W.S.W. and we could never gain the Isle Rupert, with the favourable tide. The frigate sailed very ill; drove to leeward beyond measure; and the Etoile had an incredible

credible advantage over us. We plyed all day between Rupert iſland, and a head-land of the continent, which we called the Point of the Paſſage, in order to wait for the ebb; with which I hoped either to gain the anchoring-place of Bay Dauphine, upon the iſle of Louis le Grand, or that of Elizabeth bay*. But as we loſt ground by plying, I ſent a boat to ſound to the S. E. of Rupert's-iſland, intending to anchor there, till the tide became favourable. They made ſignal of an anchoring-place, and came to a grapnel there; but we were already too much fallen to leeward of it. We made one board in-ſhore, to endeavour to gain it on the other tack; the frigate miſſed ſtays twice; and it became neceſſary to wear; but at the very moment when, by the manœuvres, and by the help of our boats, ſhe began to wear, the force of the tide made her come to the wind again; a ſtrong current had already carried us within half a cable's length of the ſhore. We let go our anchor in eight fathom: the anchor, falling upon rocks, came home, and our proximity to the ſhore did not al-

* From Cape Galant to Bay Elizabeth, the coaſt runs nearly W. N. W. and the diſtance from the one to the other, is about four leagues. In this ſpace there is no anchoring-place on the main-land. The depth is too great, even cloſe to the ſhore. Bay Elizabeth is open to the S. W. Its breadth between the points is three quarters of a league; and its depth pretty near the ſame. The ſhore in the bottom of the bay is ſandy; and ſo is the S. E. ſhore. In its northern part lies a ledge, ſtretching a good way to the offing. The good anchoring in this bay is nine fathom, bottom of ſand, gravel, and coral; and has the following marks: the E. point of the bay bears S. S. E. ½ E. its W point, W. b. N. The E. point of the iſle of Louis le Grand, S. S. W. ½ S. the ledge N. W. b. N.

low us to veer away cable. We had now no more than three fathom and a half of water a-ſtern; and were only thrice the length of the ſhip from the ſhore, when a little breeze ſprung up from thence, we immediately filled our ſails, and the frigate fell to leeward: all our boats, and thoſe of the Etoile, which came to our aſſiſtance, were a-head, towing her We veered away our cable, upon which we had put a buoy; and near half of it was out, when it got foul between decks, and ſtopt the frigate, which then ran the greateſt danger. We cut the cable, and by the prompt execution of this manœuvre, we ſaved the ſhip. The breeze at length freſhened; and, after having made two or three unprofitable boards, I returned to Port Galant, where we anchored again in twenty fathom oozy bottom. Our boats, which I left to weigh our anchor, returned towards night with it and the cable. Thus this appearance of fine weather ſerved only to give us cruel alarms.

Violent hurricane.

The day following was more ſtormy than all the preceding ones. The wind raiſed a mountainous ſea in the channel; and we often ſaw ſeveral waves run in contrary directions. The ſtorm appeared to abate towards ten o'clock; but at noon a clap of thunder, the only one we ever heard in this ſtrait, was as it were the ſignal at which the wind again began to blow with more violence than in the morning. We dragged our

anchor

anchor, and were obliged to let go our ſheet-anchor, and ſtrike our lower-yards and top-maſts. Notwithſtanding this, the ſhrubs and plants were now in flower, and the trees afforded a very brilliant verdure, which however was not ſufficient to diſpel that ſadneſs which the repeated ſight of this unlucky ſpot had caſt over us. The moſt lively temper would be overcome in this dreadful climate, which is ſhunned by animals of every element, and where a handful of people lead a languid life, after having been rendered ſtill more unfortunate by their intercourſe with us.

On the 18th and 19th there were ſome intervals between the bad weather: we weighed our ſheet anchor, ſquared our yards, and ſet up our top-maſts; and I ſent the Etoile's barge, which was in ſo good a condition as to be able to go out in almoſt any weather, to view the channel of *Sainte Barbe*. According to the extract M. Frezier gives of the Journal of M. Marcant, who diſcovered and paſſed through it, this channel muſt bear S. W. and S. W. by S. from Bay Elizabeth. The barge returned on the 20th, and M. Landais, who commanded it, informed me, that having followed the track and marks taken notice of by M. Marcant, he had not found the true mouth, but only a narrow channel, cloſed by ſhoals of ice and the land, which it is the more dangerous to follow, as it has not a ſingle good anchoring place, and as it is

Aſſertion concerning the channel of Sainte Barbe diſcuſſed.

 croſſed

croffed in the middle by a fand covered with mufcles. He then went all round the ifle of Louis le Grand to the fouthward, and re-entered the channel of Magalhaens, without having found any other. He only faw a fine bay on the coaft of Terra del Fuego, which is certainly the fame with that which Beauchefne calls Nativity Bay. Upon the whole, by going S. W. and S. W. by S. from Bay Elizabeth, as Frezier fays that Marcant did, you muft cut through the middle of the ifle of Louis le Grand.

This information gave me room to believe that the channel of Sainte Barbe was oppofite the very bay where we now lay. From the top of the mountains which furround Port Galant, we had often difcovered fouthward of the ifles Charles and Monmouth, a vaft channel, full of little iflands, and terminated by no land to the fouthward; but, as at the fame time we perceived another inlet fouthward of the ifle of Louis le Grand, we took that for the channel of St. Barbe, as being more conformable to Marcant's account. As foon as we were fure that this inlet was no more than a deep bay, we no longer doubted that the channel of Sainte Barbe was oppofite Port Galant, fouthward of Charles and Monmouth Iflands. Indeed, reading over again the paffage in Frezier, and comparing it with his chart of the ftrait, we faw that Frezier, according to

 Mar-

Marcant's report, places Elizabeth Bay, from whence the latter set sail, in order to enter into his channel about ten or twelve leagues from Cape Forward. Marcant therefore must have mistaken Bay De Cordes for Bay Elizabeth, the former lying actually eleven leagues from Cape Forward, being a league eastward of Port Galant: setting sail from this bay, and standing S. E. and S. E. by S. he came along the westermost point of Charles and Monmouth isles, the whole of which he took for the isle of Louis le Grand; an error into which every good navigator may easily fall, unless he is well provided with good directions: and then he stood into the channel full of isles, of which we had a prospect from the top of the mountains.

Utility whic[h] would accru[e] from the knowledge [of] the channel of Sainte Barbe.

The perfect knowledge of the channel of Sainte Barbe would be so much the more interesting, as it would considerably shorten the passing of the straits of Magalhaens. It does not take much time to come to Port Galant; the greatest difficulty before you come there, being to double Cape Forward, which is now rendered pretty easy, by the discovery of three ports upon Terra del Fuego: when you are once got to Port Galant, should the winds prevent your taking the ordinary channel, if they be ever so little upon the northerly points, the channel is open to you, opposite to this port; in twenty-four hours you can then be in the South Seas.

 I in-

I intended to have ſent two barges into this channel which I firmly believe to be that of Sainte Barbe; they would have completely ſolved this problem, but the bad weather prevented their going out.

Exceeding violent ſquall.

The 21ſt, 22d, and 23d, ſqualls, ſnow, and rain, were continual. In the night between the 21ſt and 22d, there was a calm interval; it ſeemed that the wind afforded us that momentary repoſe, only in order to fall harder upon us afterwards. A dreadful hurricane came ſuddenly from S. S. W. and blew with ſuch fury as to aſtoniſh the oldeſt ſeamen. Both our ſhips had their anchors come home, and were obliged to let go their ſheet-anchor, lower the lower yards, and hand the top-maſts: our mizen was carried away in the brails. Happily this hurricane did not laſt long. On the 24th the ſtorm abated, we got calm weather and ſun-ſhine, and put ourſelves in a condition to proceed. Since our re-entering Port Galant, we took ſeveral ton weight of ballaſt, and altered our ſtowage, endeavouring by this means to make the frigate ſail well again; and we ſucceeded in part. Upon the whole, whenever it is neceſſary to navigate in the midſt of currents, it will always be found very difficult to manage ſuch long veſſels as our frigates generally are.

On the 25th, at one o'clock in the morning, we unmoored, and hove a peek; at three o'clock we weighed,

and

and were towed by our boats; the breeze was northerly; at half paſt five it ſettled in the eaſt, and we ſet all our top-gallant and ſtudding-ſails, which are very ſeldom made uſe of here. We kept the middle of the ſtrait, following its windings, for which Narborough juſtly calls it Crooked Reach. Between the Royal Iſles and the continent, the ſtrait is about two leagues wide; the channel between Rupert Iſle and Point Paſſage, is not above a league broad; then there is the breadth of a league and a half between the iſle of Louis le Grand and Bay Elizabeth, on the eaſterly point of which, there is a ledge covered with ſea weeds, extending a quarter of a league into the ſea.

We leave Bay Forteſcue.

From Bay Elizabeth the coaſt runs W. N. W. for about two leagues, till you come to the river which Narborough calls Bachelor, and Beaucheſne, du Maſſacre; at the mouth of which, is an anchoring-place. This river is eaſily known; it comes from a deep valley; on the weſt, it has a high mountain; its weſterly point is low, wooded, and the coaſt ſandy. From the river Bachelor, to the entrance of the falſe ſtrait or St. Jerom's channel, I reckon three leagues, and the bearing is N. W. by W. The entrance of this channel ſeems to be half a league broad, and in the bottom of it, the lands are ſeen cloſing in to the northward. When you are oppoſite the river du Maſſacre, or Bachelor, you can only ſee this

Deſcription of the ſtrai[t] from Cape Galant to t[he] open ſea.

this false strait, and it is very easy to take it for the true one, which happened even to us, because the coast then runs W. by S. and W. S. W. till Cape *Quade*, which stretching very far, seems to close in with the westerly point of the isle of Louis le Grand, and leave no outlet. Upon the whole, the safest way not to miss the true channel, is to keep the coast of Louis le Grand island on board, which may be done without any danger. The distance of St. Jerom's channel to Cape Quade, is about four leagues, and this cape bears E. 9 N. and W. 9° S. with the westerly point of the isle of Louis le Grand.

That island is about four leagues long, its north side runs W. N. W. as far as Bay Dauphine, the depth of which, is about two miles, and the breadth at the entrance, half a league; it then runs W. to its most westerly extremity, called Cape St. Louis. As, after finding out our error concerning the false strait, we run within a mile of the shore of Louis le Grand island, we distinctly saw Port Phelippeaux, which appeared to be a very convenient and well situated creek. At noon Cape Quade bore W. 13° S. two leagues distant, and Cape St. Louis, E. by N. about two leagues and a half off. The fair weather continued all day, and we bore away with all our sails set.

From

From Cape Quade the ſtrait runs W. N. W. and N. W. by W. without any conſiderable turnings, from which it has got the name of Long-Lane, or Long-Reach, (*Longue Rue*). The figure of Cape Quade is remarkable. It conſiſts of craggy rocks, of which, thoſe forming its higheſt ſummits, do not look unlike ancient ruins. As far as this cape, the coaſts are every where wooded, and the verdure of the trees ſoftens the aſpect of the frozen tops of the mountains. Having doubled Cape Quade, the nature of the country is quite altered. The ſtrait is incloſed on both ſides by barren rocks, on which there is no appearance of any ſoil. Their high ſummits are always covered with ſnow, and the deep vallies are filled with immenſe maſſes of ice, the colour of which bears the mark of antiquity. Narborough, ſtruck with this horrid aſpect, called this part, Deſolation of the South, nor can any thing more dreadful be imagined.

Being oppoſite Cape Quade, the coaſt of Terra del Fuego ſeems terminated by an advanced cape, which is Cape Monday, and which I reckon is about fifteen leagues from Cape Quade. On the coaſt of the main land, are three capes, to which we gave names. The firſt, which from its figure, we called *Cap Fendu*, or Split Cape, is about five leagues from Cape Quade, between two fine bays, in which the anchorage is ſafe, and

and the bottom as good as the ſheltered ſituation. The other two capes received the names of our ſhips, Cap de l'Etoile, three leagues weſt of Cap Fendu, and Cap de la Boudeuſe, in the ſame ſituation, and about the ſame diſtance from the Cape of the Etoile. All theſe lands are high and ſteep; both coaſts appear clear, and ſeem to have good anchoring places, but happily, the wind being fair for our courſe, did not give us time to ſound them. The ſtrait in this part, called *Longue Rue*, is about two leagues broad; it grows more narrow towards Cape Monday, where it is not above four miles broad.

Dangerous night.

A nine o clock in the evening, we were about three leagues E. by S. and E. S. E. off Cape Monday. It always blew very freſh from eaſt, and the weather being fine, I reſolved to continue my courſe during the night, making little ſail. We handed the ſtudding ſails, and cloſe-reefed the top-ſails. Towards ten o'clock at night the weather became foggy, and the wind encreaſed ſo much, that we were obliged to haul our boats on board. It rained much, and the weather became ſo black at eleven, that we loſt all ſight of land. About half an hour after, reckoning myſelf a-breaſt of Cape Monday, I made ſignal to bring-to on the ſtar-board tack, and thus we paſſed the reſt of the night, filling or backing, according as we reckoned ourſelves to be too near one

or

or the other ſhore. This night we have been in one of the moſt critical ſituations during the whole voyage.

At half an hour paſt three, by the dawn of day, we had ſight of the land, and I gave orders to fill. We ſtood W. by N. till eight o'clock, and from eight till noon, between W. by N. and W. N. W. The wind was always eaſt, a little breeze, and very miſty. From time to time we ſaw ſome parts of the coaſt, but often we entirely loſt ſight of it. At laſt, at noon, we ſaw Cape Pillar, and the Evangeliſts. The latter could only be ſeen from the maſt-head. As we advanced towards the ſide of Cape Pillar, we diſcovered, with joy, an immenſe horizon, no longer bounded by lands, and a great ſea from the weſt, which announced a vaſt ocean to us. The wind did not continue E. it ſhifted to W. S. W. and we ran N. W. till half an hour paſt two, when Cape Victory bore N. W. and Cape Pillar, S. 3° W.

End of the ſtrait, and deſcription of that part.

After paſſing Cape Monday, the north coaſt bends like a bow, and the ſtrait opens to four, five, and ſix leagues in breadth. I reckon about ſixteen leagues from Cape Monday to Cape Pillar, which terminates the ſouth coaſt of the ſtraits. The direction of the channel between theſe two capes, is W. by N. The ſouthern coaſt is here high and ſteep, the northern one is bordered with iſlands and rocks, which make it dangerous to come near it: it is more prudent to keep the ſouth coaſt

 on

on board. I can ſay no more concerning theſe laſt lands: I have hardly ſeen them, except at ſome ſhort intervals, when the fogs allowed our perceiving but ſmall parts of them. The laſt land you ſee upon the north coaſt, is Cape Victory (*Cap des Victoires*), which ſeems to be of middling height, as is Cape Deſeado (*Deſiré*), which is without the ſtraits, upon Terra del Fuego, about two leagues S. W. of Cape Pillar. The coaſt between theſe two capes is bounded for near a league into the ſea, by ſeveral little iſles or breakers, known by the name of the Twelve Apoſtles.

Cape Pillar is a very high land, or rather a great maſs of rocks, which terminates in two great cliffs, formed in the ſhape of towers, inclining to N. W. and making the extremity of the cape. About ſix or ſeven leagues N. W. of this cape, you ſee four little iſles, called the Evangeliſts; three of them are low, the fourth, which looks like a hay-ſtack, is at ſome diſtance from the reſt. They ly S. S. W. about four or five leagues off Cape Victory In order to come out of the ſtrait, it is indifferent whether you leave them to the ſouth or northward; in order to go in, I would adviſe that they ſhould be left to the northward. It is then likewiſe neceſſary to range along the ſouthern coaſt; the northern one is bordered with little iſles, and ſeems cut by large bays, which might occaſion dangerous miſ-

miſtakes. From two o'clock in the afternoon, the winds were variable, between W. S. W. and W. N. W. and blew very freſh; we plyed till ſun-ſetting, with all our ſails ſet, in order to double the Twelve Apoſtles. We were for a long while afraid we ſhould not be able to do it, but be forced to paſs the night ſtill in the ſtraits, by which means we might have been obliged to ſtay there more than one day. But about ſix o'clock in the evening we gave over plying; at ſeven, Cape Pillar was doubled, and at eight we were quite clear of the land, and advancing, all ſails ſet, and with a fine northerly wind, into the weſterly ocean. We then laid down the bearings whence I took my departure, in 52° 50′ S. lat. and 79° 9′ W. long. from Paris.

Departure taken from the ſtrait of Magalhaens.

Thus, after conſtant bad and contrary weather at Port Galant, for twenty-ſix days together, thirty-ſix hours of fair wind, ſuch as we never expected, were ſufficient to bring us into the Pacific Ocean; an example, which I believe is the only one, of a navigation without anchoring from Port Galant to the open ſea.

General obſervations on this navigation.

I reckon the whole length of the ſtrait, from Cape Virgin (Mary) to Cape Pillar, at about one hundred and fourteen leagues. We employed fifty-two days to make them. I muſt repeat here, that from Cape Virgin to Cape Noir, we have conſtantly found the flood tide to ſet to the eaſtward, and the ebb to the weſtward,

 and

and that the tides are very ſtrong; that they are not by much ſo rapid from Cape Noir to Port Galant, and that their direction is irregular there; that laſtly, from Port Galant to Cape Quade, the tides are violent; that we have not found them very conſiderable from this cape to Cape Pillar, but that in all this part from Port Galant, the water is ſubject to the ſame laws which put them in motion from Cape Virgin; viz. that the flood runs towards the eaſterly, and the ebb towards the weſterly ſeas. I muſt at the ſame time mention, that this aſſertion concerning the direction of the tides in the ſtrait of Magalhaens, is abſolutely contrary to what other navigators ſay they have obſerved there on this head. However, it would not be well if every one gave another account.

Upon the whole, how often have we regretted that we had not got the Journals of Narborough and Beaucheſne, ſuch as they came from their own hands, and that we were obliged to conſult disfigured extracts of them: beſides the affectation of the authors of ſuch extracts, of curtailing every thing which is uſeful merely in navigation; likewiſe, when ſome details eſcape them that have a relation to that ſcience, their ignorance of the ſea-phraſes makes them miſtake neceſſary and uſual expreſſions for vicious words, and they replace them by abſurdities. All their

their aim is to compile a work agreeable to the effeminate people of both ſexes, and their labour ends in compoſing a book that tires every body's patience, and is uſeful to nobody *.

Conclusions drawn from hence.

Notwithſtanding the difficulties which we have met with in our paſſage of the ſtrait of Magalhaens, I would always adviſe to prefer this courſe to that of doubling Cape Horn, from the month of September to the end of March. During the other months of the year, when the nights are ſixteen, ſeventeen, and eighteen hours long, I would paſs through the open ſea. The wind a-head, and a high ſea, are not dangerous; whereas, it is not ſafe to be under a neceſſity of ſailing blindfold between the ſhores. Certainly there will be ſome obſtacles in paſſing the ſtraits, but this retardment is not entirely time loſt. There is water, wood, and ſhells in abundance, ſometimes there are likewiſe very good fiſh; and I make no doubt but the ſcurvy would make more havock among a crew, who ſhould

* This complaint of our author is applicable only to the French publications, for it is well known that the Engliſh voyages, chiefly when publiſhed by authority, are remarkable both for the fine language, and the ſtrict keeping of the marine phraſes, ſo neceſſary to make theſe publications uſeful to future navigators, and which are underſtood by the greater part of this nation, ſo much uſed to the ſea and its phraſes, that our romances and plays are full of them, and that they have even a run in common life. F.

come

come into the South Seas by the way of Cape Horn, than among thoſe who ſhould enter the ſame Seas through the ſtraits of Magalhaens: when we left it, we had no ſick perſon on board.

END OF THE FIRST PART.

A VOYAGE ROUND THE WORLD.

PART the SECOND.

From our entrance into the Western Sea, to our return to France.

Et nos jam *tertia* portat
Omnibus errantes terris and fluctibus æstas. VIRG. Lib. I.

CHAP. I.

The run from the straits of Magalhaens to our arrival at the Isle of Taiti; discoveries which precede it.

January. 1768. Direction of our track, after leaving the strait.

FROM our entrance into the Western Sea, after some days of variable winds, between S. W. and N. W. we soon got S. and S. S. E. winds. I did not expect to meet with them so soon; the west winds generally last to about 30°; and I intended to go to the isle of Juan Fernandez, in order to make good astronomical observations there. I intended by this means to fix a sure point

point of departure, in order to crofs this immenfe ocean the extent of which is differently laid down, by differently navigators. The early meeting with the S. and S. E. winds, obliged me to lay afide this fcheme of putting in there, which would have prolonged my voyage.

Obfervation on the pofition of the coafts of Chili.

During the firft days, I ftood as near weft as poffible; as well to keep my wind, as to get off from the coaft; the bearings of which are not laid down with any certainty in the charts: however, as the winds were then always in the weftern board, we fhould have fallen in with the land, if the charts of Don George Juan, and Don Antonio de Ulloa had been exact. Thefe Spanifh officers have corrected the old maps of North America *: they make the coaft run N. E. and S. E. between Cape Corfo and Chiloe; and that upon conjectures, which they have certainly thought well-founded. This correction happily deferves another; it was not a very comfortable one for thofe navigators, who after coming out of the ftrait, endeavour to get to the northward, with winds which conftantly vary from S. W. to N. W. by W. Sir John Narborough, after leaving the ftraits of Magalhaens, in 1669, run along the coaft of Chili, examining all the inlets and creeks, as far as the river of Baldivia, into which he entered; he fays

* It muft be fuppofed, that the author means South America. F.

 expref sly,

expreſsly, that the courſe from Cape Deſire to Baldivia is N. 5° E. This is ſomething more certain than the conjectural aſſertion of Don George and Don Antonio. If, upon the whole, their conjecture had been true, by the courſe which we were obliged to take we muſt have fallen in with the land.

When we were got into the Pacific Ocean, I agreed with the commander of the Etoile, that, in order to diſcover a greater ſpace of the ſea, he ſhould go every morning ſouthward, as far from me as the weather would allow, keeping within ſight; and that every evening we ſhould join; and that then he ſhould keep in our wake, at about half a league's diſtance. By this means, if the Boudeuſe had met with any ſudden danger, the Etoile was enabled to give us all the aſſiſtance which the caſe might require. This order of ſailing has been followed throughout the whole voyage. Order of ſailing of the Boudeuſe and Etoile.

On the 30th of January, a ſailor fell into the ſea; our efforts were uſeleſs; and we were unable to ſave him: it blew very freſh, and we had a great ſea. Loſs of a ſailor fallen into the ſea.

I directed my courſe for making the land, which Davis *, an Engliſh privateer, ſaw in 1686, between 27° and 28° ſouth latitude; and which Roggewein, a Dutchman †, ſought for in vain, in 1722. I continu- Fruitleſs ſearch for Davis's land.

* Mr. Boungainville writes *David:* indeed, he and moſt writers of his nation, mutilate all foreign names; not only inadvertently, but often on purpoſe, through mere caprice. F.

† A Mecklenburger, who, with his father, had been in the Dutch ſervice. F.

 ed

1768. February.

ed to ftand in fearch of it till the 17th of February. According to M. de Bellin's chart, I muft have failed over this land on the 14th. I did not chufe to go in fearch of Eafter ifland, as its latitude is not laid down with certainty. Many geographers agree in placing it in 27° or 28° S. M. Buache, alone, puts it in 31° However, on the 14th, being in 27° 7′ of latitude obferved, and 104° 12′ computed weft longitude, we faw two birds very like *Equerrets* *, which generally do not go further than 60 or 80 leagues from land; we likewife faw a tuft of that green plant, which faftens on fhips' bottoms; and, for thefe reafons, I continued to ftand on the fame courfe till the 17th. Upon the whole, I think, from the account which Davis gives of the land he faw, that it is no other than the ifles of St. Ambrofe and St. Felix, which are two hundred leagues from the coaft of Chili.

Incertitude on the latitude of Eafter ifland.

From the 23d of February, to the 3d of March, we had wefterly winds, conftantly varying between S. W. and N. W. with calms and rain: every day, either a little before noon, or foon after, we had fudden gufts of rain, accompanied with thunder. It was ftrange to us to meet with this extraordinary wind, under the tropic, and in that ocean, fo much renowned above all other feas, for the uniformity and the frefhnefs of the E. and S. E.

Meteorological Obfervations.

* A kind of fea-fowl; probably of the gull or tern kind. F.

trade-

trade-winds; which are ſaid to reign in it all the year round. We ſhall find more than one opportunity to make the ſame obſervation.

Aſtronomical obſervations, compared with the ſhip' reckoning.

During the month of February, M. Verron communicated to me the reſult of four obſervations, towards determining our longitude. The firſt, which was made on the 6th at noon, differed from my reckoning only 31′; which I was more to the weſtward than his obſervation. The ſecond, taken at noon on the 11th, differed from my eſtimated longitude 37′ 45″, which I was to the eaſtward of him. By the third obſervation, made on the 22d, reduced to noon, I was more weſtward than he, by 42′ 30″; and I had 1° 25′ of difference weſt, from the longitude determined by the obſervations of the 27th. Then we met with calms and contrary winds. The thermometer, till we came into 45° lat. always kept between 5° and 8° above the freezing-point: it then roſe ſucceſſively; and when we ran between 27° and 24° of lat. it varied from 17° to 19°

There was an almoſt epidemical ſore-throat among the crew of my frigate, as ſoon as we had left the ſtraits. As it was attributed to the ſnow waters of the ſtraits, I ordered every day, that a pint of vinegar, and red hot bullets ſhould be put into the ſcuttled caſk, containing the water for the crew to drink, on the upper deck. Happily theſe ſore throats yielded to the ſimpleſt reme-

 dies;

dies; and, at the end of March, we had no-body upon the ſick-liſt. Only four ſailors were attacked by the ſcurvy. About this time we got plenty of Bonitos and Great-ears *(Grandes-Oreilles)*; and, during eight or ten days, ſufficient were taken to afford one meal a-day for the crews of both ſhips.

Meeting with the firſt iſles.

1768. March.

During March, we ran on the parallel of the firſt lands and iſles marked on the chart of M. Bellin, by the name of Quiros's Iſles. On the 21ſt we caught a tunny, in whoſe belly we found ſome little fiſh, not yet digeſted, of ſuch ſpecies as never go to any diſtance from the ſhore. This was a ſign of the vicinity of land. Indeed, the 22d, at ſix in the morning, we ſaw at once four little iſles, bearing S. S. E. $\frac{1}{2}$E. and a little iſle about four leagues weſt. The four iſles I called *les quatre Facardins*; and as they were too far to windward, I ſtood for the little iſle a-head of us. As we approached it, we diſcovered that it is ſurrounded with a very level ſand, and that all the interior parts of it are covered with thick woods, above which the cocoa-trees raiſe their fertile heads. The ſea broke much to the N. and S. and a great ſwell beating all along the eaſtern ſide, prevented our acceſs to this iſle in that part. However, the verdure charmed our eyes, and the cocoa-trees every where expoſed their fruits to our ſight, and over-ſhadowed a graſs-plot adorned with flowers; thouſands of birds

Obſervations on one of theſe iſles.

were

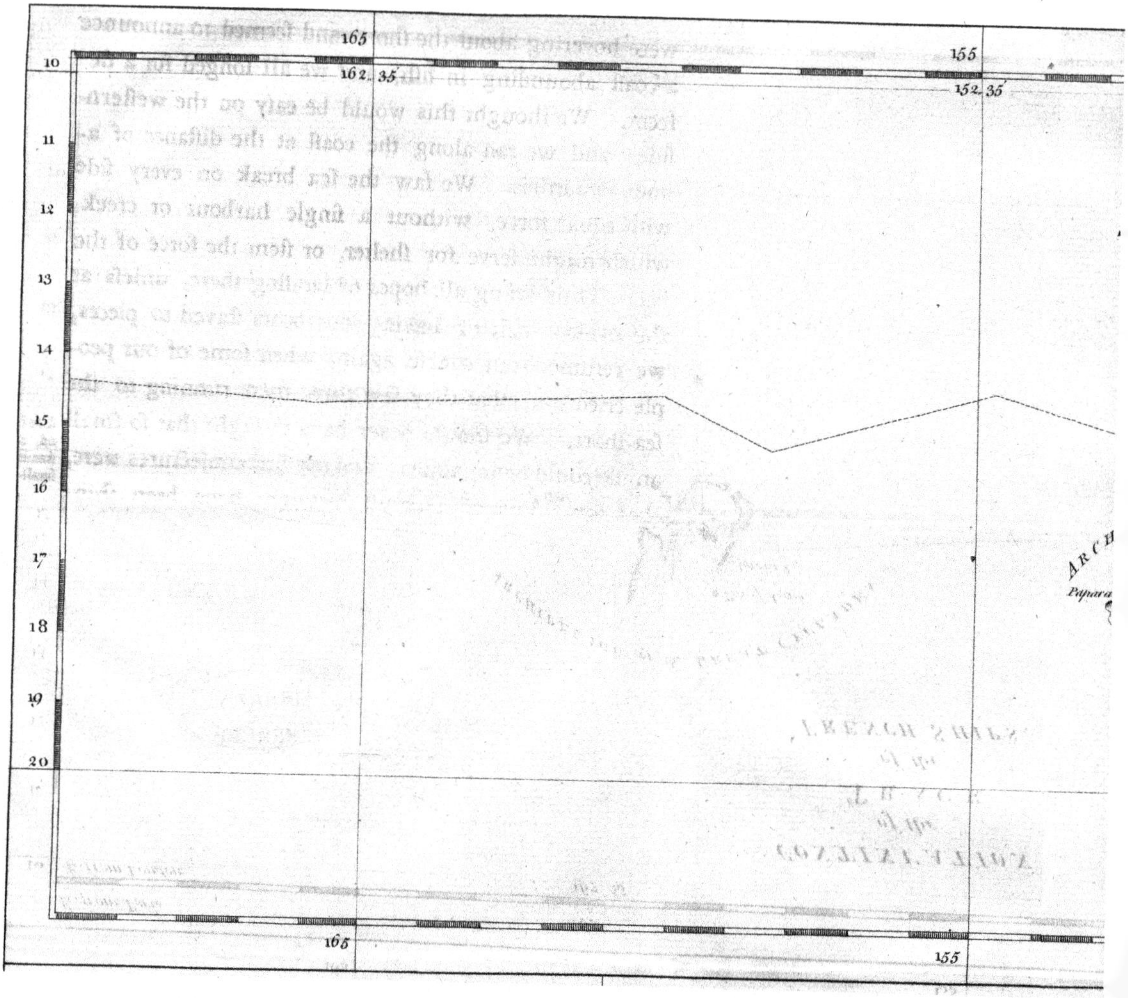

165
155
162. 35
152. 35
10
11
12
13
14
15
16
17
18
19
20
ARCH
Papara
165
155

to face Page 205

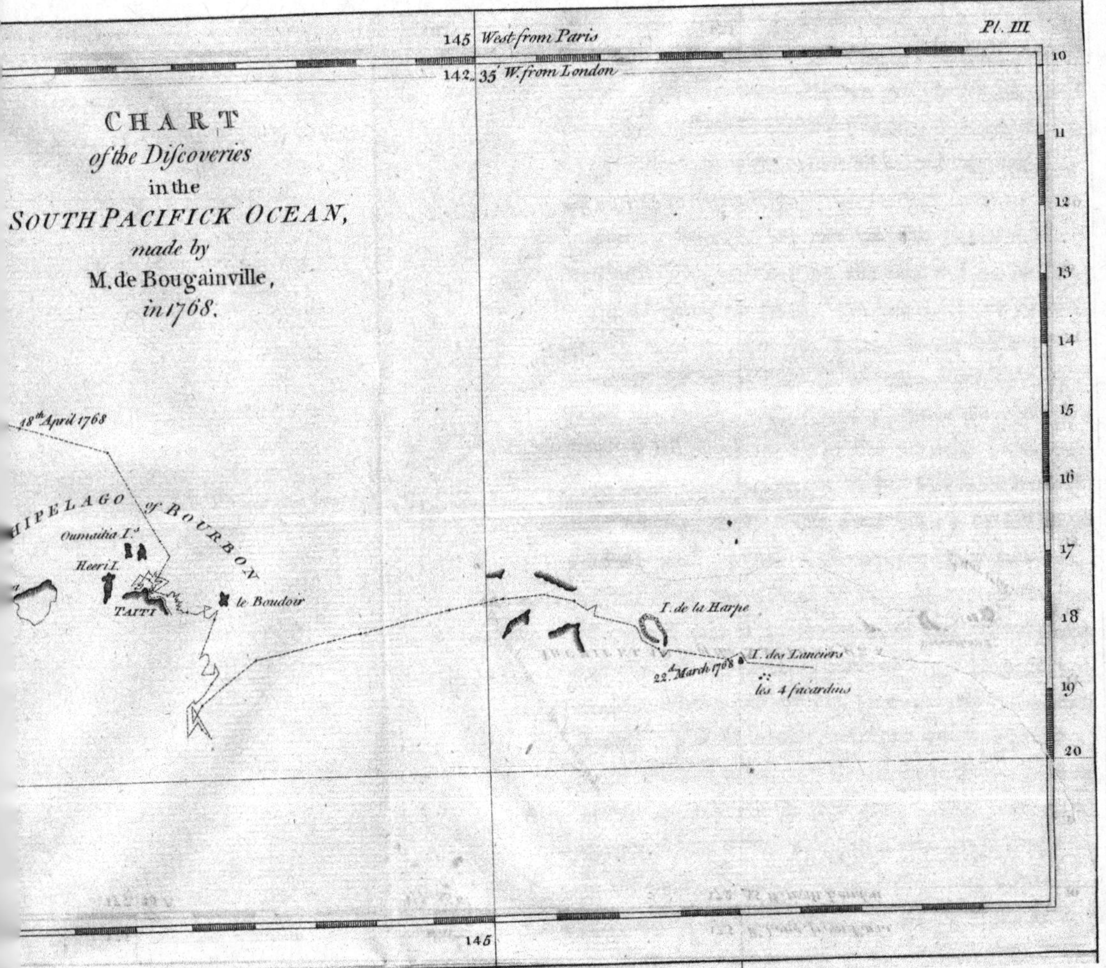

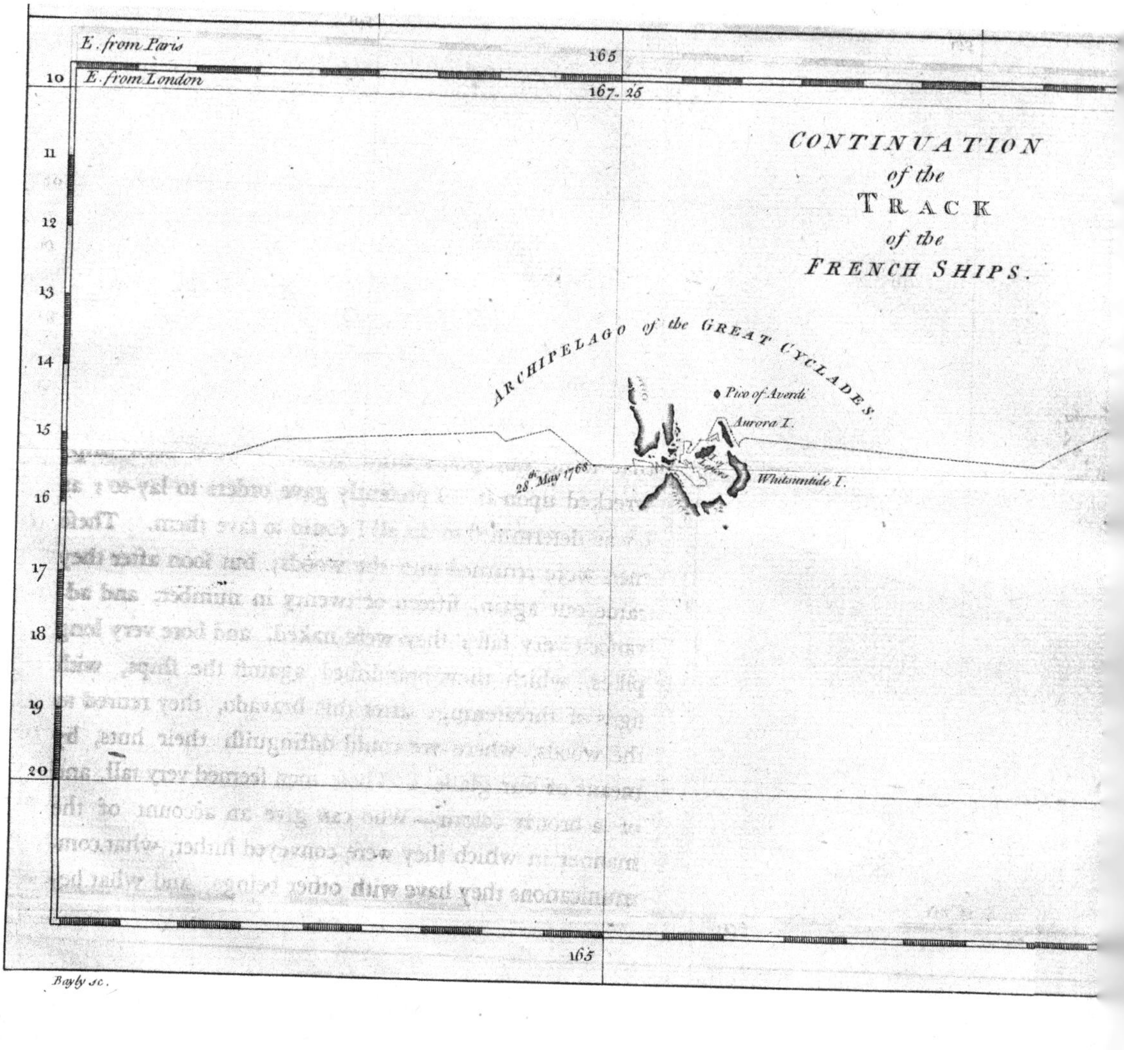

CONTINUATION
of the
TRACK
of the
FRENCH SHIPS.

E. from Paris 165
10 E. from London 167. 25

11
12
13
14
15
16
17
18
19
20

ARCHIPELAGO of the GREAT CYCLADES.

Pico of Averdi
Aurora I.
Whitsuntide I.
28th May 1768

165

Bayly sc.

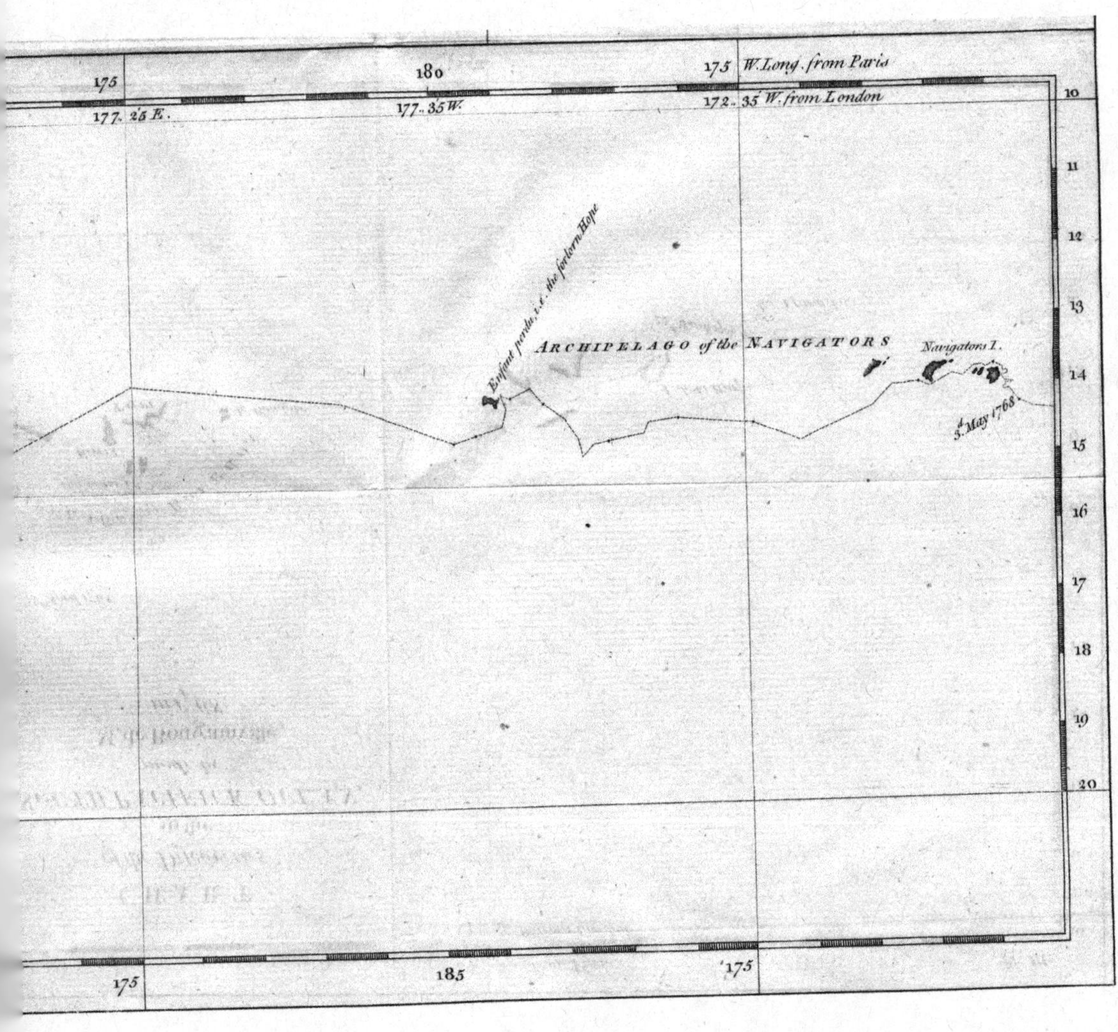

175
180
175 W. Long. from Paris
177. 25 E.
177. 35 W.
172. 35 W. from London
10
11
12
13
14
15
16
17
18
19
20
Enfant perdu, i.e. the forlorn Hope
ARCHIPELAGO of the NAVIGATORS
Navigators I.
3.d May 1768
175
185
175

were hovering about the ſhore, and ſeemed to announce a coaſt abounding in fiſh, and we all longed for a deſcent. We thought this would be eaſy on the weſtern-ſide; and we ran along the coaſt at the diſtance of about two miles. We ſaw the ſea break on every ſide with equal force, without a ſingle harbour or creek, which might ſerve for ſhelter, or ſtem the force of the ſea. Thus loſing all hopes of landing there, unleſs at the evident riſk of having our boats ſtaved to pieces, we reſumed our courſe again, when ſome of our people cried out, that they ſaw three men running to the ſea-ſhore. We ſhould never have thought that ſo ſmall an iſle could be inhabited; and my firſt conjectures were, that ſome Europeans muſt certainly have been ſhipwrecked upon it. I preſently gave orders to lay-to; as I was determined to do all I could to ſave them. Theſe men were returned into the woods; but ſoon after they came out again, fifteen or twenty in number, and advanced very faſt; they were naked, and bore very long pikes, which they brandiſhed againſt the ſhips, with ſigns of threatening; after this bravado, they retired to the woods, where we could diſtinguiſh their huts, by means of our glaſſes. Theſe men ſeemed very tall, and of a bronze colour—Who can give an account of the manner in which they were conveyed hither, what communications they have with other beings, and what becomes

It is inhabited, notwithſtanding its ſmall ſize.

comes of them when they multiply on an iſle, which has no more than a league in diameter? I called it *Iſle des Lanciers* *. Being leſs than a league to the N. E. of this iſle, I made the ſignal to the Etoile to ſound; ſhe did ſo with a line of two hundred fathom, without finding any bottom.

From that day we always ſhortened ſail at night, fearing to meet all at once ſome of theſe low-lands, to which it is ſo dangerous to come near. We were obliged to bring-to †, during a part of the night, between the 22d and 23d; as we had a ſtorm, with very high wind, rain, and thunder. At day-break we ſaw land, bearing from us, from N. E. b. N. to N. N. W. We ſtood for it, and at eight o'clock were about three leagues from its eaſtermoſt point. Then, though it was ſomewhat hazy, we perceived breakers along this coaſt, which appeared very low, and covered with trees: therefore, we ſtood out to ſea again, waiting for the fair weather to allow us to come nearer the coaſt at a leſs riſk; this we were able to do towards ten o'clock. Being only one league off the iſland, we ran along it, endeavouring to find a proper landing-place; we could not find bottom with 120 fathom. A bar, over which the ſea broke with great violence, lay along the whole coaſt; and we

Farther meeting with iſlands.

* Iſle of Lancebearers.

† *Reſter en travers.*

ſoon

ſoon diſcovered, that this iſland is formed by two very narrow ſlips of land, which join at the N. W. end, and leave an opening to the S. E. between their extremities. The middle of this iſle is therefore occupied by the ſea, in all its length, which is about ten or twelve leagues S. E. and N. W. ſo that it appears like a very oblong horſe-ſhoe, whereof the opening or entrance is at S. E.

Deſcription of the largeſt of theſe iſles.

The two necks of land are ſo very narrow, that we could perceive the ſea beyond the moſt northerly one. They ſeem compoſed of nothing but ſandy downs, interſperſed with low grounds, without either trees or verdure. The higher downs are covered with cocoa-nut and other leſſer trees, which were very ſhady After noon we ſaw periaguas in the kind of lake which this iſland forms; ſome ſailing, others paddling. The ſavages in them were naked. In the evening we ſaw a great number of theſe iſlanders along the coaſt. They likewiſe ſeemed to have ſuch long lances as the inhabitants of the firſt iſland threatened us with. We had not yet found any place where our canoes could land. The ſea foamed every where with equal violence. Night interrupted our reſearches; we paſſed it plying under our top-ſails; and not diſcovering any landing-place, on the 24th in the morning, we continued our courſe, and left this inacceſſible iſland; which, on account of its

 figure,

figure I called *Harp Island.* I question whether this extraordinary land is rising and encreasing, or whether it is decaying? How was it peopled? Its inhabitants appeared to us tall and well proportioned. I admire their courage, if they live unconcerned on these little slips of sand, which are exposed to be buried in the sea every moment by a hurricane

First division: *Dangerous Archipelago.*

The same day, at five in the afternoon, we saw another land, about seven or eight leagues distant; the uncertainty of its position, the inconstant squally and tempestuous weather, and the obscurity of the night obliged us to stand off and on. The 25th, in the morning, we came near the land, which we found to be another very low island, extending S. E. and N. W. about twenty-four leagues. We continued till the 27th to sail between low and partly overflowed islands, four of which we examined, all of the same nature, and all inaccessible, and not deserving that we should lose our time in visiting them. I gave the name of *Dangerous Archipelago* to this cluster of islands; of which we saw eleven, and which are probably more numerous. It is very dangerous sailing amidst these low isles, surrounded with breakers and shoals; where it is necessary, especially at night, to use the utmost precaution.

Error in the chart of this part of the Pacific Ocean.

I determined to stand more southerly, in order to get clear of these dangerous parts. Indeed, on the 28th,

we

we ceaſed to ſee the land. Quiros, diſcovered firſt, in 1606, the ſouth end of this chain of iſlands, which extend W. N. W. and among which admiral Roggewein found himſelf engaged in 1722, in about 15° lat. he called them the *Labyrinth.* Upon the whole, I know not on what grounds our geographers lay down after theſe iſles, a beginning of land ſeen, as they ſay, by Quiros, and to which they give ſeventy leagues of extent. All that can be inferred from the Journal of this navigator is, that the firſt place he *landed* at, after his departure from Peru, was eight leagues in extent. But far from conſidering it as a conſiderable coaſt, he ſays, that the ſavages who inhabit it, gave him to underſtand, that he ſhould find great countries in his way*. If any conſiderable land exiſted hereabouts, we could not fail meeting with it; as the leaſt latitude we were hitherto arrived at, was 17° 40′ S. which is the ſame that Quiros obſerved on this very coaſt, whereof the geographers have been pleaſed to make a great continent.

I agree, that it is difficult to conceive ſuch a number of low iſlands, and almoſt drowned lands, without ſuppoſing a continent near it. But Geography is a ſcience

* The continent, which the geographers place in theſe parts, ought to have been laid down only as a ſign of land, which Quiros ſays he met with the 27th of January 1606. But theſe ſigns of continent Quiros found before he came to the iſle of Sagittaria, which is the firſt he landed at, after ſailing from Peru. See Mr. *Dalrymple's Hiſtorical Collection of Voyages in the South Pacific Ocean*, part i. 107, 108. and the chart of the South Seas annexed. F

of facts; in studying it, authors must by no means give way to any system, formed in their studies, unless they would run the risk of being subject to very great errors, which can be rectified only at the expence of navigators.

Astronomical observations, compared with my reckoning.

Mr. Verron, in March, gave me three observations of longitude. The first, taken by Hadley's octant, on the 3d in the afternoon, was only 21′ 30″ different from my reckoning, I being so much to the westward of the observed longitude. The second, made by the megameter, and reduced to noon of the 10th, differed considerably from my reckoning, as my computed longitude was 3° 6′ more westward than that taken by observation. On the contrary, from the result of the third observation, taken with the octant on the 27th, my reckoning agreed within 39′ 15″, which he found I was more eastward than his longitude. It must be observed, that since my leaving the straits of Magalhaens, I have always followed the longitude of my departure, without correcting it in the least, or making use of the observations.

Meteorological observations.

The thermometer * constantly kept between 19° and 20°, during this month, and even near the land. Towards the end of the month, we had five days west winds, with squalls and storms, which succeeded each

* Reaumur's.

other

other almoſt without interruption. It rained continually; and the ſcurvy made its appearance on eight or ten perſons of the crew. Moiſtneſs is one of the moſt powerful cauſes of this diſeaſe. Each ſailor got daily a pint of lemonade, prepared with a kind of powder, called powder of *faciot*; which we made great uſe of, during the courſe of this voyage On the third of March I had likewiſe begun to make uſe of the diſtilling apparatus of M. Poiſſonier; and we continued till we arrived at New Britain to make uſe of the ſea-water, which was by this means deprived of its ſalt; employing it in broth, and in boiling meat and legumes. The ſupply of water it procured us, during this long run, was a very great reſource. We lighted our fire at five in the evening, and put it out by five or ſix in the morning, making above a barrel of water every night. By way of ſparing our freſh water, we always kneaded our bread with ſalt water.

Advantageous uſe of lemonade-powder at ſea.

Water deprived of it ſalt.

1768. April.

The ſecond of April, at ten in the morning, we perceived to the N. N. E. a high and very ſteep mountain, ſeemingly ſurrounded by the ſea. I called it the *Boudoir*, or the *Peak of the Boudeuſe*. We ſtood to the northward, in order to make it plain, when we ſaw another land, bearing W. by N. the coaſt of which was not ſo high, but afforded an indeterminate extent to our eyes. We had a very urgent neceſſity for touching at ſome place

Second diviſion of lands. Archipelago of Bourbon.

Sight of Taiti.

where we might get refreſhments and wood, and we flattered ourſelves to find them on this land. It was a calm almoſt the whole day. In the evening a breeze ſprung up, and we ſtood towards the land till two in the morning, when we ſtood off ſhore again, for three hours together. The ſun roſe obſcured by clouds and haze; and it was nine o'clock in the morning before we could ſee the land again, its ſouthermoſt point then bearing W. by N. We could no longer ſee the peak of the Boudeuſe, but from the maſt-head. The wind blew N. and N. N. E. and we ſtood as cloſe upon it as we could, in order to fall in to windward of the iſland. As we came nearer we ſaw, beyond its northermoſt point a diſtant land, ſtill further to northward, without our being able at that time to diſtinguiſh whether it joined to the firſt iſle, or whether it formed a ſecond.

Manœuvres in order to land there.

During the night, between the third and fourth, we turned to windward, in order to get more to the north ward. With joy we ſaw fires burning on every part of the coaſt, and from thence concluded that it was inhabited.

The 4th, at day-break, we diſcovered that the two lands, which before appeared ſeparate, were united together by a low land, which was bent like a bow, and formed a bay open to the N. E. We run with all ſails ſet towards the land, ſtanding to windward of this bay, when we perceived a periagua coming from the offing, and ſtanding for the land, and making uſe of her ſail and paddles. She paſſed athwart us, and joined a number of others, which ſailed a-head of us, from all parts of the iſland. One of them went before all the reſt; it was manned by twelve naked men, who preſented us with branches of bananas; and their demonſtrations ſigniſied that this was their olive-branch. We anſwered them with all the ſigns of friendſhip we could imagine; they then came along ſide of our ſhip; and one of them, remarkable for his prodigious growth of hair, which ſtood like briſtles divergent on his head, offered us, together with his branch of peace, a little pig, and a cluſter of bananas. We accepted his preſent, which he faſtened to a rope that was thrown over to him; we gave him caps and handkerchiefs; and theſe firſt preſents were the pledges of our alliance with theſe people

First traffic with theſe iſlanders.

The two ſhips were ſoon ſurrounded with more than an hundred periaguas of different ſizes, all which had outriggers. They were laden with cocoa-nuts, bananas, and other fruits of the country. The exchange of theſe fruits, which were delicious to us, was made very honeſtly for all ſorts of trifles; but without any of the iſlanders venturing to come aboard. We were obliged either to come into their periaguas, or ſhew them at a diſtance

diſtance what we offered in exchange; when both parties were agreed, a baſket or a net was let down by a rope; they put their goods in it, and ſo we did ours; giving before they had received, or receiving before they gave indifferently, with a kind of confidence, which made us conceive a good opinion of their character. We further ſaw no kind of arms in their periaguas, in which there were no women at this firſt interview. The periaguas kept along-ſide of the ſhips, till the approach of night obliged us to ſtand off ſhore, when they all retired.

We endeavoured, during night, to go to the northward, never ſtanding further than three leagues from the land. All the ſhore was, till near midnight, covered as the night before, with little fires at a ſhort diſtance from each other: it ſeemed as if it was an illumination made on purpoſe, and we accompanied it with ſeveral ſky-rockets from both our ſhips.

The 5th we ſpent in plying, in order to work to windward of the iſland, and in letting the boats ſound for an anchoring-place. The aſpect of this coaſt, elevated like an amphitheatre, offered us the moſt enchanting proſpect. Notwithſtanding the great height of the mountains, none of the rocks has the appearance of barrenneſs; every part is covered with woods. We hardly believed our eyes, when we ſaw a peak covered with trees,

Deſcription of the coaſt as ſeen from the offing.

trees, up to its ſolitary ſummit, which riſes above the level of the mountains, in the interior parts of the ſouthermoſt quarter of the iſland. Its apparent ſize ſeemed to be no more than of thirty toiſes in diameter, and grew leſs in breadth as it roſe higher. At a diſtance it might have been taken for a pyramid of immenſe height, which the hand of an able ſculptor had adorned with garlands and foliage. The leſs elevated lands are interſperſed with meadows and little woods; and all along the coaſt there runs a piece of low and level land, covered with plantations, touching on one ſide the ſea, and on the other bordering the mountainous parts of the country. Here we ſaw the houſes of the iſlanders amidſt bananas, cocoa-nut, and other trees loaded with fruit.

As we ran along the coaſt, our eyes were ſtruck with the ſight of a beautiful caſcade, which came from the tops of the mountains, and poured its foaming waters into the ſea. A village was ſituated at the foot of this caſcade, and there appeared to be no breakers in this part of the coaſt. We all wiſhed to be able to anchor within reach of this beautiful ſpot; we were conſtantly ſounding aboard the ſhips, and our boats took ſoundings cloſe under the ſhore; but we found a bottom of nothing but rocks in this port, and were forced to go in ſearch of another anchorage.

The

Continuation of the traffic with the iſlanders.

The periaguas returned to the ſhip at ſun-riſing, and continued to make exchanges all the day. We likewiſe opened new branches of commerce; for, beſides the fruits, which they brought the day before, and other refreſhments, ſuch as fowls and pigeons, the iſlanders brought with them ſeveral inſtruments for fiſhing; ſtone chiſels, *(herminettes de pierre)* ſtrange kinds of cloth, ſhells, &c. They wanted iron and ear-rings in exchange This bartering trade was carried on very honeſtly, as the day before: this time ſome pretty and almoſt naked women came in the periaguas. One of the iſlanders went on board the Etoile, and ſtayed there all night, without being in the leaſt uneaſy.

This night was likewiſe ſpent in plying; and on the 6th in the morning we were got to the moſt northerly extremity of the iſland. Another iſle now came within ſight but ſeeing ſeveral breakers that ſeemed to obſtruct the paſſage between the two iſles, I determined to return in ſearch of anchorage in the firſt bay, which we ſaw on the day of our land-fall. Our boats which ſounded a head of us towards ſhore, found the north ſide of the bay every where ſurrounded, at a quarter of a league's diſtance, by a reef which appears at low water. However, about a league from the north point, they diſcovered a gap in the reef, of the width of twice a cable's length at moſt, where there was 30 and 35 fathom

fathom of water, and within it a pretty extenſive road, where the bottom varied from nine to thirty fathom. This road was bounded to the ſouth by a reef, which, proceeding from the land, joined that which ſurrounded the ſhore. Our boats had conſtantly found a ſandy bottom, and diſcovered ſeveral little rivers fit for watering at. Upon the reef, on the north ſide, there are three little iſlands.

Anchorage at Taiti.

This account determined me to come to an anchor in the road, and we immediately made ſail to enter into it. We ranged the point of the ſtarboard reef in entering; and as ſoon as we were got within it, we let go our beſt bower in 34 fathom, bottom of grey ſand, ſhells, and gravel; and we immediately carried out the ſtream-anchor to the north-weſt, in order to let go our ſmall bower there. The Etoile went to windward, and came to an anchor a cable's length to the northward of us. As ſoon as we were moored, we ſtruck yards and top-maſts.

Difficulty of mooring the ſhips.

As we came nearer the ſhore, the number of iſlanders ſurrounding our ſhips encreaſed. The periaguas were ſo numerous all about the ſhips, that we had much to do to warp in amidſt the croud of boats and the noiſe. All theſe people came crying out *tayo*, which means friend, and gave a thouſand ſigns of friendſhip; they all aſked nails and ear-rings of us. The periaguas were

 full

full of females; who, for agreeable features, are not inferior to moſt European women; and who in point of beauty of the body might, with much reaſon, vie with them all. Moſt of theſe fair females were naked; for the men and the old women that accompanied them, had ſtripped them of the garments which they generally dreſs themſelves in. The glances which they gave us from their periaguas, ſeemed to diſcover ſome degree of uneaſineſs, notwithſtanding the innocent manner in which they were given; perhaps, becauſe nature has every where embelliſhed their ſex with a natural timidity; or becauſe even in thoſe countries, where the eaſe of the golden age is ſtill in uſe, women ſeem leaſt to deſire what they moſt wiſh for. The men, who were more plain, or rather more free, ſoon explained their meaning very clearly. They preſſed us to chooſe a woman, and to come on ſhore with her; and their geſtures, which were nothing leſs than equivocal, denoted in what manner we ſhould form an acquaintance with her. It was very difficult, amidſt ſuch a ſight, to keep at their work four hundred young French ſailors, who had ſeen no women for ſix months. In ſpite of all our precautions, a young girl came on board, and placed herſelf upon the quarter-deck, near one of the hatchways, which was open, in order to give air to thoſe who were heaving at the capſtern below it. The girl

care-

carelessly dropt a cloth, which covered her, and appeared to the eyes of all beholders, such as Venus shewed herself to the Phrygian shepherd, having, indeed, the celestial form of that goddess. Both sailors and soldiers endeavoured to come to the hatch-way; and the capstern was never hove with more alacrity than on this occasion.

At last our cares succeeded in keeping these bewitched fellows in order, though it was no less difficult to keep the command of ourselves. One single Frenchman, who was my cook, having found means to escape against my orders, soon returned more dead than alive. He had hardly set his feet on shore, with the fair whom he had chosen, when he was immediately surrounded by a croud of Indians, who undressed him from head to feet. He thought he was utterly lost, not knowing where the exclamations of those people would end, who were tumultuously examining every part of his body. After having considered him well, they returned him his clothes, put into his pockets whatever they had taken out of them, and brought the girl to him, desiring him to content those desires which had brought him on shore with her. All their persuasive arguments had no effect; they were obliged to bring the poor cook on board, who told me, that I might reprimand him as much as I pleased, but that I could never frighten him so much, as he had just now been frightened on shore.

 CHAP.

CHAP. II.

Stay at Taiti; account of the good and evil which befel us there.

Landing.

I Have pointed out the obſtacles which we met with in coming to an anchor. When we were moored, I went on ſhore with ſeveral officers, to ſurvey the watering-place. An immenſe croud of men and women received us there, and could not be tired with looking at us; the boldeſt among them came to touch us; they even puſhed aſide our clothes with their hands, in order to ſee whether we were made exactly like them: none of them wore any arms, not ſo much as a ſtick. They ſufficiently expreſſed their joy at our arrival. The chief of this diſtrict conducted and introduced us into his houſe, in which we found five or ſix women, and a venerable old man. The women ſaluted us, by laying their hands on their breaſts, and ſaying ſeveral times *tayo*. The old man was the father of our hoſt. He had no other character of old age, than that reſpectable one which is imprinted on a fine figure. His head adorned with white hair, and a long beard; all his body, nervous and fleſhy, had neither wrinkles, nor ſhewed any marks of decrepitude. This venerable man ſeemed to be rather diſpleaſed with our arrival; he even retired

Viſit paid to the chief of the diſtrict.

retired without anſwering our civilities, without giving any ſigns of fear, aſtoniſhment, or curioſity; very far from taking part in the raptures all this people was in at our ſight, his thoughtful and ſuſpicious air ſeemed to ſhew that he feared the arrival of a new race of men would trouble thoſe happy days which he had ſpent in peace.

Deſcription of his houſe

We were at liberty to examine the interior parts of the houſe. It had no furniture, no ornament to diſtinguiſh it from the common huts, except its extent. It was about eighty feet long and twenty feet wide. In it we obſerved a cylinder of ozier, three or four feet long, ſet with black feathers, which was ſuſpended from the thatch; and beſides it, there were two wooden figures which we took for idols. One, which was their god *, ſtood upright againſt one of the pillars; the goddeſs was oppoſite, leaned againſt the wall, which ſhe ſurpaſſed in height, and was faſtened to the reeds, of which their walls are made. Theſe figures, which were ill made, and without any proportion, were about

* The people of *Otahitee*, or as our author wrongly calls it, *Taiti*, are not idolaters, according to the laſt publiſhed account, and therefore it is certain, that Mr. de B. took ſome ornamental figures for thoſe of their divinities. Had this circumnavigator made a longer ſtay in this iſland, had he thoroughly ſtudied the language of the country, and looked upon many things with a more philoſophical, or leſs prejudiced eye, his account would have proved leſs ſubject to the miſtakes it abounds with. The Engliſh, more uſed to philoſophical enquiries, will give more faithful accounts in the work that is going to be publiſhed, of the great diſcoveries made by the Britiſh nation in thoſe ſeas. F.

three

three feet high, but ſtood on a cylindrical pedeſtal, hollow within, and carved quite through. This pedeſtal was made in the ſhape of a tower, was ſix or ſeven feet high, and about a foot in diameter. The whole was made of a black and very hard wood.

Reception which he gives us.

The chief then propoſed that we ſhould ſit down upon the graſs before his houſe, where he ordered ſome fruit, broiled fiſh and water to be ſet before us: during the meal he ſent for ſome pieces of cloth, and for two great collars or gorgets of oziers, covered with black feathers and ſhark's teeth. They are pretty like in form to the immenſe ruffs, worn in the time of Francis the firſt. One of theſe he put upon the neck of the Chevalier d'Oraiſon, another upon mine, and diſtributed the cloths. We were juſt going to return on board when the Chevalier de Suzannet miſſed a piſtol, which had been very dexterouſly ſtolen out of his pocket. We informed the chief of it, who immediately was for ſearching all the people who ſurrounded us, and even treated ſome of them very harſhly. We ſtopt his reſearches, endeavouring only to make him underſtand, that the thief would fall a victim to his own crime, and that what he had ſtolen could kill him.

The chief and all his people accompanied us to our boats. We were almoſt come to them when we were ſtopped by an iſlander, of a fine figure, who lying under

under a tree, invited us to ſit down by him on the graſs. We accepted his offer: he then leaned towards us, and with a tender air he ſlowly ſung a ſong, without doubt of the Anacreontic kind, to the tune of a flute, which another Indian blew with his noſe: this was a charming ſcene, and worthy the pencil of a Boucher. Four iſlanders came with great confidence to ſup and lye on board. We let them hear the muſic of our flutes, baſe-viols, and violins, and we entertained them with a fire-work of ſky-rockets and fire-ſnakes. This ſight cauſed a mixture of ſurprize and of horror in them.

On the 7th in the morning, the chief, whoſe name was Ereti, came on board. He brought us a hog, ſome fowls, and the piſtol which had been ſtolen at his houſe the day before. This act of juſtice gave us a good opinion of him. However, we made every thing ready in the morning, for landing our ſick people, and our water caſks, and leaving a guard for their defence. In the afternoon I went on ſhore with arms and implements, and we began to make a camp on the banks of a little brook, where we were to fill our water. Ereti ſaw the men under arms, and the preparations for the encampment, without appearing at firſt ſurpriſed or diſcontented. However, ſome hours after he came to me, accompanied by his father and the principal people of the

Project of a camp for our ſick on ſhore.

Oppoſition on the part of the iſlanders.

the diſtrict, who had made remonſtrances to him on this occaſion, and gave me to underſtand that our ſtay on ſhore diſpleaſed them, that we might ſtay there during day-time as long as we pleaſed, but that we ſhould ly on board our ſhips at night. I inſiſted upon eſtabliſhing the camp, making him comprehend that it was neceſſary to us, in order to get wood and water, and to facilitate the exchanges between both nations. They then held a ſecond council, the reſult of which, was, that Ereti came to aſk me whether we intended to ſtay here for ever, or whether we intended to go away again, and how ſoon that would be. I told him that we ſhould ſet ſail in eighteen days, in ſign of which, I gave him eighteen little ſtones. Upon this they held a new conference, at which they deſired I would be preſent. A grave man, who ſeemed to have much weight with the members of the council, wanted to reduce the number of days of our encamping to nine; but as I inſiſted on the number I had at firſt required, they at laſt gave their conſent.

They conſent to it on ſome conditions.

From that moment their joy returned; Ereti himſelf offered us an extenſive building like a ſhed, cloſe to the river, under which were ſome periaguas, which he immediately got taken away. Under this ſhed we raiſed the tents for thoſe who were ill of the ſcurvy, being thirty-four in number, twelve from the Boudeuſe, and

[E]ſtabliſhment [o]f a camp for [o]ur ſick and [ar]tificers.

twenty

twenty-two from the Etoile, and for ſome neceſſary hands. The guard conſiſted of thirty ſoldiers and I likewiſe landed muſkets enough to arm the workmen and the ſick. I ſtaid on ſhore the firſt night, which Ereti likewiſe choſe to paſs under our tents. He ordered his ſupper to be brought, and joined it to ours, driving away the crowd which ſurrounded the camp, and retaining only five or ſix of his friends. After ſupper he deſired to ſee ſome ſky-rockets played off, and they frightened him at leaſt as much as they gave him pleaſure. Towards the end of night he ſent for one of his wives, whom he ſent to ſleep in prince Naſſau's tent. She was old and ugly.

Precautions taken. Conduct of the natives.

The next day was ſpent in completing our camp. The ſhed was well made, and entirely covered over by a kind of mats. We left only one entrance to it, which we provided with a barrier, and placed a guard there. Ereti, his wives and his friends alone were allowed to come in; the croud kept on the outſide of the ſhed, and only a ſingle man of our people with a ſwitch in his hand was ſufficient to clear the way. Hither the natives from all ſides brought fruits, fowls, hogs, fiſh, and pieces of cloth, which they exchanged for nails, tools, beads, buttons, and numberleſs other trifles, which were treaſures to them They were, upon the whole, very attentive to learn what would give us pleaſure;

ſure; they ſaw us gathering antiſcorbutic plants, and ſearching for ſhells: their women and children ſoon vied with each other in bringing us bundles of the ſame plants, which they had ſeen us collecting, and baſkets full of ſhells of all ſorts. Their trouble was paid at a ſmall expence.

Aſſiſtance they give us.

This ſame day I deſired the chief to ſhew me where I might cut wood. The low country where we were, was covered only with fruit trees, and a kind of wood full of gum, and of little conſiſtence; the hard wood grows upon the mountains. Ereti pointed out to me the trees which I might cut down, and even ſhewed towards which ſide I ſhould fell them. The natives aſſiſted us greatly in our works; our workmen cut down the trees and made them into faggots, which the iſlanders brought to the boats; they likewiſe gave us their aſſiſtance in making our proviſion of water, filling the caſks, and bringing them to the boats. Their labour was paid in nails, of which, the number was proportionate to the work they had done. The only conſtraint which their preſence put upon us, was, that they obliged us to have our eyes upon every thing that was brought on ſhore, and even to look to our pockets; for even in Europe itſelf, one cannot ſee more expert filchers than the people of this country.

6 However,

Precautions taken againſt thieves.

However, it does not appear that ſtealing is uſual among themſelves. Nothing is ſhut up in their houſes, every piece of furniture lies on the ground, or is hung up, without being under locks, or under any perſon's care. Doubtleſs their curioſity for new objects excited violent deſires in them; and beſides that, there are always baſe-minded people every where. During the two firſt nights we had ſome things ſtolen from us, notwithſtanding our guards and patroles, at whom the thieves had even thrown ſtones. Theſe thieves hid themſelves in a marſh full of graſs and reeds, extending behind our camp. This marſh was partly cleared by my orders, and I commanded the officer upon duty to fire upon any thieves who ſhould come for the future. Ereti himſelf told me to do it, but took great care to ſhew me ſeveral times the ſpot where his houſe was ſituated, earneſtly recommending it to me, to fire towards the oppoſite quarter. I likewiſe ſent every evening three of our boats, armed with pedereroes and ſwivel guns, to ly at anchor before the camp.

All our tranſactions were carried on in as friendly a manner as poſſible, if we except thieving. Our people were daily walking in the iſle without arms, either quite alone, or in little companies. They were invited to enter the houſes, where the people gave them to eat; nor did the civility of their landlords ſtop at a ſlight collation, they offered them young girls; the hut was immediately filled with a curious croud of men and women, who made a circle round the gueſt, and the young victim of hoſpitality. The ground was ſpread with leaves and flowers, and their muſicians ſung an hymeneal ſong to the tune of their flutes. Here Venus is the goddeſs of hoſpitality, her worſhip does not admit of any myſteries, and every tribute paid to her is a feaſt for the whole nation. They were ſurpriſed at the confuſion which our people appeared to be in, as our cuſtoms do not admit of theſe public proceedings However, I would not anſwer for it, that every one of our men had found it impoſſible to conquer his repugnance, and conform to the cuſtoms of the country.

Beauty of the interior parts of the country.

I have often, in company with only one or two of our people, been out walking in the interior parts of the iſle. I thought I was tranſported into the garden of Eden; we croſſed a turf, covered with fine fruit-trees, and interſected by little rivulets, which keep up a pleaſant coolneſs in the air, without any of thoſe inconveniences which humidity occaſions. A numerous people there enjoy the bleſſings which nature ſhowers liberally down upon them. We found companies of men and women ſitting under the ſhade of their fruit-trees: they all greeted us with ſigns of friendſhip: thoſe who met us upon the road ſtood aſide to let us paſs by;

by; every where we found hoſpitality, eaſe, innocent joy, and every appearance of happineſs amongſt them.

Preſents of European fowls and ſeeds made to the chief.

I preſented the chief of the diſtrict in which we were with a couple of turkies, and ſome ducks and drakes; they were to be conſidered as the mites of the widow. I likewiſe deſired him to make a garden in our way, and to ſow various ſorts of ſeeds in them, and this propoſal was received with joy. In a ſhort time, Ereti prepared a piece of ground, which had been choſen by our gardeners, and got it incloſed. I ordered it to be dug; they admired our gardening inſtruments. They have likewiſe around their houſes a kind of kitchen gardens, in which they plant an eatable hibiſcus or okra, potatoes, yams, and other roots. We ſowed for their uſe ſome wheat, barley, oats, rice, maize, onions, and pot herbs of all kinds. We have reaſon to believe that theſe plantations will be taken care of; for this nation appeared to love agriculture, and would I believe be eaſily accuſtomed to make advantage of their ſoil, which is the moſt fertile in the univerſe.

Viſit of the chief of a neighbouring diſtrict.

During the firſt days of our arrival, I had a viſit from the chief of a neighbouring diſtrict, who came on board with a preſent of fruits, hogs, fowls, and cloth. This lord, named *Toutaa*, has a fine ſhape, and is prodigiouſly tall. He was accompanied by ſome of his relations, who were almoſt all of them ſix feet (French mea-

meaſure) high. I made them preſents of nails, ſome tools, beads, and ſilk ſtuffs. We were obliged to repay this viſit at his houſe, where we were very well received, and where the good-natured Toutaa offered me one of his wives, who was very young and pretty handſome. The aſſembly was very numerous, and the muſicians had already began the hymenean. Such is their manner of receiving viſits of ceremony.

On the 10th, an iſlander was killed, and the natives came to complain of this murder. I ſent ſome people to the houſe, whither they had brought the dead body; it appeared very plain that the man had been killed by a fire-arm. However, none of our people had been ſuffered to go out of the camp, or to come from the ſhips with fire-arms. The moſt exact enquiries which I made to find out the author of this villainous action proved unſucceſsful. The natives doubtleſs believed that their countryman had been in the wrong; for they continued to come to our quarters with their uſual confidence. However, I received intelligence that many of the people had been ſeen carrying off their effects to the mountains, and that even Ereti's houſe was quite unfurniſhed. I made him ſome more preſents, and this good chief continued to teſtify the ſinncereſt friendſhip for us

ofs of our nchors, daners which e meet with.

I haſtened in the mean while the completing of our works of all kinds; for though this was an excellent

place

place to ſupply our wants at, yet I knew that we were very ill moored. Indeed, though we under-run the cables almoſt every day with the long boat, and had not yet found them chafed *, yet we had found the bottom was ſtrewed with large coral; and beſides, in caſe of a high wind from the offing, we had no room to drive. Neceſſity had obliged us to take this anchorage, without leaving us the liberty of chooſing, and we ſoon found that our fears were but too well grounded.

Account of the manoeuvres which ſaved us.

The 12th, at five in the morning, the wind being ſouth, our S. E. cable, and the hawſer of the ſtream-anchor, which by way of precaution we had extended to the E. S. E. parted at the bottom. We immediately let go our ſheet-anchor, but before it had reached the bottom, the frigate ſwung off to her N. W. anchor, and we fell aboard the Etoile on the larboard ſide. We hove upon our anchor, and the Etoile veered out cable as faſt as poſſible, ſo that we were ſeparated before any damage was done. The ſtore ſhip then ſent us the end of a hawſer, which ſhe had extended to the eaſtward, and upon which we hove, in order to get farther from her. We then weighed our ſheet-anchor, and hove in our hawſer and cable, which parted at the bottom. The latter had been cut about thirty fathom from the clinch;

* Rayés.

we

we ſhifted it end for end, and bent it to a ſpare anchor of two thouſand ſeven hundred weight, which the Etoile had ſtowed in her hold, and which we ſent for. Our S. E. anchor, which we had let go without any buoy-rope, on account of the great depth, was entirely loſt; and we endeavoured, without ſucceſs, to ſave the ſtream-anchor, whoſe buoy was ſunk, and for which it was impoſſible to ſweep the bottom. We preſently ſwayed up our fore-top-maſt and fore-yard, in order to be ready for ſailing as ſoon as the wind ſhould permit.

In the afternoon the wind abated and ſhifted to the eaſtward. We then carried out to the S. E. a ſtream-anchor, and the anchor we had got from the Etoile, and I ſent a boat to ſound to the northward, in order to know whether there was a paſſage that way, by which means we might have got out almoſt with any wind. One misfortune never comes alone; as we were occupied with a piece of work on which our ſafety depended, I was informed that three of the natives had been killed or wounded with bayonets in their huts, that the alarm was ſpread in the country, that the old men, the women and the children fled towards the mountains with their goods, and even the bodies of the dead, and that we ſhould perhaps be attacked by an army of theſe enraged men. Thus our ſituation gave us room to fear a war on ſhore, at the very moment when both

Another murder of ſome iſlanders.

ſhips

ſhips were upon the point of being ſtranded. I went aſhore, and came into the camp, where, in preſence of the chief, I put four ſoldiers in irons, who were ſuſpected to be the authors of this crime: theſe proceedings ſeemed to content the natives.

Precautions againſt the conſequences which it might have had.

I paſſed a part of the night on ſhore, and reinforced the watches, fearing that the inhabitants might revenge their countrymen. We occupied a moſt excellent poſt, between two rivers, diſtant from each other at moſt only a quarter of a league; the front of the camp was covered by a marſh, and on the remaining ſide was the ſea, of which we certainly were the maſters. We had a fair chance to defend this poſt againſt the united forces of the whole iſland; but happily the night paſſed very quietly in the camp, excepting ſome alarms occaſioned by thieves.

Continuati[on] of the dangers which the ſhips r[un].

It was not from this part that I dreaded the worſt that could happen; the fear of ſeeing the ſhips loſt upon the coaſt, gave me infinitely more concern. From ten o'clock in the evening, the wind freſhened very much from the eaſt; and was attended with a great ſwell; rain, tempeſt, and all the ſad appearances which augment the horror of theſe dreadful ſituations.

Towards two o'clock in the morning, a ſquall drove the ſhips towards the coaſt: I came on board; the ſquall happily was not of long duration; and as ſoon as it was

 blown

blown over, the wind blew off ſhore. At day-break we encountered new misfortunes; our N. W. cable parted; the hawſer, which the Etoile had given us, and which held us by her ſtream-anchor, had the ſame fate a few minutes after. The frigate then ſwinging off to her S. E. anchor and hawſer, was no more than a cable's length off ſhore, upon which the ſea broke with great violence. In proportion, as the danger became more preſſing, our reſources failed us; the two anchors of which the cable's had juſt parted, were entirely loſt to us; their buoys diſappeared, being either ſunk, or taken away, during the night, by the Indians. Thus we had loſt already four anchors, in four and twenty hours, and had yet ſeveral loſſes to ſuſtain.

At ten o'clock in the morning, the new cable we had bent to the anchor of two thouſand ſeven hundred weight from the Etoile, which held us to the S. E. parted, and the frigate, riding by a ſingle hawſer, began to drive upon the coaſt. We immediately let go our ſheet-anchor under foot; it being the only one which we had remaining at our bow but of what uſe could it be to us? We were ſo cloſe to the breakers, that we muſt have been upon them before we had veered out cable ſufficient to make the anchor catch hold in the ground. We expected every moment the ſad concluſion of this adventure, when a S. W. breeze gave us ſome hopes of

ſetting

ſetting ſail. Our jib and ſtay-ſails were ſoon hoiſted; the ſhip began to ſhoot a-head, and we were endeavouring to make ſail, in order to veer away cable and hawſer, and get out; but the wind almoſt immediately ſhifted to the eaſtward again. This interval had, however, given us time to take on board the end of a hawſer, from a ſecond ſtream-anchor of the Etoile, which ſhe had juſt carried out to the eaſtward, and which ſaved us for this time. We hove in upon both hawſers, and got ſomewhat further from the ſhore. We then ſent our long-boat aboard the Etoile, to help her in mooring her ſecurely; her anchors happily lay in a bottom leſs covered with coral than that where we had let ours go. This being done, our long-boat went to weigh the anchor of 2700 weight by its buoy-rope; we bent another cable to it, and carried it out to the N. E. we then weighed the ſtream-anchor belonging to the Etoile, and returned it to her. During theſe two days M. de la Giraudais, captain of that ſtore-ſhip, had a very great ſhare in the preſervation of the frigate, by the aſſiſtance which he gave me: it is with pleaſure that I pay this tribute of gratitude to an officer, who has already been my companion on former voyages, and whoſe zeal equals his talents.

Peace ma[de] with the iſlanders.

However, when the day appeared, no Indian was come near the camp, not a ſingle periagua was ſeen ſail-

 ing,

ing, all the neighbouring houſes were abandoned, and the whole country appeared as a deſert. The prince of Naſſau, who with only four or five men was gone out a little further, in order to ſcearch for ſome of the natives, and to inſpire them with confidence again, found a great number of them with Ereti, about a league from the camp. As ſoon as that chief knew the prince again, he came up to him with an air of conſternation.

The women, who were all in tears, fell at his feet, kiſſed his hands, weeping and repeating ſeveral times, *Tayo, maté*, you are our friends, and you kill us. By his careſſes and demonſtrations of friendſhip, he at laſt ſucceeded in regaining their confidence. I ſaw from on board a croud of people running to our quarters: fowls, cocoa-nuts, and branches full of bananas, embelliſhed this proceſſion, and promiſed a peace. I immediately went aſhore with an aſſortment of ſilk ſtuffs, and tools of all ſorts; I diſtributed them among the chiefs, expreſſing my concern to them on account of the diſaſter which had happened the day before, and aſſuring them, that I would puniſh the perpetrators. The good iſlanders loaded me with careſſes; the people applauded the reunion, and, in a ſhort time, the uſual croud and the thieves returned to our quarters, which looked like a fair. This day, and the following, they brought more refreſhments than ever. They likewiſe deſired to have

have ſeveral muſkets fired in their preſence, which frightened them very much, as all the creatures which we ſhot at were killed immediately.

The Etoile ſets ſail.

The boat, which I had ſent to ſound to the northward, was returned with the good news of having found a very fine paſſage. It was then too late to profit of it the ſame day; for night was coming on. Happily it paſſed quietly, both on ſhore and at ſea. The 14th in the morning, wind at eaſt, I ordered the Etoile, who had got her water and all her men on board, to weigh and go out by the new north paſſage. We could not go out by that paſſage before the ſtore-ſhip, ſhe being moored to the northward of us. At eleven ſhe came to ſail, from a hawſer, which ſhe had carried on board of us. I kept her long-boat and two ſmall anchors; I likewiſe took on board, as ſoon as ſhe was got under ſail, the end of the cable of her S. E. anchor, which lay in a good bottom. We now weighed our ſheet-anchor, carried the two ſtream anchors further out; and were by this means moored by two great, and three ſmall anchors. At two o'clock in the afternoon, we had the ſatisfaction of ſeeing the Etoile without the reefs. Our ſituation by this means became leſs terrifying; we had at leaſt ſecured to ourſelves the means of returning to our country, by putting one of the ſhips out of danger. When M. de la Giraudais was got out into

into the offing, he ſent back his boat to me, with Mr. Lavari Leroi, who had been employed to ſurvey the paſſage

Inſcription buried.

We laboured all day, and a part of the night, to complete our water, and to remove the hoſpital and the camp. I buried near the ſhed, an act of taking poſſeſſion, inſcribed on an oak plank, and a bottle well corked and glued, containing the names of the officers of both ſhips. I have followed the ſame method in regard to all the lands diſcovered during the courſe of this voyage. It was two o'clock in the morning, before every one of our people were on board: the night was ſtill ſtormy enough to give us ſome diſturbance, notwithſtanding the number of anchors we had moored.

The Boudeuſe ſets ſail; runs new dangers.

On the 15th, at ſix o'clock in the morning, the wind blowing off ſhore, and the ſky looking ſtormy, we weighed our anchor, veered away the cable of that which belonged to the Etoile, cut one of the hawſers, and veered out the other two, ſetting ſail under our foreſail and top ſails, in order to go out by the eaſtern paſſage. We left the two long-boats to weigh the anchors; and as ſoon as we were got out of the reefs, I ſent the two barges armed, under the command of enſign the chevalier de Suzannet, to protect the work of the long boats. We were about a quarter of a league off ſhore, and began to give ourſelves joy of having ſo

happily

happily left an anchorage, that had given us ſuch terrible alarms, when the wind ceaſing all at once, the tide and a great ſwell from the eaſtward, began to drive us towards the reefs to leeward of the paſſage. The worſt conſequences of the ſhipwreck, with which we had hitherto been threatened, would have been to paſs the remainder of our days on an iſle adorned with all the gifts of nature, and to exchange the ſweets of the mother-country, for a peaceable life, exempted from cares. But now ſhipwreck appeared with a more cruel aſpect; the ſhip being rapidly carried upon the rocks, could not have reſiſted the violence of the ſea two minutes, and hardly ſome of the beſt ſwimmers could have ſaved their lives. At the beginning of the danger, I had made ſignal for the long boats and barges to return and tow us. They came at the very moment, when we being only 35 or 36 fathom (50 toiſes) from the reef, our ſituation was become quite deſperate; the more ſo as we could not let go an anchor. A weſterly breeze, ſpringing up that inſtant, brought hope along with it; it actually freſhened by degrees; and at nine o'clock in the morning, we were quite clear of all dangers.

Departur from Tai loſies whi we ſuſtai there.

I immediately ſent the boats back in queſt of the anchors, and I remained plying to wait for them. In the afternoon we joined the Etoile. At five in the evening our long-boat came on board with the beſt bower, and the cable of the Etoile, which ſhe carried to her: our barge, that of the Etoile, and her long-boat returned ſoon after; the latter bringing us our ſtream-anchor and a hawſer. As to the other two ſtream-anchors, the night coming on, and the ſailors being extremely fatigued, they could not weigh them that day. I at firſt intended to keep plying off and on during night, and to ſend them out for them the next morning; but at mid-night a ſtrong gale ſprung at E. N. E. obliging me to hoiſt in the boats, and make ſail, in order to get clear of the coaſt.

Thus an anchorage of nine days coſt us ſix anchors; which we ſhould not have loſt, had we been provided with ſome iron chains. This is a precaution which no navigator ought to forget, if he is going upon ſuch a voyage as this.

Regret of the iſlanders at our departing.

Now that the ſhips are in ſafety, let us ſtop a moment to receive the farewel of the iſlanders. At day-break, when they perceived us ſetting ſail, Ereti leaped alone into the firſt periagua he could find on ſhore, and came on board. There he embraced all of us, held us ſome moments in his arms, ſhedding tears, and appearing much affected at our departure. Soon after, his great periagua came on board, laden with refreſhments of all kinds; his wives were in the periagua; and with them the ſame iſlander, who, on the firſt day of our

One of them embarks with us, at his own and his nation's request.

our land-fall, had lodged on board the Etoile. Ereti took him by the hand, and, presenting him to me, gave me to understand, that this man, whose name was Aotourou, desired to go with us, and begged that I would consent to it. He then presented him to each of the officers in particular; telling them that it was one of his friends, whom he entrusted with those who were likewise his friends, and recommending him to us with the greatest signs of concern. We made Ereti more presents of all sorts; after which he took leave of us, and returned to his wives, who did not cease to weep all the time of the periagua's being along-side of us. In it there was likewise a young and handsome girl, whom the islander that stayed along with us went to embrace. He gave her three pearls which he had in his ears, kissed her once more; and, notwithstanding the tears of this young wife or mistress, he tore himself from her, and came aboard the ship. Thus we quitted this good people; and I was no less surprised at the sorrow they testified on our departure, than at their affectionate confidence on our arrival.

CHAP. III.

Description of the new island; manners and character of its inhabitants.

Lucis habitamus opacis,
Riparumque toros & prata recentia rivis
Incolimus.
VIRG. Æneid. Lib. VI.

Geographical position of Taiti.

THE isle which at first was called New Cythera, is known by the name of Taiti amongst its inhabitants. Its latitude has been determined in our camp from several meridian altitudes of the sun, observed on shore with a quadrant. Its longitude has been ascertained by eleven observations of the moon, according to the method of the horary angles. M. Verron had made many others on shore, during four days and four nights, to determine the same longitude; but the paper on which he wrote them having been stolen, he has only kept the last observations, made the day before our departure. He believes their result exact enough, though their extremes differ among themselves 7° or 8°. The loss of our anchors, and all the accidents I have mentioned before, obliged us to leave this place much sooner than we intended, and have made it impossible for us to survey its coasts. The southern part of it is

entirely

entirely unknown to us; that which we have obſerved from the S. E. to the N. W. point, ſeems to be fifteen or twenty leagues in extent, and the poſition of its principal points, is between N. W. and W. N. W.

Better anchorage than that where we were.

Between the S. E. point and another great cape advancing to the northward about ſeven or eight leagues from the former, you ſee a bay open to the N. E. which has three or four leagues depth. Its ſhores gradually deſcend towards the bottom of the bay, where they have but little height, and ſeem to form the fineſt and beſt peopled diſtrict of the whole iſland. It ſeems it would be eaſy to find ſeveral good anchoring-places in this bay. We were very ill ſerved by fortune in meeting with our anchorage. In entering into it by the paſſage where the Etoile came out at, M. de la Giraudais aſſured me, that between the two moſt northerly iſles, there was a very ſafe anchorage for at leaſt thirty ſhips; that there was from twenty-three to between twelve and ten fathom of water, grey ſand and ooze; that there was a birth of a league in extent, and never any ſea. The reſt of the ſhore is high, and ſeems in general to be quite ſurrounded by a reef, unequally covered by the ſea, and forming little iſles in ſome parts, on which the iſlanders keep up fires at night on account of their fiſhery, and for the ſafety of their navigation; ſome gaps from ſpace to ſpace form entrances to the part

 within

within the reefs, but the bottom muſt not be too much relied upon. The lead never brings up any thing but a grey ſand; this ſand covers great maſſes of hard and ſharp coral, which can cut through a cable in one night, as fatal experience taught us.

Beyond the north point of this bay, the coaſt forms no creek, nor no remarkable cape. The moſt weſterly point is terminated by a low ground, from which to the N. W. and at about a league's diſtance, you ſee a low iſle, extending two or three leagues to the N. W.

Aſpect of the country.

The height of the mountains in the interior parts of Taiti, is ſurpriſing in reſpect to the extent of the iſland. Far from making its aſpect gloomy and wild, they ſerve only to embelliſh it, offering to the eye many proſpects and beautiful landſcapes, covered with the richeſt productions of nature, in that beautiful diſorder which it was never in the power of art to imitate. From thence ſpring a vaſt number of little rivulets, which greatly contribute to the fertility of the country, and ſerve no leſs to ſupply the wants of the inhabitants than to adorn and heighten the charms of the plains. All the flat country, from the ſea-ſhore to the foot of the mountains, is deſtined for the fruit-trees, under which, as I have already obſerved before, the houſes of the people of Taiti are built, without order, and without forming any villages. One would think himſelf in the

Elyſian

Elysian fields: Public paths, very judiciously laid out, and carefully kept in a good condition form the most easy communication with every part of the country.

The chief productions of the isle are * cocoa-nuts, plantains or bananas, the bread-fruit, yams, curassol, okras, and several other roots and fruits peculiar to the country: plenty of sugar-canes which are not cultivated, a species of wild indigo, a very fine red and a yellow substance for dying, of which I cannot say from whence they get them. In general, M. de Commercon has found the same kinds of vegetables there as are common in India. Aotourou, whilst he was amongst us, knew and named several of our fruits and legumes, and a considerable number of plants, cultivated by the curious, in hot-houses. The wood which is fit for carpenters work grew on the mountains, and the islanders make little use of it; they only employ it for their

Its productions.

* The cocoa-nuts, or the fruit of the *cooes nutifera*, Linn. is too well known to want any description. The plantains, or fruit of the *musa parasidiaca*, Linn. is likewise well known to all navigators, as the produce of hot countries. The bread-fruit is a production of a tree not yet described by Dr. Linnæus; Lord Anson found it upon the isle of Tinian; Dampier and the great Ray take notice of this very useful and curious tree. Yams are the roots of a plant known under the name of *dioscorea alata*. The okra is the fruit of the *hibiscus esculentus*, Linn. The curassol is one of the *annonas* or custard-apples. In general it must be observed that the botanical knowledge of our author is very superficial, and though he enumerates these fruits as the growth of the isle of Otahitee, it cannot be ascertained with any degree of precision, whether our author is right or wrong; and the new light in which, by the indefatigable industry of our philosophers, the natural history of these countries will be placed, makes us the more ardently wish for the publication of their great discoveries. F.

great

great periaguas, which they make of cedar wood. We have likewiſe ſeen pikes of a black, hard and heavy wood among them, very like iron-wood. For building their common periaguas, they make uſe of the tree which bears the bread-fruit This is a wood which will not ſplit, but is ſo ſoft and full of gum, that it is only as it were bruiſed when worked with a tool.

It does not appear that there are any mines

This iſle, though abounding with very high mountains, does not ſeem to contain any minerals, ſince the hills are every where covered with trees and other plants *. At leaſt it is certain that the iſlanders do not know any metals. They give the ſame name of *aouri*, by which they aſked us for iron, to all the kinds of metals we could ſhew them. But in what manner they became acquainted with iron, is not eaſily underſtood; however, I ſhall ſoon mention what I think on this ſubject. I know of only a ſingle rich article of commerce, viz. very fine pearls. The wives and children of the chief people wear them at their ears; but they hid them during our ſtay amongſt them. They

There are fine pearls.

* This aſſertion of Mr. de B. proves him to be little acquainted with mining; ſince our beſt writers on that ſubject give a gently ſloping ridge of mountains, with a fine turf, covered with groves of trees, and well ſupplied with water, amongſt many more, as the characteriſtics of a place where it is probable to find minerals in: See *Lehman's Art des Mines Metalliques*, vol. i. p. 17. But the whole iſle of Otahitee ſeems to be produced by a Volcano, and the rocks on it are chiefly lava, conſequently there are very little hopes of finding any regular veins with minerals on it, except ſome iron-ſtone, which has been liberally ſcattered by the benevolent hand of nature all over the various parts of the globe. F.

make

make a kind of caſtanets of the ſhells of the pearl-oyſter, and this is one of the inſtruments employed by their dancers.

Animals of the country

We have ſeen no other quadrupeds than hogs, a ſmall but pretty ſort of dogs, and rats in abundance. The inhabitants have domeſtic cocks and hens, exactly like ours We have likewiſe ſeen beautiful green turtle doves, large pigeons of a deep blue plumage and excellent taſte, and a very ſmall ſort of perrokeets, very ſingular on account of the various mixture of blue and red in their feathers. The people feed their hogs and their fowls with nothing but plantains. Taking together what has been conſumed by us on ſhore, and what we have embarked in both ſhips, we have in all got by our exchanges, upwards of eight hundred fowls, and near one hundred and fifty hogs; and if it had not been for the troubleſome work on the laſt days, we ſhould have got much more, for the inhabitants brought every day a greater quantity of them.

Meteorological obſervations.

We have not obſerved great heat in this iſland. During our ſtay, Reaumur's thermometer never roſe above 22°, and was ſometimes at 18°, but it may be obſerved that the ſun was already eight or nine degrees on the other ſide of the equator. However, this iſle has another ineſtimable advantage, which is that of not being infeſted by thoſe myriads of troubleſome inſects that are the

the plague of other tropical countries: neither have we observed any venomous animals in it. The climate upon the whole is so healthy, that notwithstanding the hard work we have done in this island, though our men were continually in the water, and exposed to the meridian sun, though they slept upon the bare soil and in the open air, none of them fell sick there. Those of our men who were sent on shore because they were afflicted with the scurvy, have not passed one night there quietly, yet they regained their strength, and were so far recovered, in the short space of time they staid on shore, that some of them were afterwards perfectly cured on board. In short, what better proofs can we desire of the salubrity of the air, and the good regimen which the inhabitants observe, than the health and strength of these same islanders, who inhabit huts exposed to all the winds, and hardly cover the earth which serves them as a bed with a few leaves; the happy old age to which they attain without feeling any of its inconveniences; the acuteness of all their senses; and lastly, the singular beauty of their teeth, which they keep even in the most advanced age?

Goodness of the climate: vigour of the inhabitants.

Vegetables and fish are their principal food; they seldom eat flesh, their children and young girls never eat any; and this doubtless serves to keep them free from almost all our diseases. I must say the same of their

Their nourishment.

their drink; they know of no other beverage than water. The very ſmell of wine or brandy diſguſted them; they likewiſe ſhewed their averſion to tobacco, ſpices, and in general to every thing ſtrong.

The inhabitants of Taiti conſiſt of two races of men, very different from each other, but ſpeaking the ſame language, having the ſame cuſtoms, and ſeemingly mixing without diſtinction. The firſt, which is the moſt numerous one, produces men of the greateſt ſize; it is very common to ſee them meaſure ſix (Paris) feet and upwards in height. I never ſaw men better made, and whoſe limbs were more proportionate: in order to paint a Hercules or a Mars, one could no where find ſuch beautiful models. Nothing diſtinguiſhes their features from thoſe of the Europeans: and if they were cloathed; if they lived leſs in the open air, and were leſs expoſed to the ſun at noon, they would be as white as ourſelves: their hair in general is black. The ſecond race are of a middle ſize, have frizzled hair as hard as briſtles, and both in colour and features they differ but little from mulattoes. The Taiti man who embarked with us, is of this ſecond race, though his father is chief of a diſtrict: but he poſſeſſes in underſtanding what he wants in beauty.

There are two races of men in the iſle.

Both races let the lower part of the beard grow, but they all have their whiſkers, and the upper part of the

Account of ſome of their cuſtoms.

the cheeks ſhaved. They likewiſe let all their nails grow, except that on the middle finger of the right hand. Some of them cut their hair very ſhort, others let it grow, and wear it faſtened on the top of the head. They have all got the cuſtom of anointing or oiling it and their beard with cocoa-nut oil. I have met with only a ſingle cripple amongſt them; and he ſeemed to have been maimed by a fall. Our ſurgeon aſſured me, that he had on ſeveral of them obſerved marks of the ſmall-pox; and I took all poſſible meaſures to prevent our people's communicating the other ſort to them; as I could not ſuppoſe that they were already infected with it.

Their dreſs.

The inhabitants of Taiti are often ſeen quite naked, having no other clothes than a ſaſh, which covers their natural parts. However, the chief people among them generally wrap themſelves in a great piece of cloth, which hangs down to their knees. This is likewiſe the only dreſs of the women; and they know how to place it ſo artfully, as to make this ſimple dreſs ſuſceptible of coquetry. As the women of Taiti never go out into the ſun, without being covered, and always have a little hat, made of canes, and adorned with flowers, to defend their faces againſt its rays; their complexions are, of courſe, much fairer than thoſe of the men. Their features are very delicate; but what diſtinguiſhes them,

6 is

is the beauty of their bodies, of which the *contour* has not been disfigured by a torture of fifteen years duration.

Custom of puncturing the skin.

Whilst the women in Europe paint their cheeks red, those of Taiti dye their loins and buttocks of a deep blue. This is an ornament, and at the same time a mark of distinction. The men are subject to the same fashion. I cannot say how they do to impress these indelible marks, unless it is by puncturing the skin, and pouring the juice of certain herbs upon it, as I have seen it practised by the natives of Canada. It is remarkable, that this custom of painting has always been found to be received among nations who bordered upon a state of nature. When Cæsar made his first descent upon England, he found this fashion established there; *omnes vero Britanni se vitro inficiunt, quod cæruleum efficit Colorem.* The learned and ingenious author of the *Recherches philosophiques sur les Americains* *, thinks this general custom owes its rise to the necessity of defending the body from the puncture of insects, multiplying beyond conception in uncultivated countries. This cause, however, does not exist at Taiti, since, as we have already said above, the people there are not troubled with such insupportable insects. The custom of painting is accordingly a mere fashion, the same as at Paris. Another

* Supposed to be the marquis de Pau. F.

 custom

custom at Taiti, common to men and women, is, to pierce their ears and to wear in them pearls or flowers of all sorts. The greatest degree of cleanliness further adorns this amiable nation; they constantly bathe, and never eat nor drink without washing before and after it.

Interior policy.

The character of the nation has appeared mild and beneficent to us. Though the isle is divided into many little districts, each of which has its own master, yet there does not seem to be any civil war, or any private hatred in the isle. It is probable, that the people of Taiti deal amongst each other with unquestioned sincerity. Whether they be at home or no, by day or by night, their houses are always open. Every one gathers fruits from the first tree he meets with, or takes some in any house into which he enters. It should seem as if, in regard to things absolutely necessary for the maintainance of life, there was no personal property amongst them, and that they all had an equal right to those articles. In regard to us, they were expert thieves; but so fearful, as to run away at the least menace. It likewise appeared, that the chiefs disapproved of their thefts, and that they desired us to kill those who committed them. Ereti, however, did not himself employ that severity which he recommended to us. When we pointed out a thief to him, he himself pursued him as fast as possible; the

man

man fled; and if he was overtaken, which was commonly the caſe, for Ereti was indefatigable in the purſuit, ſome laſhes, and a forced reſtitution of the ſtolen goods, was all the puniſhment inflicted on the guilty. I at firſt believed they knew of no greater puniſhment; for when they ſaw that ſome of our people were put in irons, they expreſſed great concern for them; but I have ſince learnt, that they have undoubtedly the cuſtom of hanging thieves upon trees, as it is practiſed in our armies.

They are at war with the neighbouring iſlands.

They are almoſt conſtantly at war with the inhabitants of the neighbouring iſles. We have ſeen the great periaguas, which they make uſe of to make deſcents, and even in ſea-fights. Their arms are the bow, the ſling, and a kind of pike of a very hard wood. They make war in a very cruel manner. According to Aotourou's information, they kill all the men and male children taken in battle; they ſtrip the ſkins, with the beards from the chins, and carry them off as trophies of their victory, only preſerving the wives and daughters of their enemies, whom the conquerors do not diſdain to admit to their bed. Aotourou himſelf is the ſon of a chief of Taiti, and of a captive woman from the iſle of Oopoa, which is near Taiti, and often at war with its inhabitants. To this mixture I attribute the difference of the races we have obſerved among them. I am not acquainted

quainted with their method of healing wounds: our surgeons admired the ſcars which they ſaw.

I ſhall, towards the end of this chapter, give an account of what I have been able to diſcover, concerning their form of government, the extent of the power of their petty ſovereigns, the kind of diſtinction exiſting between the men of note and the common people; and, laſtly, the ties which unite together, under the ſame authority, this multitude of robuſt men, whoſe wants are ſo few. I ſhall only obſerve here, that in matters of conſequence, the lord of the diſtrict does not give his deciſion without the advice of a council. I have mentioned above, that a deliberation of the people of note in the nation was required on the ſubject of our eſtabliſhing a camp on ſhore. I muſt add too, that the chief ſeems to be implicitly obeyed by every body; and that the men of note have likewiſe people to ſerve them, and over whom they have an authority.

Important cuſtom.

Cuſtoms on the ſubject of their dead.

It is very difficult to give an account of their religion. We have ſeen wooden ſtatues among them, which we took for idols; but how did they worſhip them? The only religious ceremony, which we have been witneſſes to, concerns the dead. They preſerve their corpſes a long while, extended on a kind of ſcaffold, covered by a ſhed. The infection which they ſpread does not prevent the women from going to weep around the corpſe, dur-

during part of the day, and from anointing the cold relicks of their affection with cocoa-nut oil. Those women, with whom we were acquainted, would sometimes allow us to come near these places, which are consecrated to the manes of the deceased; they told us *emoé*, he sleeps. When nothing but the skeletons remain, they carry them into their houses, and I do not know how long they keep them there. I only know, because I have seen it, that then a man of consideration among the people comes to exercise his sacred rites there; and that in these awful ceremonies, he wears ornaments which are much in request.

Superstition of the islanders.

We have asked Aotourou many questions concerning his religion; and believe, we understood that, in general his countrymen are very superstitious; that the priests have the highest authority amongst them; that besides a superior being, named *Eri-t-Era*, king of the sun or of light, and whom they do not represent by any material image, they have several divinities; some beneficent, others mischievous; that the name of these divinities or genii is *Eatoua*; that they suppose, that at each important action of human life, there presides a good and an evil genius; and that they decide its good or bad success. What we understand with certainty is, that when the moon has a certain aspect, which they call *Malama Tamai*, or moon in state of war *, (an aspect in which

* *Lune en état de Guerre.*

which we have not been able to diſtinguiſh any characteriſtic mark, by which it could be defined) they ſacrifice human victims. Of all their cuſtoms, one which moſt ſurpriſed me, is that of ſaluting thoſe who ſneeze by ſaying, *Evaroua-t-eatoua*, that the good *eatoua* may awaken thee, or that the evil *eatoua* may not lull thee aſleep. Theſe are marks which prove, that they have the ſame origin with the people of the old continent. Upon the whole, ſcepticiſm is reaſonable, eſpecially when we treat of the religion of different nations; as there is no ſubject in which it is more eaſy to be deceived by appearances.

Polygamy. Polygamy ſeems eſtabliſhed amongſt them; at leaſt it is ſo amongſt the chief people. As love is their only paſſion, the great number of women is the only luxury of the opulent. Their children are taken care of, both by their fathers and their mothers. It is not the cuſtom at Taiti, that the men occupied only with their fiſhery and their wars, leave to the weaker ſex the toilſome works of huſbandry and agriculture. Here a gentle indolence falls to the ſhare of the women; and the endeavours to pleaſe, are their moſt ſerious occupation. I cannot ſay whether their marriage is a civil contract, or whether it is conſecrated by religion; whether it is indiſſoluble, or ſubject to the laws of divorce. Be this as it will, the wives owe their huſbands a blind ſubmiſſion; they

would

would waſh with their blood any infidelity committed without their huſbands' conſent. That, it is true, is eaſily obtained; and jealouſy is ſo unknown a paſſion here, that the huſband is commonly the firſt who perſuades his wife to yield to another. An unmarried woman ſuffers no conſtraint on that account; every thing invites her to follow the inclination of her heart, or the inſtinct of her ſenſuality; and public applauſe honours her defeat: nor does it appear, that how great ſoever the number of her previous lovers may have been, it ſhould prove an obſtacle to her meeting with a huſband afterwards. Then wherefore ſhould ſhe reſiſt the influence of the climate, or the ſeduction of examples? The very air which the people breathe their ſongs, their dances, almoſt conſtantly attended with indecent poſtures; all conſpire to call to mind the ſweets of love, all engage to give themſelves up to them. They dance to the ſound of a kind of drum, and when they ſing, they accompany their voices with a very ſoft kind of flute, with three or four holes, which, as I have obſerved above, they blow with their noſe. They likewiſe practiſe a kind of wreſtling; which, at the ſame time, is both exerciſe and play to them.

Thus accuſtomed to live continually immerſed in pleaſure, the people of Taiti have acquired a witty and humorous temper, which is the offspring of eaſe and of joy. Character the iſlande joy. They likewiſe contracted from the ſame ſource a character of ficklineſs, which conſtantly amazed us. Every thing ſtrikes them, yet nothing fixes their attention: amidſt all the new objects, which we preſented to them, we could never ſucceed in making them attend for two minutes together to any one. It ſeems as if the leaſt reflection is a toilſome labour for them, and that they are ſtill more averſe to the exerciſes of the mind, than to thoſe of the body.

Account of ſome of their works.

I ſhall not, however, accuſe them of want of underſtanding Their ſkill and ingenuity in the few neceſſary inſtances of induſtry, which notwithſtanding the abundance of the country, and the temperature of the climate they cannot diſpenſe with, would be ſufficient to deſtroy ſuch aſſertion. It is amazing with how much art their fiſhing tackle is contrived; their hooks are made of mother-of pearl, as neatly wrought as if they were made by the help of our tools; their nets are exactly like ours; and knit with threads, taken from the great American *Aloes*. We admired the conſtruction of their extenſive houſes, and the diſpoſition of the leaves of the *Thatch-palm*, with which they are covered.

Conſtruction of their boats.

They have two ſorts of periaguas; ſome are little, and without much ornament; being made of a ſingle ſtem of a tree hollowed out; the others are much larger, and wrought with much art. A hollow tree forms the bot-

tom

Canoe, of the isle of Navigators, under sail.

Indian Canoe, of the isle of Choiseul.

Indian Canoe, of the isle of Choiseul.

to face Page 259.

Bayly, sculp.

Canoe, of the isle of Taiti, under sail.

tom of the periagua; from the head, to two-thirds of the intended length, another tree forms the back part, which is bent, and greatly elevated; in so much, that the extremity of the stern rises five or six feet above the water. These two pieces are joined together, as an arch of a circle; and as they have no nails to fasten them together with, they pierce the extremity of both pieces in several places, and by the means of strings, (made of the filaments which surround cocoa-nuts) they tie them together. The sides of the periagua are raised by two boards, about one foot broad, sewed to the bottom, and to each other, with the preceding sort of strings. They fill the seams with the fibrous substance round cocoa-nuts; but do not cover or pay them with any coating. A plank, which covers the head of the periagua, and projects about five or six feet beyond it, prevents its plunging entirely into the water, when there is a great sea. To make these light boats less subject to overset, they fix an out-rigger to one of its sides. This is nothing more than a pretty long piece of wood, supported by two cross pieces, of about four or five feet in length; the other end of which is fastened to the periagua. When she is sailing, a plank projects along the side, opposite to the out-rigger; a rope is fastened to it, which supports the mast, and it likewise makes the periagua stiffer, by placing a man or a weight at the end of the plank.

Their ingenuity appears ſtill more to advantage in the means they employ to render theſe veſſels proper to tranſport them to the neighbouring iſles, with which they have a communication, having no other guides than the ſtars on ſuch navigations. They faſten two great periaguas together alongſide of each other, (leaving about four feet diſtance between them) by means of ſome croſs pieces of wood tied very faſt to the ſtar-board of one and larboard of the other boat. Over the ſtern of theſe two veſſels thus joined, they place a hut, of a very light conſtruction, covered by a roof of reeds. This apartment ſhelters them from the ſun and rain, and at the ſame time affords them a proper place for keeping their proviſions dry. Theſe double periaguas can contain a great number of perſons, and are never in danger of overſetting. We have always ſeen the chiefs make uſe of them; they are navigated both by a ſail and by oars, as the ſingle periaguas: the ſails are compoſed of mats, extended on a ſquare frame, formed by canes, of which one of the angles is rounded.

The Taiti people have no other tool for all theſe works than a chiſſel, the blade of which is made of a very hard black ſtone *. It is exactly of the ſame form as

* The ſtone employed by the inhabitants of Otahitee for chiſſels and other tools and even for ornaments to be hung in the ears, is by all appearances a kind of *lapis nephriticus*, which when tranſparent is pale-green, very ſoft, and employed for the latter purpoſe; but when opaque, it is of a deeper hue and harder. In South

as that of our carpenters, and they uſe it with great expertneſs: they uſe very ſharp pieces of ſhells to bore holes into the wood.

Their cloths.

The manufacturing of that ſingular cloth, of which their dreſs is made up, is likewiſe one of their greateſt arts. It is prepared from the rind of a ſhrub, which all the inhabitants cultivate around their houſes. A ſquare piece of hard wood, fluted on its four ſides by furrows of different ſizes, is made uſe of in beating the bark on a ſmooth board: they ſprinkle ſome water on it during this operation, and thus they at laſt form a very equal fine cloth, of the nature of paper, but much more pliable, and leſs apt to be torn, to which they give a great breadth. They have ſeveral ſorts of it, of a greater or leſs thickneſs, but all manufactured from the ſame ſubſtance: I am not acquainted with their methods of dying them.

South America the ſame kind of ſtone is employed by the natives for ornaments; and is much valued among the *Topayos*, or Tapuyas, a nation in the interior parts of Braſil, living along the river of that name, which falls into the river of Amazons. This ſtone is called *tapuravas* by the Galibis, a nation in Guiana; the Europeans ſettled in theſe parts of the world, call it the *Amazon's-ſtone*; the European jewellers think it to be a *jade*, a kind of precious ſtone of the ſame colour brought from the eaſt. It is ſaid that ſtones of this kind are found near the river St. Jago, forty miles from Quito, in the province of las Eſmeraldas, in Peru. They grow more and more ſcarce, being much coveted by the nations of Guiana, the Tapuyas, and ſome other Indian nations, and likewiſe frequently bought up by the Europeans. BARRERE *Nouvelle Relation de la France equinoxiale*, Paris 1743, and CONDAMINE *Relation abregée d'un Voyage fait en deſcendant la Riviere des Amazones*, Paris 1746. F.

I. ſhall

Account of the Taiti-man, whom I brought to France.

I ſhall conclude this chapter in exculpating myſelf, for people oblige me to uſe this word, for having profited of the good will of Aotourou, and taken him on a voyage, which he certainly did not expect to be of ſuch a length; and likewiſe, in giving an account of the information he has given me concerning his country, during the time that he has been with me.

Reaſons for which I took him.

The zeal of this iſlander to follow us was unfeigned. The very firſt day of our arrival at Taiti, he manifeſted it to us in the moſt expreſſive manner, and the nation ſeemed to applaud his project. As we were forced to ſail through an unknown ocean, and ſure to owe all the aſſiſtance and refreſhments on which our life depended, to the humanity of the people we ſhould meet with, it was of great conſequence to us to take a man on board from one of the moſt conſiderable iſlands in this ocean. It was to be ſuppoſed that he ſpoke the ſame language as his neighbours, that his manners were the ſame, and that his credit with them would be deciſive in our favour, when he ſhould inform them of our proceedings towards his countrymen, and our behaviour to him. Beſides, ſuppoſing our country would profit of an union with a powerful people, living in the middle of the fineſt countries in the world, we could have no better pledge to cement ſuch an alliance, than the eternal obligation which we were going to confer

on

on this nation, by ſending back their fellow-countryman well treated by us, and enriched by the uſeful knowledge which he would bring them. Would to God that the neceſſity and the zeal which inſpired us, may not prove fatal to the bold Aotourou!

His ſtay at Paris.

I have ſpared neither money nor trouble to make his ſtay at Paris agreeable and uſeful to him. He has been there eleven months, during which he has not given any mark at all of being tired of his ſtay. The deſire of ſeeing him has been very violent; idle curioſity, which has ſerved only to give falſe ideas to men whoſe conſtant practice it is to traduce others, who never went beyond the capital, never examine any thing, and who being influenced by errors of all ſorts, never caſt an impartial eye upon any object, and yet pretend to decide with magiſterial ſeverity, and without appeal! How, ſaid ſome of them to me, in this man's country the people ſpeak neither French, nor Engliſh, nor Spaniſh? What could I anſwer them? I was ſtruck dumb; however, it was not on account of the ſurprize at hearing ſuch a queſtion aſked. I was uſed to them, becauſe I knew that at my arrival, many of thoſe who even paſs for people of abilities, maintained that I had not made the voyage round the world, becauſe I had not been in China. Some other ſharp critics conceived and propagated a very mean idea of the poor iſlander, becauſe,

after

after a ſtay of two years amongſt Frenchmen, he could hardly ſpeak a few words of the language. Do not we ſee every day, ſaid they, that the Italians, Engliſh, and Germans learn the French in ſo ſhort a time as one year at Paris? I could have anſwered them perhaps with ſome reaſon, that, beſides the phyſical obſtacle in the organs of ſpeech of this iſlander, (which ſhall be mentioned in the ſequel) which prevented his becoming converſant in our language, he was at leaſt thirty years old; that his memory had never been exerciſed by any kind of ſtudy, nor had his mind ever been at work; that indeed an Italian, an Engliſhman, a German could in a year's time ſpeak a French jargon tolerably well, but that was not ſtrange at all, as theſe ſtrangers had a grammar like ours, as their moral, phyſical, political, and ſocial ideas were the ſame with ours, and all expreſſed by certain words in their language as they are in French; that they had accordingly no more than a tranſlation to fix in their memory, which had been exerted from their very infancy. The Taiti-man, on the contrary, only having a ſmall number of ideas, relative on the one hand to a moſt ſimple and moſt limited ſociety, and on the other, to wants which are reduced to the ſmalleſt number poſſible; he would have been obliged, firſt of all, as I may ſay, to create a world of previous ideas, in a mind which is as indolent as his body, be-

6 fore

fore he could come ſo far as to adapt to them the words in our language, by which they are expreſſed. All this I might perhaps have anſwered: but this detail required ſome minutes of time, and I have always obſerved, that, loaded with queſtions as I was, whenever I was going to anſwer, the perſons that had honoured me with them were already far from me. But it is common in a capital to meet with people who aſk queſtions, not from an impulſe of curioſity, or from a deſire of acquiring knowledge, but as judges who are preparing to pronounce their judgment; and whether they hear the anſwer or no, it does not prevent them from giving their deciſion *.

However, though Aotourou could hardly blabber out ſome words of our language, yet he went out by himſelf every day, and paſſed through the whole town without once miſſing or loſing his way. He often made ſome purchaſes, and hardly ever paid for things beyond their real value. The only ſhew which pleaſed him, was the opera, for he was exceſſively fond of dancing. He knew perfectly well upon what days this kind of entertainment was played; he went thither

* Though our author has ſtrongly pleaded in this paragraph in behalf of Aotourou, it cannot, however, be denied that he was one of the moſt ſtupid fellows; which not only has been found by Engliſhmen who ſaw him at Paris, during his ſtay there, and whoſe teſtimony would be deciſive with the public, were I at liberty to name them; but the very countrymen of Aotourou were, without exception, all of the ſame opinion, that he had very moderate parts, if any at all. F.

by himſelf, paid at the door the ſame as every body elſe, and his favourite place was in the galleries behind the boxes *. Among the great number of perſons who have been deſirous of ſeeing him, he always diſtinguiſhed thoſe who were obliging towards him, and his grateful heart never forgot them. He was particularly attached to the ducheſs of Choiſeul, who has loaded him with favours, and eſpecially ſhewed marks of concern and friendſhip for him, to which he was infinitely more ſenſible than to preſents. Therefore, he would, of his own accord, go to viſit this generous benefactreſs as often as he heard that ſhe was come to town.

His departure from France.

He left Paris in March, 1770, and embarked at Rochelle, on board the Briſſon, which was to carry him to the Iſle de France.

Steps taken to ſend him home.

During this voyage he has been truſted to the care of a merchant, who went a paſſenger in the ſame ſhip, which he had equipped in part. The miniſtry have ſent orders to the governor and the intendant of the Iſle of France, to ſend Aotourou home to his iſle from thence. I have given a very minute account of the courſe that muſt be taken in order to go thither, and thirty-ſix thouſand francs, (about fifteen hundred pounds ſterling) which is the third part of my

* In the French theatre there is, in the door of each box, a ſmall window or hole, where people may peep through, which made it poſſible to Aotourou to enjoy even in the galleries the ſight of the dancers. F.

whole fortune, towards the equipment of the ship intended for this navigation. The duchess of Choiseul has been so humane as to consecrate a sum of money for bringing to Taiti a great number of the most necessary tools, a quantity of seeds, and a number of cattle; and the king of Spain has been pleased to permit that this ship might, if necessary, touch at the Philippines. O may Aotourou soon see his countrymen again!—I shall now give an account of what I have learnt in my conversations with him, concerning the customs of his country.

Farther accounts of the customs of Taiti.

I have already observed that the Taiti people acknowledge a supreme Being, who cannot be represented by any factitious image, and inferior divinities of two classes, represented by wooden figures. They pray at sun rise and at sun-set; but they have besides a great number of superstitious practices, in order to conciliate the influence of the evil genii. The comet, visible at Paris in 1769, and which Aotourou has very well taken notice of, has given me an opportunity of learning that the people of Taiti know this kind of stars, which do not appear again, as Aotourou said, till after a great number of moons. They call comets *evetou-eave*, and do not combine any sinister ideas with their apparition. Those meteors, however, which are here called shooting stars, are known to the people of

 Taiti

Taiti by the name of *epao*, and are by them thought to be evil genii *eatoua toa*.

The better inſtructed people of this nation (without being aſtronomers, as our gazettes have pretended) have, however, a name for every remarkable conſtellation; they know their diurnal motion, and direct their courſe at ſea by them, from iſle to iſle. In theſe navigations, which ſometimes extend three hundred leagues, they loſe all ſight of land. Their compaſs is the ſun's courſe in day-time, and the poſition of the ſtars during the nights, which are almoſt always fair between the tropics.

Neighbouring iſles.

Aotourou has mentioned ſeveral iſles to me; ſome of which are allies of, and others at war with Taiti. The friendly iſles are Aimeo, Maoroua, Aca, Oumaitia, and Tapouamaſſou. The enemies iſles are Papara, Aiatea, Otaa, Toumaraa, Oopoa. Theſe iſles are as big as Taiti.

The iſle of Pare, which is very abundant in pearls, is ſometimes in alliance, and ſometimes at war with Taiti. Enoua-motou, and Toupai, are two little uninhabited iſles, abounding with fruits, hogs, fowls, fiſh, and turtle; but the people believe, that they are the habitation of the genii; they are their domains; and unhappy are the boats which chance or curioſity has conducted to theſe ſacred iſles. Almoſt all thoſe, who endeavour to land there, muſt loſe their lives in the attempt.

tempt. These isles ly at different distances from Taiti. The greatest distance, which Aotourou mentioned to me, was fifteen days sail. It was, doubtless, about the same distance that he supposed our country was at, when he resolved to go with us.

Inequality of ranks.

I have mentioned above, that the inhabitants of Taiti seemed to live in an enviable happiness. We took them to be almost equal in rank amongst themselves; or at least enjoying a liberty, which was only subject to the laws established for their common happiness. I was mistaken; the distinction of ranks is very great at Taiti, and the disproportion very tyrannical. The kings and grandees have power of life and death over their servants and slaves, and I am inclined to believe, they have the same barbarous prerogative with regard to the common people, whom they call *Tata-einou*, vile men; so much is certain, that the victims for human sacrifices are taken from this class of people. Flesh and fish are reserved for the tables of the great; the commonalty live upon mere fruits and pulse. Even the very manner of being lighted at night, shews the difference in the ranks; for the kind of wood, which is burnt for people of distinction, is not the same with that which the common people are allowed to make use of. Their kings, alone, are allowed to plant before their houses, the tree which we call the *Weeping-willow*, or *Babylonian-willow* *. It

* *Arbre du Grand Seigneur.*

is

is known, that by bending the branches of this tree, and planting them in the ground, you can extend its ſhadow as far as you will, and in what direction you pleaſe; at Taiti, their ſhade affords the dining-hall of their kings.

The grandees have liveries for their ſervants. In proportion as the maſter's rank is more or leſs elevated, their ſervants wear their ſaſhes more or leſs high. This ſaſh is faſtened cloſe under the arms, in the ſervants of the chiefs, and only covers the loins in thoſe belonging to the loweſt claſs of nobility. The ordinary hours of repaſt, are when the ſun paſſes the meridian, and when he is ſet. The men do not eat with the women; the latter ſerving up the diſhes, which the ſervants have prepared.

Cuſtom of going into mourning.

At Taiti they wear mourning regularly, and call it *eeva*. The whole nation wear mourning for their kings. The mourning for the fathers is very long. The women mourn for their huſbands; but the latter do not do the ſame for them. The marks of mourning, are a head dreſs of feathers; the colour of which is conſecrated to death, and a veil over the face. When the people in mourning go out of their houſes, they are preceded by ſeveral ſlaves, who beat the caſtanets in a certain cadence; their doleful ſound gives every body notice to clear the way, whether out of reſpect for the grief of the perſons in mourning, or becauſe meeting

them

them is feared as an unlucky and ominous accident. However at Taiti, as in every other part of the world, the most respectable customs are abused; Aotourou told me, that this practice of mourning was favourable to the private meetings; doubtless, as I believe, of lovers with wives, whose husbands are not very complaisant. The instrument, whose sound disperses every body, and the veil which covers the face, secure to the lovers both secrecy and impunity.

Reciprocal assistance in their diseas.

In all diseases, which are any way dangerous, all the near relations assemble in the sick person's house. They eat and sleep there as long as the danger lasts; every one nurses him, and watches by him in his turn. They have likewise the custom of letting blood; but this operation is never performed at the foot or arm. A *Taoua*, i. e. a doctor, or inferior priest, strikes with a sharp piece of wood on the cranium of the patient; by this means he opens the *sagittal* vein; and when a sufficient quantity of blood is run out, he surrounds the head with a bandage, which shuts up the opening; the next day he washes the wound with water.

This is all that I have learnt concerning the customs of this interesting country, both upon the spot, and from my conversations with Aotourou. At the end of this work I shall add a Vocabulary of as many Taiti words as I could collect. When we arrived at this island, we observed

obſerved that ſome of the words pronounced by the iſlanders ſtood in the vocabulary at the end of Le Maire's Voyage, under the name of Vocabulary of Cocos iſland. Indeed thoſe iſlands, according to Le Maire and Schouten's reckoning, cannot be far from Taiti, and perhaps may be ſome of thoſe which Aotourou named to me. The language of Taiti is ſoft, harmonius, and eaſy to be pronounced; its words are compoſed of almoſt mere vowels, without aſpirates *. You meet with no naſal, nor no mute and half ſounded ſyllables, nor that quantity of conſonants, and of articulations which render ſome languages ſo difficult. Therefore our Taitiman could never learn to pronounce the French. The ſame reaſons for which our language is accuſed of not being very muſical, rendered it inacceſſible to his organs. It would have been eaſier to make him pronounce Spaniſh or Italian.

M. Pereire, celebrated for his art of teaching people, who are born deaf and dumb, to ſpeak and articulate words, has examined Aotourou ſeveral times, and has found that he could not naturally pronounce moſt of our conſonants, nor any of our naſal vowels. M. Pereire has been ſo obliging as to communicate to me a me-

* The contrary, of the laſt mentioned circumſtance, has been obſerved by our Engliſh navigators; and it is therefore highly probable Mr. de B. picked up his vocabulary of words from Aotourou, who had an impediment in his ſpeech. F.

moir on this ſubject. Upon the whole, the language of this iſland is abundant enough; I think ſo, becauſe Aotourou, during the courſe of the voyage, pronounced every thing that ſtruck him in rhythmic ſtanzas. It was a kind of blank verſe, which he ſpoke extempore. Theſe were his annals; and it ſeems as if his language furniſhed him with expreſſions ſufficient to deſcribe a number of objects unknown to him. We further heard him pronounce every day ſuch words as we were not yet acquainted with; and he likewiſe ſpoke a long prayer, which he calls the prayer of the kings, and of all the words that compoſe it, I do not underſtand ten.

I learnt from Aotourou, that about eight months before our arrival at his iſland, an Engliſh ſhip had touched there. It is the ſame which was commanded by Mr. Wallace. The ſame chance by which we have diſcovered this iſle, has likewiſe conducted the Engliſh thither, whilſt we lay in Rio de la Plata. They ſtayed there a month; and, excepting one attack of the iſlanders, who had conceived hopes of taking the ſhip, every thing has paſſed very friendly between them. From hence, doubtleſs, proceeds the knowledge of iron, which we found among the natives of Taiti, and the name of *aouri*, by which they call it, and which ſounds pretty like the Engliſh word *iron*. I am yet ignorant, whether the people of Taiti, as they owe the first knowledge of iron to the English, may not likewise be indebted to them for the venereal disease, which we found had been naturalized amongst them, as will appear in the sequel.

CHAP. IV.

Departure from Taiti; discovery of other islands; navigation to our clearing the great Cyclades.

OUR touching at Taiti has been productive of good, and of disagreeable consequences; danger and alarms followed all our steps to the very last moments of our stay; yet we considered this country as a friend, whom we must love with all his faults. On the 16th of April, at eight o'clock in the morning, we were about ten leagues N. E. by N. of the north point of the island, and from hence I took my departure. At ten o'clock we perceived land to leeward, seeming to form three isles, and we were still in sight of Taiti. At noon we plainly saw, that what we had taken for three isles, was no more than a single one, whose eminences had appeared as separate isles at a distance. Beyond this new land, we thought we saw another at a greater distance.

1768. April.

ight of Ou-aitia.

tance. This iſle is of a middling height, and covered with trees; it may be ſeen at ſea, about eight or ten leagues diſtant. Aotourou called it Oumaitia. He gave us to underſtand, in a manner which admitted of no doubt, that it was inhabited by a nation allied to his, that he had been there ſeveral times, that he had a miſtreſs there, and that we ſhould meet with the ſame reception and refreſhments there as at Taiti.

Direction of the courſe.

We loſt ſight of Oumaitia this day, and I directed my courſe ſo as to avoid the Pernicious Iſles, which we were taught to ſhun, by the diſaſters of admiral Roggewein. Two days afterwards, we had an inconteſtable proof, that the inhabitants of the iſles in the Pacific Ocean communicate with each other, even at conſiderable diſtances. The night was very fair, without a ſingle cloud, and all the ſtars ſhone very bright. Aotourou. after attentively obſerving them, pointed at the bright ſtar in Orion's ſhoulder, ſaying, we ſhould direct our courſe upon it; and that in two days time we ſhould find an abundant country, which he well knew, and where he had friends: we even believed his geſtures meant that he had a child there. As I did not alter my courſe, he repeated ſeveral times, that there were cocoa-nuts, plantains, fowls, hogs, and above all, women, whom by many expreſſive geſtures he deſcribed as very complaiſant. Being vexed that theſe reaſons did not

 make

make any impreſſion upon me, he ran to get hold of the wheel of the helm, the uſe of which he had already found out, and endeavoured in ſpite of the helm's-man to change it, and ſteer directly upon the ſtar, which he pointed at. We had much ado to quiet him, and he was greatly vexed at our refuſal. The next morning, by break of day, he climbed up to the top of the maſt, and ſtayed there all the morning, always looking towards that part where the land lay, whither he intended to conduct us, as if he had any hopes of getting ſight of it. He had likewiſe told us that night, without any heſitation, all the names which the bright ſtars that we pointed at, bear in his language. We have ſince been aſſured with certainty, that he knows the phaſes of the moon perfectly well, and is well acquainted with different prognoſticks, which often give notice to navigators of the changes of weather that are to happen at ſea ſome time after. One of the opinions of theſe people, which Aotourou made very intelligible to us is, that they poſitively believe that the ſun and moon are inhabited. What Fontenelle taught them the plurality of worlds?

During the latter end of April we had very fine weather, but not much wind, and the eaſterly winds kept more to the northward than ſouthward. On the night between the 26th and 27th, our pilot of the coaſt of

France died ſuddenly of an apoplexy. Theſe pilots generally are called coaſting pilots *, and all the king's ſhips have a pilot of the coaſt of France †. They differ from thoſe of the crew who are called pilots, and under-pilots, or pilot's boys ‡. The world has a very inaccurate idea of the functions which theſe pilots exerciſe on board our ſhips. They are generally thought to be the perſons who direct the courſe, and who ſerve as a ſtaff and ſupport to the blind. I know not whether there is ſtill any nation where they leave the art of piloting, that eſſential part of navigation, to thoſe ſubaltern people. In our ſhips, the buſineſs of the pilot is to take care that the helmſman exactly follows the courſe, for which the captain alone gives the orders, to mark down all the alterations of the courſe that happen, either in conſequence of the changes of winds, or of the orders of the commander, and likewiſe to obſerve the ſignals; nor have they the care of all theſe particulars, but under the direction of the officer of the watch. The ſuperior officers of the king's navy certainly know more of geometry, even at leaving ſchool, than is neceſſary to have a perfect knowledge of all the laws of pilotage. The claſs of pilots, properly ſo called, are moreover charged with the care of the common

* *Pilotes-côtiers.* † *Pilote-pratique de la côte de France.* ‡ *Pilotes, Aide-pilotes, ou Pilotins.*

and

and azimuth compaſſes, of the log and ſounding lines, of the lanthorns, the colours, &c. and it is plain, that theſe particulars require nothing more than exactneſs. Nor was my maſter-pilot above twenty years old, the ſecond pilot was of the ſame age, and the pilots boys * were making their firſt voyage.

Aſtronomical obſervations.

My reckoning compared twice during this month, with M. Verron's aſtronomical obſervations, differed, the firſt time, and that was at Taiti, only 13′ 10″, which I was more to the weſtward. The ſecond time, which was the 27th at noon, 1° 13′ 37″, which I was to the eaſtward of the obſervation. The different iſles diſcovered during this month, form the ſecond diviſion of iſles in this vaſt ocean; I named them Archipelago of Bourbon.

Second diviſion of iſles.

May.

The third of May, almoſt at day-break, we diſcovered more land to the north weſt, about ten or twelve leagues off. The wind was north eaſterly, and I gave orders to ſtand to windward of the north point of the land, which was very high, intending to reconnoitre it. The nautical knowledge of Aotourou did not extend to theſe places, for his firſt idea when he ſaw this land, was, that it was our country. During the day we had ſome ſqualls, followed by calms, rain, and weſterly breezes, ſuch as are obſerved in this ocean at the approach of

Sight of new iſlands.

* *Aides-pilotes.*

the

the leaſt land. Before ſun-ſet we diſtinguiſhed three iſles, one of which was much more conſiderable than the others. During the night, which happened to be moon-light, we kept ſight of the land; we ſtood in for it the next day, and ranged the eaſtern ſhore of the greater iſle, from its ſouth to its north point; that was its longeſt ſide, being about three leagues long. The iſle extends two leagues eaſt and weſt. Its ſhores are every where ſteep, and the whole iſle is as it were nothing more than a high mountain, covered with trees up to its ſummit, without either vallies or plains. The ſea broke very violently upon the ſhore. We ſaw fires on the iſland, ſome huts covered with reeds, and terminating in a point, built under the ſhadow of cocoa-trees, and about thirty men running along the ſea ſhore. The two little iſles bear W. N. W. corrected, and one league diſtant from the great one, and have likewiſe the ſame ſituation among themſelves. A narrow arm of the ſea ſeparates them, and at the W. point of the weſtermoſt iſle, there is a ſmall iſle or key. Each of the above two is not more than half a league long, and their ſhores are equally high and ſteep.

Exchanges made with the iſlande

At noon I made ſail to paſs between the little iſles and the great one, when the ſight of a periagua coming towards us made me bring to. She approached within piſtol ſhot of the ſhip, but none of her people would come

come on board, notwithſtanding all the ſigns of friendſhip which we could poſſibly invent and give to five men who conducted her. They were naked, excepting their natural parts, and ſhewed us cocoa-nuts and roots. Our Taiti-man ſtripped naked as they were, and ſpoke his language to them, but they did not underſtand him: they are no more of the ſame nation here. Being tired to ſee that they did not venture to come nearer, notwithſtanding the deſire they expreſſed of having ſeveral trifles which were diſplayed to them, I hoiſted out the pinnace. As ſoon as they ſaw her, they made all the haſte they could to get off, and I would not purſue them. Soon after we ſaw ſeveral other periaguas arrive, ſome of them under ſail. They ſeemed leſs miſtruſtful than the former one, and came near enough to make exchanges, though none of them would come on board. We got from them yams, cocoa-nuts, a water hen of a ſuperb plumage, and ſome pieces of a very fine ſhell. One of them had a cock which he would never exchange. They likewiſe brought ſtuffs of the ſame make as thoſe of Taiti, but much coarſer, and died with ugly red, brown, and black colours; bad fiſh hooks, made of the bones of fiſh, ſome mats, and ſome lances, ſix feet long, made of a kind of wood which was hardened in the fire. They did not chooſe to have any iron: they preferred little bits of red ſtuffs

to

to nails, knives, and ear-rings, which had had ſo great a ſucceſs at Taiti. I do not believe that theſe men are ſo gentle as thoſe of Taiti; their features were more ſavage, and we were always obliged to be upon our guard againſt their cunning tricks to cheat us by their barter.

Deſcription of theſe iſlanders.

Theſe iſlanders appeared to be of a middle ſize, but active and nimble. They paint their breaſt and their thighs, almoſt down to the knee, of a dark blue; their colour is bronzed; but we obſerved one man among them who was much whiter than the reſt. They ſhave or tear out their beards, and only one of them wore a pretty long one. They all had black hair, which ſtood upright on the head. Their periaguas are made with a good deal of ſkill, and have an out-rigger. Neither the head nor the ſtern is raiſed, but there is a kind of deck over each of them, and in the middle of theſe decks is a row of wooden pegs, ending in form of large nails, but their heads are covered with a fine ſhell, which is of a clear white. The ſail of their periaguas is of a triangular ſhape, compoſed of ſeveral mats. Two of its ſides are bent to two ſticks, one of which ſupported it up along the maſt; and the other, which is fixed in the outer clew, anſwers the purpoſe of a boom. Theſe periaguas followed us pretty far out to ſea, when we filled the ſails; ſome came likewiſe from the

Deſcription of their periaguas.

two little iſles, and in one of them was an ugly old woman. Aotourou expreſſed the greateſt contempt for theſe iſlanders.

We met with ſome calms, being to leeward of the larger iſland, which made me give up the ſcheme of paſſing between it and the little ones. The channel between them is a league and a half in breadth, and it ſeems as if there was ſome anchorage to be found. At ſix in the evening we diſcovered from the maſts more land to W. S. W. appearing as three detached hummocks. We ſteered S. W. and two hours after mid-night we ſaw the ſame land again, in W. 2° S. The firſt iſlands, which by the help of the moon-ſhine we ſtill could perceive, then bore N. E. of us.

Continuation of iſlands.

On the 5th in the morning we ſaw that this new land was a very fine iſle, of which we had only ſeen the ſummits the day before. It was interſperſed with mountains and vaſt plains, covered with cocoa-nut and many other trees. We ranged its ſouthern coaſt, at one or two leagues diſtance, without ſeeing any appearances of anchorage, the ſea breaking upon the ſhore very violently. There are even breakers to the weſtward of its weſtermoſt point, which runs about two leagues into the ſea. We have from ſeveral bearings got the exact poſition of this coaſt. A great number of periaguas ſailing, and ſimilar to thoſe of the laſt iſles, came around the ſhips, without however venturing to come close to us; a single one came alongside of the Etoile. The Indians seemed to invite us by signs to come on shore: but the breakers prevented it. Though we ran seven or eight knots at this time, yet the periaguas sailed round us with the same ease as if we had been at anchor. Several of them were seen from the masts sailing to the southward.

At six o'clock in the morning we had got sight of another land to westward; some clouds then intercepted it from our sight, and it appeared again at ten. Its shore ran S.W. and appeared to be at least as high, and of as great extent as the former ones, with which it lies nearly E. and W. about twelve leagues asunder. A thick fog which rose in the afternoon, and continued all the next night and ensuing day, prevented our viewing it more particularly. We only distinguished at its N. E. extremity two little isles, of unequal sizes.

Position of these isles which form the second division.

The longitude of these isles is nearly the same in which Abel Tasman was, by his reckoning, when he discovered the isles of Amsterdam, Rotterdam, Pylstaart, those of Prince William, and the shoals of Fleemskerk *. It is likewise the same which, within a very little, is assigned to the Solomon's isles. Besides, the periaguas, which we saw sailing to the southward, seem to shew that there are other isles in that part. Thus these isles

* Valentyn and others say *Heemskirk*. See Dalrymple's Historical Collection of Voyages in the South Pacific Ocean, p. 83.

 seem

ſeem to form a chain under the ſame meridian; they make the third diviſion, which we have named *l'Archipel des Navigateurs*, or Archipelago of the Navigators *.

On the 11th in the morning, having ſteered W. by S. ſince we got ſight of the laſt iſles, we diſcovered a land bearing W. S. W. ſeven or eight leagues diſtant. At firſt it was thought they were two ſeparate iſles, and we were kept at a diſtance from them all day by a calm. On the 12th we found that it was only one iſle, of which, the two elevated parts were connected by a low land, ſeemingly bending like a bow, and forming a bay open to the N. E. The high land lies N. N. W. A head wind prevented our approaching nearer than ſix or ſeven leagues of this iſland, which I named *l'Enfant Perdu*, or the Forlorn Hope.

Meteorological obſervations.

The bad weather which began already on the 6th of this month, continued almoſt uninterrupted to the 20th, and during all that time we had calms, rains, and weſt winds to encounter. In general, in this ocean which is

* Tobia, the man who went away from Otahitee, on board the Endeavour, according to the publiſhed *Journal of a Voyage round the World*, gave our circumnavigators accounts of many more iſlands in theſe ſeas, ſome of which were really found by our people; but many more were known only from his narrative of an expedition of theſe iſlanders to the weſt. As the number of theſe iſles ſeems to be ſo conſiderable, it would certainly deſerve another expedition to diſcover them all, and though at preſent the advantages ſeem to be of no great conſequence, which might be reaped from an intercourſe with theſe iſlanders; it is however certain, that the ſame objection might have been made to the firſt diſcoverers of America; and every body is at preſent ſenſible of the benefit accruing to theſe kingdoms from its American ſettlements. F.

called Pacific, the approach to lands is attended with tempeſts, which are ſtill more frequent during the decreaſe of the moon. When the weather proves ſqually, and there are thick clouds fixed upon the horizon, they are almoſt certain ſigns of ſome iſles, and give timely notice to be upon guard againſt them. It cannot be comprehended with what precautions and what apprehenſions, theſe unknown ſeas muſt be navigated, as you are there on all ſides threatened with the unexpected appearance of lands and ſhoals, and theſe apprehenſions are heightened by the length of the nights in the torrid zone. We were obliged to make way as it were blindfold, altering our courſe when the horizon appeared too black before us. The ſcarcity of water, the want of proviſions, and the neceſſity of making advantage of the wind whenever it blew, would not allow us to proceed with the ſlowneſs of a prudent navigation, and to bring to, or ſtand on our boards, whenever it was dark.

Critical ſituation we are in.

The ſcurvy in the mean while made its appearance again. A great part of the crew, and almoſt all the officers, had their gums affected, and the mouth inflamed with it. We had no refreſhments left, except for the ſick, and it is difficult to uſe one's ſelf to eat nothing but ſalt fleſh and dried pulſe. At the ſame time there appeared in both ſhips ſeveral venereal complaints,

plaints, contracted at Taiti. They had all the symptoms known in Europe. I ordered Aotourou to be searched; he was quite ruined by it; but it seems in his country this disease is but little minded; however, he consented to be taken care of by the surgeons. Columbus brought this disease from America; here it is in an isle in the midst of the greatest ocean. Have the English brought it thither? Or ought the physician to win, who laid a wager, that if four healthy stout men were shut up with one healthy woman, the venereal complaint would be the consequence of their commerce?

The 22d at day break, as we stood to the westward we saw a long high land a-head. When the sun rose we discovered two isles; the most southerly one bore from S. by E. to S. W. by S. and seemed to run N. N. W. corrected, being about twelve leagues long in that direction. It received the name of the day, *Isle de la Pentecôte*, Whitsuntide isle. The second bore from S. W. ¼ S. to W. N. W. the time when it first appeared to us was the occasion of our giving it the name of Aurora. We immediately stood as near as possible on the larboard tack, in order to pass between the two isles. The wind failed us, and we were obliged to bear away in order to pass to the leeward of the isle of Aurora. As we advanced to the northward, along its eastern shore, we saw a little isle rising like

like a ſugar-loaf, bearing N. by W. which we called Peak of the Etoile (*Pic de l'Etoile*). We continued to range the iſle of Aurora a league and a half diſtant. It runs N. and S. corrected from its ſouthermoſt point to about the middle of its length, which in the whole is ten leagues. It then declines to the N. N. W. it is very narrow, being to the utmoſt two leagues broad. Its ſhores are ſteep, and covered with woods. At two o'clock in the afternoon we perceived the ſummits of high mountains over this iſland, and about ten leagues beyond it. They belonged to a land, of which at half paſt three we ſaw the S. W. point, bearing S. S. W. by the compaſs, above the northern extremity of Aurora iſland. After doubling the latter we ſteered S. S. W. when at ſun-ſetting a new elevated coaſt, of conſiderable extent, came in ſight. It extended from W. S. W. to N. W. by N. about fifteen leagues diſtant.

We made ſeveral boards during night to get to the S. E. in order to diſcover whether the land which lay to S. S. W. of us joined to Whitſuntide iſle, or whether it formed a third iſle. This we verified on the 23d at day-break. We diſcovered the ſeparation of the three iſlands. The iſles of Whitſuntide and Aurora are nearly under the ſame meridian, two leagues diſtant from each other. The third iſle lies S. W. of Aurora, and in the neareſt part, they are three or four leagues aſunder

ſunder. Its north-weſt coaſt has at leaſt twelve leagues in extent, and is high, ſteep, and woody. We coaſted it during part of the morning on the 23d. Several periaguas appeared along the ſhore, but none ſeemed deſirous to come near us. We could ſee no huts, only a great number of ſmokes riſing out of the woods, from the ſea-ſhore, up to the tops of the mountains. We ſounded ſeveral times very near the ſhore; but found no bottom with fifty fathom of line.

Landing up on one of the iſles.

About nine o'clock the ſight of a coaſt, where it ſeemed landing would prove eaſy, determined me to ſend on ſhore, in order to take in ſome wood, which we were much in need of, to gain intelligence concerning the country, and to endeavour to get refreſhments from thence for our ſick. I ſent off three armed boats, under the command of enſign * the chevalier de Kerué, and we ſtood off and on, ready to ſend them any aſſiſtance, and to ſupport them by the artillery from both ſhips, if neceſſary. We ſaw them land, without the iſlanders ſeeming to have oppoſed their landing. In the afternoon, I and ſome other perſons went in a yawl to join them. We found our people employed in cutting wood, and the natives helping them to carry it to the boats. The officer who commanded our party, told me, that when he arrived, a numerous troop of iſlanders were

* *Enſeigne de la Marine.*

come

come to receive them on the beach, with bows and arrows in hand, making signs that they should not come ashore; but that when, notwithstanding their threats, he had given orders for landing, they had drawn back several yards; that in proportion as our people advanced, the savages retired; but always in the attitude of being ready to let go their arrows, without suffering our people to come nearer them; that at last, having given his people orders to stop, and the prince of Nassau having desired to advance alone towards them, the islanders had ceased to retire, seeing only one man come to them; that some pieces of red cloth being distributed amongst them, had brought about a kind of confidence between them. The chevalier de Kerué immediately posted himself at the entrance of the wood, made the workmen cut down trees, under the protection of the troops he had with him, and sent a detachment in search of fruits. Insensibly the islanders approached in a more friendly manner to all appearance; they even let our people have some fruits. They would not have any nails or other iron, and likewise constantly refused to exchange their bows and their clubs, only giving us some arrows. They always kept in great numbers around our people, without ever quitting their arms; and those who had no bows, held stones ready to throw at our men. They gave us to understand, that they were at war with the inha-

inhabitants of a neighbouring diſtrict. There actually appeared an armed troop of them, coming in good order from the weſt part of the iſland; and thoſe who were near us ſeemed diſpoſed to give them a warm reception; but no attack was made.

In this ſituation we found things when we came aſhore. We ſtaid there till our boats were laden with fruits and wood. I likewiſe buried at the foot of a tree, the act of taking poſſeſſion of theſe iſles, engraved on an oak plank, and after that we embarked in our boats again. This early departure, doubtleſs, ruined the project of the iſlanders to attack us, becauſe they had not yet diſpoſed every thing for that purpoſe; at leaſt we were inclined to think ſo, by ſeeing them advance to the ſea-ſhore, and ſend a ſhower of ſtones and arrows after us. Some muſkets fired off into the air, were not ſufficient to rid us of them; many advanced into the water, in order to attack us with more advantage; another diſcharge of muſkets, better directed, immediately abated their ardour, and they fled to the woods with great cries. One of our ſailors was ſlightly wounded by a ſtone.

They attack the French.

Deſcription of the iſlanders.

Theſe iſlanders are of two colours, black and mulattoes. Their lips are thick, their hair woolly, and ſometimes of a yellowiſh colour. They are ſhort, ugly, ill-proportioned, and moſt of them infected with leproſy; a circumſtance from which we called the iſland they inhabit,

habit, Isle of Lepers (*Isle des Lepreux*). There appeared but few women; and they were not less disagreeable than the men; the latter are naked, and hardly cover their natural parts; the women wear some bandages to carry their children on their backs; we saw some of the cloths, of which they are made, on which were very pretty drawings, made with a fine crimson colour. I observed that none of the men had a beard; they pierce their nose, in order to fix some ornaments to it. They likewise wear on the arm, in form of a bracelet, the tooth of a babyroussa, or a ring of a substance which I take to be ivory; on the neck they hang pieces of tortoise-shells, which they signified to us to be very common on their shores.

Their arms are bows and arrows, clubs of iron-wood, and stones, which they use without slings. The arrows are reeds, armed with a long and very sharp point made of a bone. Some of these points are square, and armed on the edges with little prickles in such a manner as to prevent the arrow's being drawn out of a wound. They have likewise sabres of iron-wood. Their periaguas did not come near us; at a distance they seemed built and rigged like those in the Isles of Navigators. Their weapons.

The beach where we landed was of very little extent. About twenty yards from the sea, you are at the foot of a moun- Descriptio[n] of the pla[ce] we lande[d] a mountain, which is covered with trees, notwithſtanding its great declivity. The ſoil is very light, and of no great depth: accordingly the fruits, though of the ſame ſpecies with thoſe at Taiti, are not ſo fine and not ſo good here. We found a particular ſpecies of figs here. There are many paths through the woods, and ſpots encloſed by palliſadoes three feet high. We could not determine whether they are intrenchments, or merely limits of different poſſeſſions. We ſaw no more than five or ſix little huts, into which one could not enter otherwiſe than by creeping on all-fours; and we were however ſurrounded by a numerous people; I believe they are very wretched, on account of the inteſtine war, of which we were witneſſes, and which brings great hardſhips upon them. We repeatedly heard the harſh ſound of a kind of drum, coming from the interior parts of the wood, towards the ſummit of the mountain. This certainly gives the ſignal to rally; for at the moment when the diſcharge of our muſkets had diſperſed them, it began to beat It likewiſe redoubled its ſound, when that body of enemies appeared, whom we ſaw ſeveral times. Our Taiti-man, who deſired to go on ſhore with us, ſeemed to think this ſet of men very ugly; he did not underſtand a ſingle word of their language.

Continuation of our courſe among the [illegible]ands.

When we came on board, we hoiſted in our boats. and made ſail ſtanding to the S. W. for a long coaſt which

which we diſcovered, extending as far as the eye could reach from S. W. to W. N. W. During night there was but little wind, and it conſtantly veered about; ſo that we were left to the mercy of the currents, which carried us to the N. E. This weather continued all the 24th, and the night following; and we could hardly get three leagues off the Iſle of Lepers. On the 25th, at five in the morning, we had a very fine breeze at E. S. E. but the Etoile, being ſtill under the land, did not feel it, and remained in a calm. I advanced, however, all ſails ſet, in order to obſerve the land, which lay to weſtward. At eight o'clock we ſaw land in all parts of the horizon; and we were, as it were, ſhut up in a great gulph. The iſle of Whitſuntide extended on the ſouth-ſide towards the new coaſt we had juſt diſcovered; and we were not ſure whether it was not connected, or whether what we took to be the ſeparation was any more than a great bay. Several places in the remainder of the coaſt likewiſe ſhewed appearances of paſſages, or of great gulphs. Among the reſt there ſeemed to be a very conſiderable one to the weſtward. Some periaguas croſſed from one land to the other. At ten o'clock we were obliged to ſtand towards the Iſle of Lepers again. The Etoile, which could no longer be ſeen from the maſt-head, was ſtill becalmed there, though the E. S. E. breeze held out at ſea. We ſtood for the ſtore-ſhip till four

four o'clock in the evening; for it was not till then that ſhe felt the breeze. It was too late when ſhe joined us to think of further diſcoveries. Thus the day of the 25th was loſt, and we paſſed the night making ſhort tacks.

The bearings we took on the 26th, at ſun-riſing, ſhewed us that the currents had carried us ſeveral miles to the ſouthward, beyond our reckoning. Whitſuntide iſle ſtill appeared ſeparated from the S. W. land, but the paſſage ſeemed narrower. We diſcovered ſeveral other openings on that coaſt, but were not able to diſtinguiſh the number of iſles which compoſed the Archipelago around us. The land ſeemed to us to extend from E. S. E. to W. N. W. by the ſouth (by compaſs); and we could not ſee the termination of it. We ſteered from N. W. by W. gradually coming round to weſt, along a fine ſhore covered with trees, on which there appeared great pieces of ground, which were either actually cultivated, or ſeemed to be ſo. The country appeared fertile; and ſome of the mountains being barren, and here and there of a red colour, ſeemed to indicate that it contained minerals. As we continued our courſe we came to the great inlet, which we had obſerved to the weſtward the day before. At noon we were in the middle of it, and obſerved the ſun's height there. Its opening is five or ſix leagues wide; and it runs due E.

Aſpect of the country.

by S. and W. by N. Some men appeared on the ſouth coaſt, and ſome others came near the ſhips in a periagua; but as ſoon as they were within muſket ſhot, they would not come nearer, though we invited them; theſe men were black.

We ranged the north coaſt at the diſtance of three quarters of a league; it is not very high, and covered with trees. A number of negroes appeared on the ſhore; even ſome periaguas came towards us; but with as little confidence as that which came from the oppoſite coaſt. After having run along this, for the ſpace of two or three leagues, we ſaw a great inlet, ſeeming to form a fine bay, at the entrance of which were two iſlands. I immediately ſent our boats well armed to reconnoitre it; and during this time we ſtood on our boards, at one or two leagues diſtance off ſhore, often ſounding without finding bottom, with 200 fathom of line.

Attempts t[o] ſearch an [an]chorage.

About five o'clock we heard a diſcharge of muſkets, which gave us much concern: it came from one of our boats, which, contrary to my orders, had ſeparated from the others, and unluckily was expoſed to the attacks of the iſlanders, being got quite cloſe to the ſhore. Two arrows, which were ſhot at the boat, ſerved as a pretence for the firſt diſcharge. She then went along the coaſt, and kept up a briſk fire from her muſkets and pede-

pedereroes, directing them both upon the shore, and upon three periaguas which passed by her within reach of shot, and had shot some arrows at her. A point of land intercepted the boat from our sight, and her continual firing gave me reason to fear that she was attacked by a whole fleet of periaguas. I was just going to send the long boat to her assistance, when I saw her quite alone, doubling the point, which had concealed her. The negroes howled excessively in the woods, whither they had all retired, and where we could hear their drum beating. I immediately made signal to the boat to come on board, and I took my measures to prevent our being dishonoured for the future, by such an abuse of the superiority of our power.

What prevents our anchoring there.

The boats of the Boudeuse found that this coast, which we took to be continued, was a number of isles; so that the bay is formed by the junction of several channels, which separate them. However, they found a pretty good sandy bottom there, in 40, 30, and 20 fathom; but its continual inequality rendered this anchorage unsafe, especially for us, who had no anchors to venture. It was, besides, necessary to anchor there above half a league off shore, as the bottom was rocky nearer the coast. Thus the ships could not have protected the boats, and the country is so woody, that we would have been obliged always to have our arms in hand, in order

to

to cover the workmen againſt ſurprizes. We could not flatter ourſelves that the natives ſhould forget the bad treatment they had juſt received, and ſhould conſent to exchange refreſhments. We obſerved the ſame productions here as as upon the Iſle of Lepers. The inhabitants were likewiſe of the ſame ſpecies, almoſt all black, naked, except their nudities, wearing the ſame ornaments of collars, and bracelets, and uſing the ſame weapons.

Another attempt to put in here.

We paſſed this night on our boards. On the 27th in the morning we bore away, and ranged the coaſt at about a league's diſtance. About ten o'clock we ſaw, on a low point, a plantation of trees, laid out in walks, like thoſe of a garden. Under the trees there was a beaten track, and the ſoil ſeemed to be ſandy. A conſiderable number of inhabitants appeared about this part; on the other ſide of this point there was an appearance of an inlet, and I hoiſted the boats out. This was a fruitleſs attempt; for it was nothing but an elbow formed by the coaſt, and we followed it to the N. W. point, without finding any anchorage. Beyond that point the land returned to N. N. W. and extended as far as the eye could reach; it was of an extraordinary height, and ſhewed a chain of mountains above the clouds. The weather was dark, with ſqualls and rain at intervals. Often in day-time we thought we ſaw land

land a-head of us; mere fog banks, which disappeared when it cleared up. We passed all the night, which was a very stormy one, in plying with short boards, and the tides carried us to the southward far beyond our reckoning. We saw the high mountains all day on the 28th till sun set, when they bore from E. to N. N. E. twenty or twenty-five leagues distant.

Conjectures concerning these lands.

The 29th in the morning we saw no more of the land, having steered W. N. W. I called the lands we had now discovered, Archipelago of the great Cyclades (*Archipel des grandes Cyclades*). To judge of this Archipelago by what we have gone through, and by what we have seen of it at a distance, it contains at least three degrees of latitude, and five of longitude. I likewise readily believe that Roggewein saw its northern extremity in 11° of latitude, and called it *Thienhoven* and *Groningen*. As for ourselves, when we fell in with it, every thing conspired to persuade us that it was the *Tierra Austral del Espiritù Santo*. Appearances seemed to conform to Quiros's account, and what we daily discovered, encouraged our researches. It is singular enough, that exactly in the same latitude and longitude where Quiros places his bay of St. Philip and St. Jago, on a coast which at first sight seemed to be that of a continent, we should find a passage exactly of the same breadth which he assigns to the entrance of his

6 bay.

bay. Has this Spanish navigator seen things in a wrong light? Or, has he been willing to disguise his discoveries? Was it by guess that the geographers made this Tierra del Espiritu Santo the same continent with New Guinea? To resolve this problem, it was necessary to keep in the same latitude for the space of three hundred and fifty leagues further. I resolved to do it, though the condition and the quantity of our provisions seemed to give us reason to make the best of our way to some European settlement. The event has shewn that little was wanting to make us the victims of our own perseverance.

Difference between the reckoning and the observations.

M. Verron made several observations during the month of May, and their results determined our longitude on the 5th, 9th, 13th, and 22d. We had not till now found so many differences between the observations and the ship's reckoning, all falling on one side. On the 5th at noon I was more to the east than the observed longitude, by 4° 00′ 42″; on the 9th, by 4° 23′ 4″; on the 13th, by 3° 38′ 15″; and lastly, on the 22d, by 3° 35′. All these differences shew, that from the isle of Taiti, the currents had carried us much to the westward. By this means it might be explained, why all the navigators who have crossed the Pacific ocean have fallen in with New Guinea much sooner than they ought. They have likewise given this ocean not by far

 so

ſo great an extent from eaſt to weſt as it really has. I muſt however obſerve, that whilſt the ſun was in the ſouthern hemiſphere, our reckoning has been to the weſtward of the obſervations; and that, after he paſſed to the other ſide of the line, our differences have changed. The thermometer during this month was commonly between 19° and 20°, it fell twice to 18°, and once to 15°.

Whilſt we were amidſt the great Cyclades, ſome buſineſs called me on board the Etoile, and I had an opportunity of verifying a very ſingular fact. For ſome time there was a report in both ſhips, that the ſervant of M. de Commercon, named Baré, was a woman. His ſhape, voice, beardleſs chin, and ſcrupulous attention of not changing his linen, or making the natural diſcharges in the preſence of any one, beſides ſeveral other ſigns, had given riſe to, and kept up this ſuſpicion. But how was it poſſible to diſcover the woman in the indefatigable Baré, who was already an expert botaniſt, had followed his maſter in all his botanical walks, amidſt the ſnows and frozen mountains of the ſtraits of Magalhaens, and had even on ſuch troubleſome excurſions carried proviſions, arms, and herbals, with ſo much courage and ſtrength, that the naturaliſt had called him his beaſt of burden? A ſcene which paſſed at Taiti changed this ſuſpicion into certainty. M. de Commercon went on ſhore to botanize there; Baré had hardly

hardly ſet his feet on ſhore with the herbal under his arm, when the men of Taiti ſurrounded him, cried out, It is a woman, and wanted to give her the honours cuſtomary in the iſle. The Chevalier de Bournand, who was upon guard on ſhore, was obliged to come to her aſſiſtance, and eſcort her to the boat. After that period it was difficult to prevent the ſailors from alarming her modeſty. When I came on board the Etoile, Baré, with her face bathed in tears, owned to me that ſhe was a woman; ſhe ſaid that ſhe had deceived her maſter at Rochefort, by offering to ſerve him in mens cloaths at the very moment when he was embarking; that ſhe had already before ſerved a Geneva gentleman at Paris, in quality of a valet; that being born in Burgundy, and become an orphan, the loſs of a law-ſuit had brought her to a diſtreſſed ſituation, and inſpired her with the reſolution to diſguiſe her ſex; that ſhe well knew when ſhe embarked that we were going round the world, and that ſuch a voyage had raiſed her curioſity. She will be the firſt woman that ever made it, and I muſt do her the juſtice to affirm that ſhe has always behaved on board with the moſt ſcrupulous modeſty. She is neither ugly nor handſome, and is no more than twenty-ſix or twenty-ſeven years of age. It muſt be owned, that if the two ſhips had been wrecked on any deſart iſle in the ocean, Baré's fate would have been a very ſingular one.

CHAP. V.

Run from the great Cyclades; discovery of the gulph of Louisiade; extremity to which we are reduced there; discovery of new isles; putting into a port on New Britain.

Direction of our course after leaving the Cyclades.

FROM the 29th of May, when we lost sight of the land, I sailed westward with a very fresh east, or south east wind. The Etoile considerably retarded our sailing. We sounded every four and twenty hours, finding no bottom with a line of two hundred and forty fathom. In day time we made all the sail we could, at night we ran under reefed top-sails, and hauling upon a wind when the weather was too dark. The night between the 4th and 5th of June, we were standing to the westward under our top-sails by moon-shine, when at eleven o'clock we perceived some breakers, and a very low sand bank, to the southward, half a league from us. We immediately got the other tacks on board, at the same time making a signal of danger to the Etoile. Thus we ran till near five in the morning, and then we resumed our former course to W. S. W. in order to view this land. We saw it again at eight o'clock, at about a league and a half distance. It is a little sandy isle, which

1768. June.

Meeting with breakers.

Pl. IIII. to face Page 303.

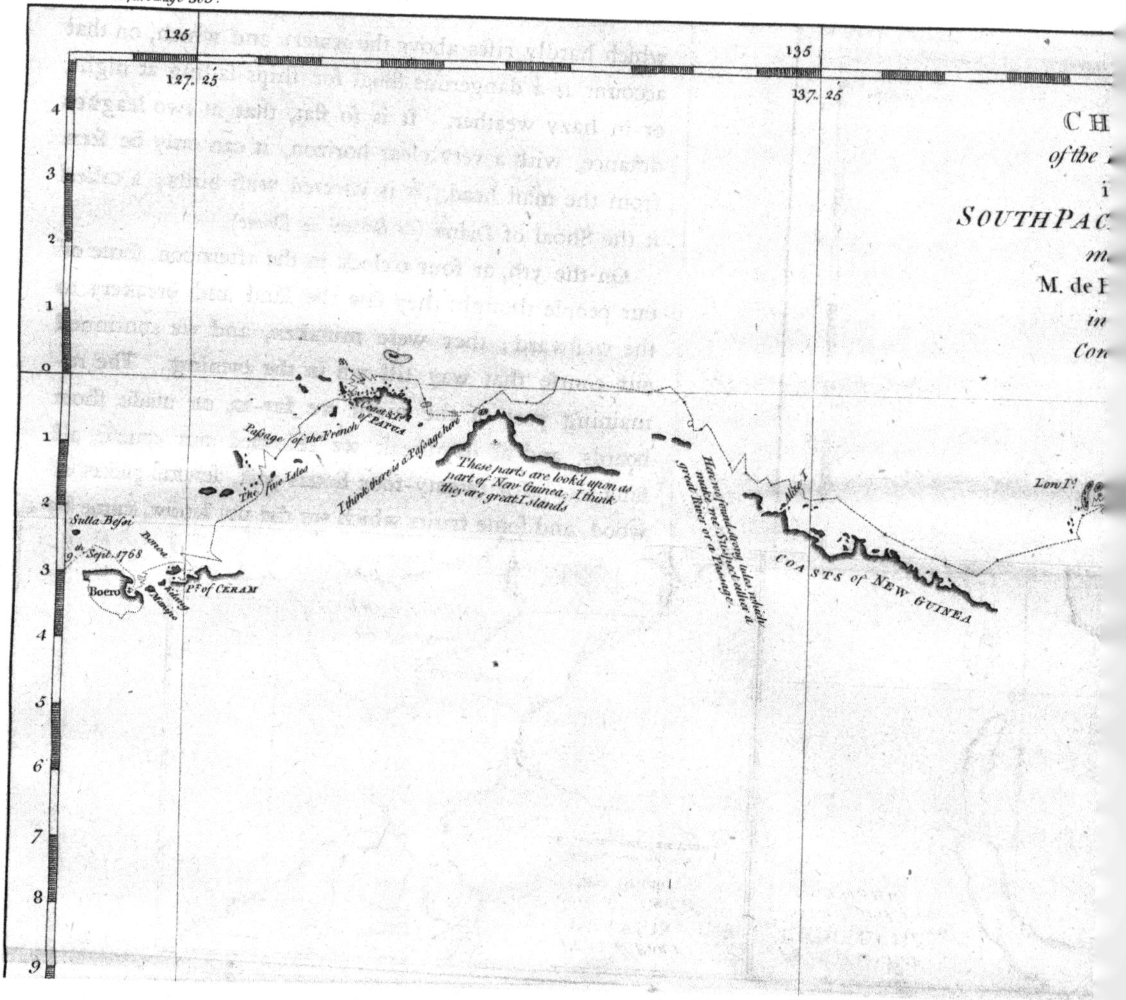

CH
of the
i
SOUTH PAC
m
M. de B
in
Con

125 127. 25 135 137. 25

Sulla Bessi
9th Sept. 1768
Boero
Bonoa
Kilang
Manipa
Pt. of CERAM
The five Isles
Passage of the French
Coasts of PAPUA
I think there is a Passage here
These parts are look'd upon as part of New Guinea. I think they are great Islands
Have found strong tides which make me suspect either a great River or a Passage.
COASTS of NEW GUINEA
Low I.s

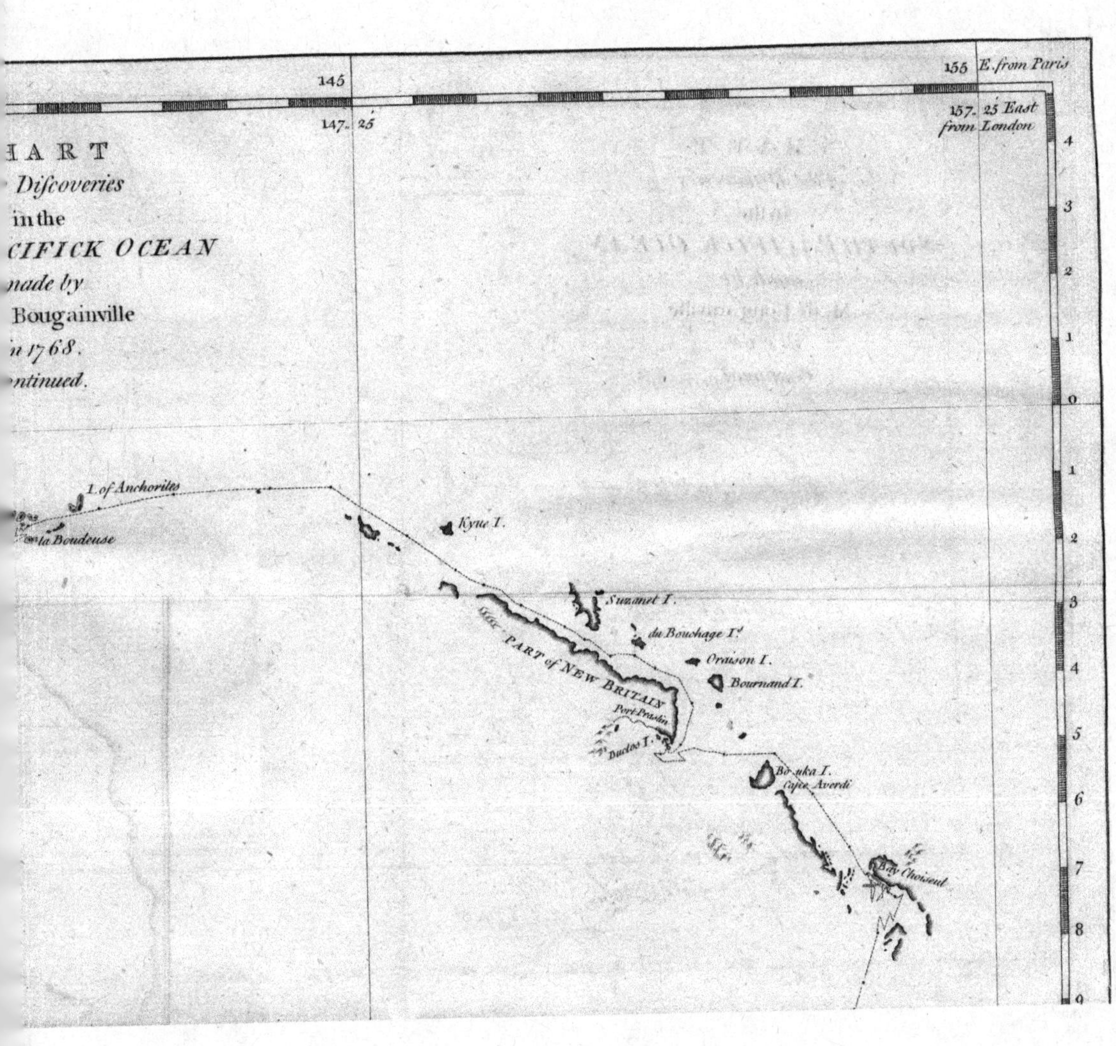

145
155 E. from Paris
147. 25
157. 25 East from London
HART
Discoveries
in the
CIFICK OCEAN
nade by
Bougainville
n 1768.
ntinued.
I. of Anchorites
la Boudeuse
Kyne I.
Suzanet I.
du Bouchage I.
PART of NEW BRITAIN
Oraison I.
Bournand I.
Port Praslin
Duclos I.
Bouka I.
Cape Averdi
Bay Choiseul

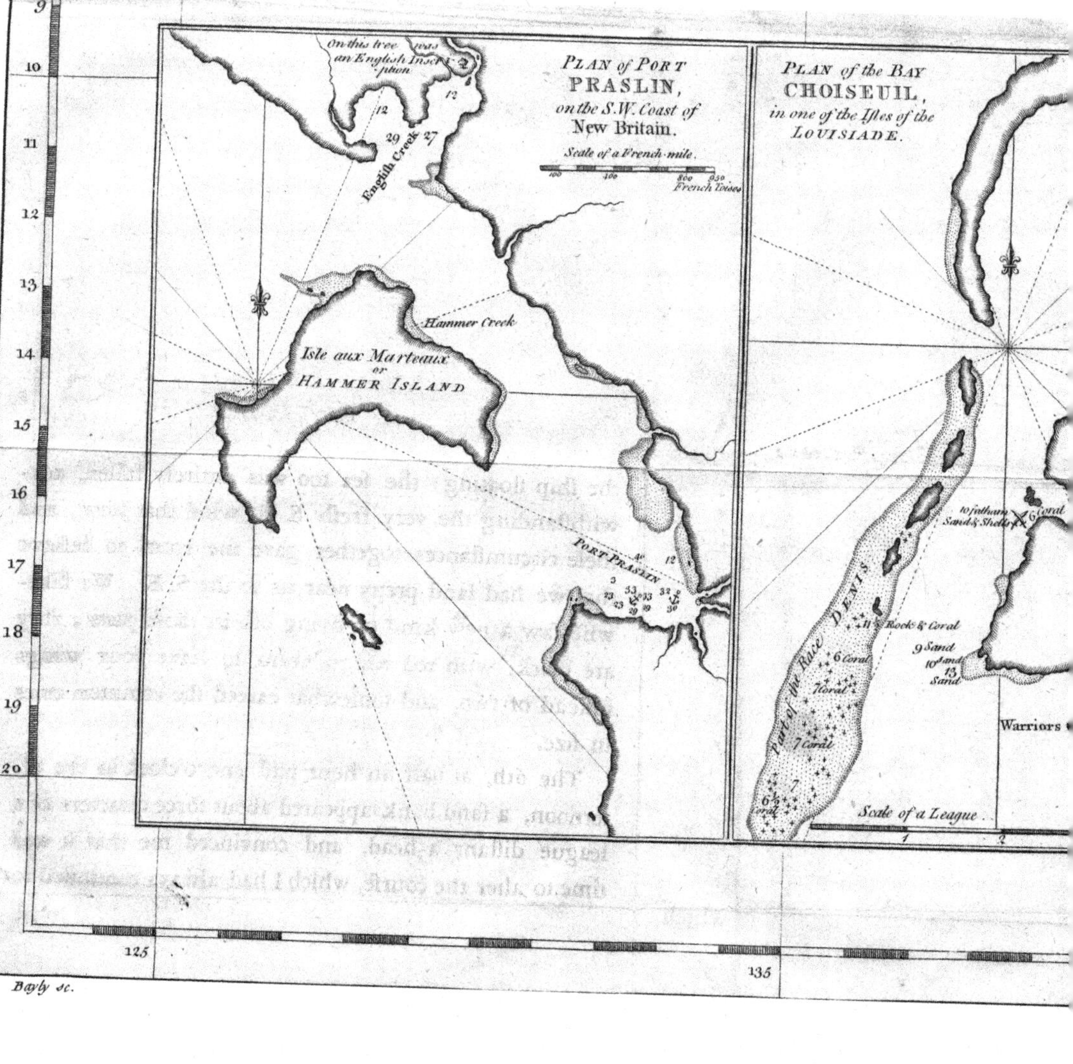
PLAN of PORT PRASLIN, on the S.W. Coast of New Britain.
Scale of a French mile.
French Toises
On this tree was an English Inscription
English Creek
Hammer Creek
Isle aux Marteaux or HAMMER ISLAND
PORT PRASLIN
PLAN of the BAY CHOISEUIL, in one of the Isles of the LOUISIADE.
Part of the Race DENIS
Rock & Coral
Sand & Shells
Warriors
Scale of a League.
125
135
Bayly sc.

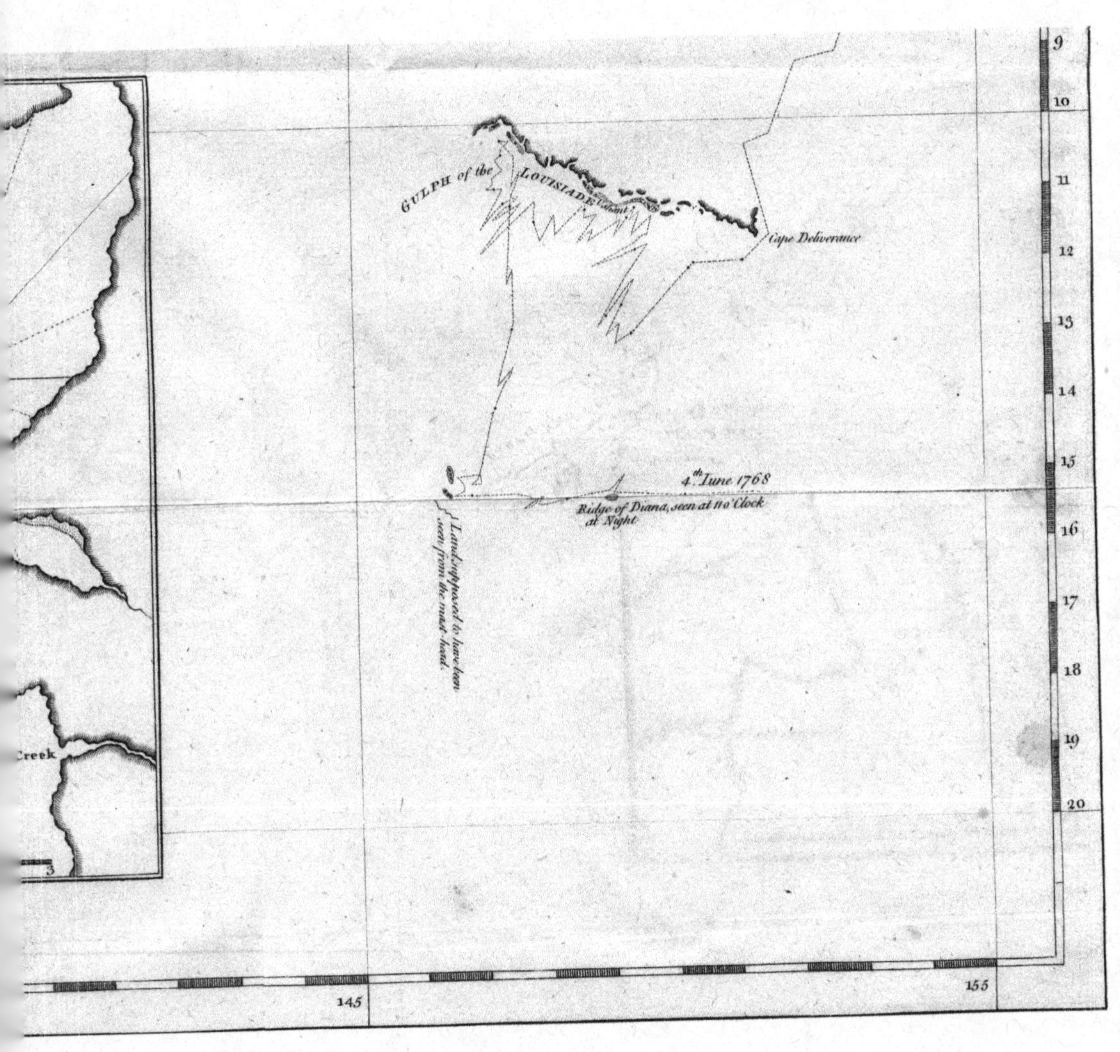

GULPH of the LOUISIADE
Cape Deliverance
4th June 1768
Ridge of Diana, seen at 11 o'Clock at Night
Land supposed to have been seen from the mast-head
Creek
145
155
9
10
11
12
13
14
15
16
17
18
19
20

which hardly riſes above the water; and which, on that account, is a dangerous ſhoal for ſhips ſailing at night, or in hazy weather. It is ſo flat, that at two leagues diſtance, with a very clear horizon, it can only be ſeen from the maſt head; it is covered with birds; I called it the Shoal of Diana (*la Bâture de Diane*).

On the 5th, at four o'clock in the afternoon, ſome of our people thought they ſaw the land and breakers to the weſtward; they were miſtaken, and we continued our courſe that way till ten in the evening. The remaining part of the night we lay-to, or made ſhort boards, and at day-break we reſumed our courſe, all ſails ſet. For twenty-four hours paſt, ſeveral pieces of wood, and ſome fruits which we did not know, came by the ſhip floating: the ſea too was entirely fallen, notwithſtanding the very freſh S. E. wind that blew, and theſe circumſtances together gave me room to believe that we had land pretty near us to the S. E. We likewiſe ſaw a new kind of flying fiſh in thoſe parts; they are black, with red wings, ſeem to have four wings inſtead of two, and ſomewhat exceed the common ones in ſize. Signs of lan

The 6th, at half an hour paſt one o'clock in the afternoon, a ſand-bank appeared about three quarters of a league diſtant a-head, and convinced me that it was time to alter the courſe, which I had always continued to weſt-

weſtward. This ſand extended at leaſt half a league from W. by S. to W. N. W. Some of our people even were of opinion they ſaw a low land to the S. W. of the breakers. We ſtood to the northward till four o'clock, and then again to the weſtward. This, however, did not laſt long; for at half paſt five o'clock, the men at the maſt-heads ſaw freſh breakers to the N. W. and N. W. by W. about a league and a half from us. We approached nearer, in order to view them better. They were ſeen to extend above two miles from N. N. E. to S. S. W. and we could not ſee an end of them. In all probability they joined thoſe which we had diſcovered three hours before. The ſea broke with great violence on theſe ſhoals, and ſome ſummits of rocks appeared above water from ſpace to ſpace. This laſt diſcovery was the voice of God, and we were obedient to it. Prudence not permitting us to purſue an uncertain courſe at night, in theſe dangerous parts, we ſpent it making ſhort boards in that ſpace, with which we had made ourſelves acquainted in the preceding day; and on the 7th, in the morning, I gave orders to ſteer N. E. by N. abandoning the ſcheme of proceeding further weſtward in the latitude of 15°.

Neceſſary alteration of the courſe.

We had certainly great reaſon to believe, that the Tierra Auſtral del Eſpiritù Santo was no more than the Archipelago of the great Cyclades, which Quiros took to be be a continent, and represented in a romantic light. When I persevered in keeping in the parallel of 15°, it was because I wanted to verify our conjectures, by getting sight of the eastern coasts of New Holland. Thus, according to the Astronomical Observations, (of which the uniformity for a month, and upwards, was a sufficient proof of their accuracy) we were already, on the 6th at noon, in 146° east latitude; that is one degree more to the westward than the Tierra del Espiritù Santo, as laid down by M. Bellin. Besides this, our repeated meeting with the breakers, which we had seen these three days; those trunks of trees, these fruits and sea-weeds, which we found at every moment; the smoothness of the sea, and the direction of the currents, all sufficiently marked the vicinity of a great land; and that it already surrounded us to the S. E. This land is nothing else than the eastern coast of New Holland. Indeed these numerous shoals, running out to sea, are signs of a low land; and when I see Dampier abandoning in our very latitude of 15° 35, the western coast of this barren region, where he did not so much as find fresh water, I conclude that the eastern coast is not much better. I should willingly believe, as he does, that this land is a cluster of isles, the approach to which is made difficult by a dangerous sea, full of shoals and sand banks. After such an explanation, it would have been rashness to

Geographi reflections.

risk running in with a coast, from whence no advantage could be expected, and which one could not get clear of, but by beating against the reigning winds. We had only bread for two months, and pulse for forty days; the salt-meat was in greater quantities; but it was noxious, and we preferred the rats to it, which we could catch. Thus it was by all means time to go to the northward, and even to deviate a little to the eastward of our course.

Unluckily the S. E. wind left us here; and when it returned, it put us into the most dangerous situation we had as yet been in. From the 7th, our course made good, was no better than N. by E. when on the 10th, at day-break, the land was discovered, bearing from east to N. W. Long before the break of day, a delicious smell announced us the vicinity of this land, which forms a great gulph open to the S. E. I have seen but few lands, which bore a finer aspect than this; a low ground, divided into plains and groves, lay along the sea-shore, and from thence it rose like an amphitheatre up to the mountains, whose summits were lost in the clouds. There were three ranges of mountains; and the highest chain was above twenty-five leagues in the interior parts of the country. The wretched condition to which we were reduced, did not allow us, either to spend some time in visiting this beautiful country, that by

Discovery of new lands.

by all appearances, was fertile and rich; nor to ſtand to weſtward in ſearch of a paſſage on the ſouth ſide of New Guinea, which might open a new and ſhort navigation to the Molucas, by the gulph of Carpentaria. Nothing, indeed, was more probable, than the exiſtence of ſuch a paſſage; it was even believed, that the land had been ſeen as far as W by S. We were now obliged to endeavour to get out of this gulph as ſoon as poſſible, and by the way which ſeemed to be moſt open: indeed we were engaged much deeper in it than we at firſt thought. Here the S. E. wind waited us, to put our patience to the greateſt trials.

During the 10th, the calm left us at the mercy of a great ſouth-eaſtern ſwell, which hove us towards the land. At four o'clock in the evening, we were no more than three quarters of a league diſtance from a little low iſle, to the eaſtern point of which lies connected a ledge, which extends two or three leagues to the eaſtward. Towards five o clock we had brought our head off, and we paſſed the night in this dreadful ſituation, making all our efforts to get off ſhore with the leaſt breezes. On the 11th, in the afternoon, we were got to about four leagues from the coaſt; at two leagues diſtance you are out of ſoundings. Several periaguas ſailed along the ſhore, on which we always ſaw great fires.

Critical ſitu[ation] in whi[ch] we are.

 fires.

fires. Here are turtles; for we found the remains of one in the belly of a ſhark.

The ſame day, at ſun ſetting, we ſet the eaſtermoſt land, bearing E. by N. 2° E. by compaſs, and the weſtermoſt bearing W. N. W. both about fifteen leagues diſtant. The following days were dreadful; every thing was againſt us; the wind conſtantly blowing very freſh at E. S. E. and S. E. the rain; a fog ſo thick, that we were obliged to fire guns, in order to keep company with the Etoile, which ſtill contained part of our proviſions; and, laſtly, a very great ſea, which hove us towards the ſhore. We could hardly keep our ground by plying, being obliged to wear, and to carry but very little ſail. Thus were we forced to make our boards; in the dark, in the midſt of a ſea, ſtrewed with ſhoals; being obliged to ſhut our eyes to all ſigns of danger. The night between the 11th and 12th, ſeven or eight of the fiſh, which are called *cornets* *, and which always keep at the bottom of the ſea, leaped upon the gang-boards. There likewiſe came ſome ſand and weeds from the bottom upon our fore-caſtle; it being left there by the waves that beat over it. I did not chooſe to ſound; it would not have leſſened the certainty of the danger, which was always the ſame, whatever expedient we could take.

Multiplied dangers which we run.

* *Cornets* are a ſpecies of ſhell-fiſh. F.

Upon

Upon the whole, we owe our ſafety to the knowledge we had of the land on the 10th in the morning, immediately before this continuance of bad and foggy weather. Indeed the winds being E. S. E. and S. E. I ſhould have thought ſteering N. E. an exceſs of precaution againſt the obſcurity of the weather. However this courſe evidently brought us into the moſt imminent danger of being loſt, as the land extended even to E. S. E.

The weather cleared up on the 16th, the wind ſtill remaining contrary; but we had at leaſt got day-light again. At ſix o'clock in the morning we ſaw the land from north to N. E. by E. by compaſs, and we plyed in order to double it. On the 17th, in the morning, we did not ſee any land at ſun-riſing; but at half paſt nine o'clock we perceived a little iſland to the N. N. E. by compaſs, five or ſix leagues diſtant, and another land to N. N. W. about nine leagues off. Soon after we diſcovered in N. E. ½ E. four or five leagues diſtant, another little iſle; which from its reſemblance to *Uſhant* *, obtained the ſame name. We continued our board to N. E. by E. hoping to double all theſe lands, when, at eleven o'clock, we diſcovered more land, bearing N. E. by E. ½ E. and breakers to E. N. E which ſeemed to join Uſhant. To the N. W. of this little iſle, we ſaw another chain of breakers, extending half a league. The firſt

* *Oueſſant.*

iſle

isle likewise seemed to be between two chains of breakers.

All the navigators, who ever came into these parts, always dreaded to fall to the southward of New Guinea, and of finding a gulph there corresponding to that of Carpentaria, which it would have proved difficult for them to clear. Consequently they have all in good time got into the latitude of New Britain, at which they touched. They all followed the same track; we opened a new one, and paid dear for the honour of the first discovery. Unhappily hunger, the most cruel of our enemies, was on board. I was obliged to make a considerable diminution in the allowance of bread and pulse. It likewise became necessary to forbid the eating of that leather, which is wrapped round the yards, and any other old leather, as it might have had the most dreadful consequences. We had a goat remaining, which had been our faithful companion since we left the Malouines, where we had taken her on board. Every day she gave us some milk. The hungry stomachs of the crew, in a capricious instant, condemned her to death; I could only pity her; and the butcher who fed her such a long time, shed tears over the victim which he thus sacrificed to our hunger. A young dog, taken in the straits of Magalhaens, shared the same fate soon after.

Extremities to which we are reduced.

On

On the 17th, in the afternoon, the currents had been ſo favourable, that we had again taken the N. N. E. board, ſtanding much to windward of Uſhant, and the ſhoals around it. But at four o'clock we were convinced, that theſe breakers extend much farther than we were at firſt aware of; ſome of them were ſeen even in E. N. E. and there was yet no end of them. We were obliged, during night, to return upon the S. S. W. tack, and in day-time the eaſtern one. On the 18th, during the whole morning, we ſaw no land; and we already gave ourſelves up to the hope of having doubled theſe iſles and breakers. Our joy was ſhort; about one o'clock in the afternoon, an iſle was ſeen in N. E. by N. by compaſs; and ſoon after it was followed by nine or ten others. Some of them bore E. N. E. and behind them a higher land extended to N. E. about ten leagues diſtant. We plyed to windward all night; the day following gave us a view of the ſame double chain of lands running nearly eaſt and weſt, viz. to the ſouthward, a number of little iſles connected by reefs, even with the ſurface of the water, to the northward of which extended the higher lands. The lands we diſcovered on the 20th ſeemed to be leſs ſouthward, and only to run E. S. E. This was an amendment in our poſition. I reſolved to run boards of four and twenty hours; we loſt too much time in putting about more frequently; the ſea being

 ex-

extremely rough, and the wind blowing very hard and constantly from the same point: we were likewise obliged to make very little sail, in order to spare our crazy masts, and damaged rigging; our ships too went very ill, being in a bad sailing trim, and not having been careened for so long a time.

We saw the land on the 25th at sun-rising, extending from N. to N. N. E. but it was now no longer low; on the contrary we saw a very high land, seemingly terminating in a large cape. It was probable that the coast after that should tend to the northward. We steered all day N. E. by E. and E. N. E. without seeing any land more easterly than the cape which we were doubling, with such a joy as I am not able to describe. On the 26th in the morning, the cape being much to leeward of us, and seeing no other lands to windward, we were at last enabled to alter our course again towards N. N. E. This cape which we had so long wished for, was named Cape Deliverance, and the gulph, of which it forms the eastermost point, Gulph of the Louisiade *(golfe de la Louisiade)*. I think we have well acquired the right of naming these parts. During the fortnight we passed in this gulph, the currents have pretty regularly carried us to the eastward. On the 26th and 27th it blew a hard gale, the sea was frightful, the weather squally and dark. It was impossible to make any way during night.

We at last double the lands of the gulph.

We

We were about ſixty leagues to the northward from Cape Deliverance, when on the 28th in the morning, we diſcovered land to the N. W. nine or ten leagues diſtant. It proved to conſiſt of two iſles, the moſt ſouthern of which, at eight o'clock, bore N. W. by W. by compaſs. Another long and high coaſt appeared at the ſame time, bearing from E. S. E. to E. N. E. This coaſt extended to the northward, and as we advanced north eaſtward, it lengthened more, and turned to N. N. W. We however diſcovered a ſpace where the coaſt was diſcontinued, either by a channel, or the opening of a large bay; for we thought we ſaw land at the bottom of it. On the 29th in the morning, the coaſt which lay to the eaſtward of us continued to extend N. W. though our horizon was not terminated by it on that ſide. I intended to come near it, and then to go along it in ſearch of an anchorage. At three o'clock in the afternoon, being near three leagues off ſhore, we found bottom in forty-eight fathoms, white ſand and broken ſhells: we then ſtood for a creek which ſeemed convenient; but we were becalmed, and thus the reſt of the day was paſſed away fruitleſsly. During night we made ſeveral ſhort boards, and on the 30th, by break of day, I ſent the boats with a detachment under the command of the chevalier Bournand, to viſit ſeveral creeks along the ſhore, which ſeemed to promiſe an anchorage, as the

We meet with new iſlands.

the bottom we had found at sea was a favourable sign. I followed him under an easy sail, ready to join him at the first signal he should give for that purpose.

Description of the islanders.

Towards ten o'clock, a dozen periaguas, of different sizes, came pretty near the ships, but would not come along-side of them. There were twenty-two men in the largest, in the middling ones eight or ten, and in the least two or three. These periaguas seemed well built; their head and stern are raised very much; they are the first we saw in these seas that had no outriggers. These islanders are as black as the negroes of Africa; their hair is curled, but long, and some of a reddish colour. They wear bracelets, and plates on the neck and forehead; I know not of what substance they were, but they seemed to be white. They are armed with bows and lances (*sagayes*); they made a great noise, and it seemed as if their disposition was far from pacific. I recalled our boats at three o'clock; the chevalier de Bournand reported that he had almost every where found good anchoring ground, from thirty, twenty-five, twenty, fifteen to eleven fathoms, oozy sand, but that it was in open road, and without any river; that he had only seen one rivulet in all that extent. The open coast is almost inaccessible, the sea breaks upon it every where, the mountains extend to the very sea shore, and the ground is entirely covered with woods.

Unsuccessful attempt to find anchorage here.

woods. In ſome little creeks there are ſome huts, but they are in very ſmall number, for the iſlanders inhabit the mountains. Our pinnace was followed by three or four periaguas, that ſeemed willing to attack her. An iſlander actually roſe ſeveral times to throw his lance (*ſagaye*); however, he did not throw it, and the boat returned on board without ſkirmiſhing.

Our ſituation was upon the whole very hazardous. We had lands, hitherto unknown, extending on one ſide from S. to N. N. W. by the E. and N. on the other ſide from W. by S. to N. W. Unhappily the horizon was ſo foggy from N. W. to N. N. W. that we could not diſtinguiſh any thing on that ſide further than two leagues off. However, I hoped in that interval to find a paſſage; we were too far advanced to return. It is true that a ſtrong tide coming from the north and ſetting to the S. E. gave us hopes of finding an opening there. The ſtrength of the tide was moſt felt from four o'clock to half an hour paſt five in the evening; the ſhips, though they had a very freſh gale, ſteered with much difficulty. The tide abated at ſix o'clock. During night we plyed from S. to S. S. W. on one tack, and from E. N. E. to N. E. on the other. The weather was ſqually, with much rain.

The 1ſt of July, at ſix in the morning, we found ourſelves at the ſame point which we left the preceding 1768. July.

 even-

evening; a proof that there was both flood and ebb. We steered N. W. and N. W. by N. At ten o'clock we entered into a passage about four or five leagues broad, between the coast which extended hither on the east side, and the land to the westward. A very strong tide, whose direction is S. E. and N. W. forms, in the middle of this passage, a race which crosses it, and where the sea rises and breaks, as if there were rocks even with the surface of the water. I called it Denis's race *(raz* * *Denis)*, from the name of the master of my ship, an old and faithful servant of the king. The Etoile, who passed it two hours after us, and more to the westward, found herself there in five fathoms of water, rocky bottom. The sea was so rough at that time, that they were obliged to lay the hatch-ways. On board the frigate we sounded forty-four fathoms, bottom of sand, gravel, shells, and coral. The eastern coast began here to lower and tend to the northward. On it we perceived, being nearly in the middle of the passage, a fine bay, which to all appearance promised a good anchorage. It was almost a calm, and the tide which then set to the N. W. carried us past it in an instant. We immediately hauled our wind, intending to visit this bay. A very violent

Dangerous shores.

* *Raz* (or *rat*, a race or whirlpool) is a place in the sea where there is some rapid and dangerous current, or where there are different tides. Such a *rat* is commonly to be met with in a strait or channel, but sometimes likewise in the high seas. See the *Dictionnaire Militaire portatif*, 12mo. 3 vols. 1758. Paris. F.

6 shower

ſhower of rain coming on at half an hour paſt eleven, prevented our ſeeing the land and the ſun, and obliged us to defer this ſcheme.

At half an hour paſt one o'clock in the afternoon, I ſent the boats, well armed, under the command of the enſign * chevalier d'Oraiſon, to ſound and viſit the bay; and during this operation, we endeavoured to keep near enough to follow his ſignals. The weather was fair, but almoſt calm. At three o'clock we ſaw the rocky bottom under us, in ten and in eight fathoms. At four our boats made ſignal of a good anchorage, and we immediately worked with all ſails ſet to gain it. It blew very little, and the tide ſet againſt us. At five we repaſſed the rocky bank in ten, nine, eight, ſeven and ſix fathoms. We likewiſe ſaw an eddy within a cable's length to the S. S. E. ſeeming to indicate that there was no more than two or three fathoms of water. By ſteering to N. W. and N. W. by N. we deepened our water. I made ſignal to the Etoile to bear away, in order to avoid this bank, and I ſent her boat to her to guide her to the anchorage. However, we did not advance, the wind being too weak to aſſiſt us in ſtemming the tide, and night coming on very faſt. In two full hours we did not gain half a league, and we were obliged to give up all thoughts of coming to this anchorage,

New attempts to find an anchorage.

* *Enſeigne de Vaiſſeau.*

chorage, as we could not go in ſearch of it in the dark, being ſurrounded by ſhoals, reefs, and rapid and irregular currents. Accordingly we ſtood W. by N. and W. N. W. in order to get off ſhore again, ſounding frequently. Having made the north point of the N. E. land, we bore away N. W. afterwards N. N. W. and then north. I now reſume the account of the expedition of our boats.

The iſlanders attack our boats.

Before they entered the bay, they had ranged its north point, which is formed by a peninſula, along which they found from nine to thirteen fathoms, ſand and coral bottom. They then entered into the bay, and about a quarter of a league from the entrance, found a very good anchorage, in nine and twelve fathoms, bottom of grey ſand and gravel, ſheltered from S. E. to S. W. by the eaſt and north. They were juſt taking ſoundings, when they all at once ſaw ten periaguas appear at the entrance of the bay, having on board about one hundred and fifty men, armed with bows, lances, and ſhields. They came out of a creek, at the bottom of which is a little river, whoſe banks are covered with huts. Theſe periaguas advanced in good order, and as faſt as poſſible towards our boats; and when they thought they were near enough, they divided very dexterouſly into two ſquadrons to ſurround them. The Indians then made horrible cries, and taking their bows

and

and lances, they began an attack, which they muſt have thought would be a mere play to them, againſt ſuch a handful of people. Our people diſcharged their arms at them; but this did not ſtop them. They continued to ſhoot their arrows and throw their lances, covering themſelves with their ſhields, which they looked upon as a defenſive weapon. A ſecond diſcharge put them to flight; ſeveral of them leaped into the ſea in order to ſwim on ſhore. Our people took two of their periaguas: they are long, well wrought, their head and ſtern very much raiſed, to ſhelter the people againſt arrows, by turning either end of the boat towards the enemy. On the head of one of theſe periaguas, they had carved the head of a man; the eyes were of mother of pearl; the ears of tortoiſe-ſhell, and the whole figure reſembled a maſk with a long beard. The lips were dyed of a bright red. In their periaguas our people found bows, arrows in great quantity, lances, ſhields, cocoa-nuts, and ſeveral other fruits, of what ſpecies we could not tell, arecca, ſeveral little utenſils employed by the Indians for various purpoſes, ſome nets with very fine meſhes, very well knit, and the jaw of a man, half broiled. Theſe iſlanders are black, and have curled hair, which they dye white, yellow or red. Their audacity in attacking us, their cuſtom of bearing offenſive and defenſive arms, and their dexterous management of them,

Deſcription of their boat

Deſcriptio[n] of the iſla[n]ders.

prove

prove that they are almoſt conſtantly at war. We have in general obſerved in the courſe of this voyage, that the black men are much more ill-natured than thoſe whoſe colour comes near to white. Theſe iſlanders are naked, excepting their privy parts, which are covered by a piece of mat. Their ſhields are oval, and made of ruſhes, twiſted above each other, and very well connected. They muſt be impenetrable by arrows. We called the river and creek from when theſe brave iſlanders came, the Warriors River (*Riviere aux Guerriers*). The whole iſle and the bay obtained the name of Iſle and Bay Choiſeul. The peninſula on the north ſide of the bay is covered all over with cocoa-nut trees.

Farther diſcoveries which we made.

It blew very little the two following days. After leaving the paſſage, we diſcovered to the weſtward a long hilly coaſt, the tops of whoſe mountains were covered with clouds. The 2d in the evening we ſtill ſaw part of the Iſle of Choiſeul. The 3d in the morning we ſaw nothing but the new coaſt, which is of a ſurpriſing height, and which lies N. W. by W. Its north part then appeared terminated by a point which inſenſibly grows lower, and forms a remarkable cape. I gave it the name of Cape *l'Averdi*. On the 3d at noon it bore about twelve leagues W. ½ N. and as we obſerved the ſun's meridian altitude, we were enabled to determine the latitude of this cape with preciſion. The

clouds

clouds which lay on the heights of the land diſperſed at ſun-ſetting, and ſhewed us mountains of a prodigious height. On the 4th, when the firſt rays of the ſun appeared, we got ſight of ſome lands to the weſtward of Cape l'Averdi. It was a new coaſt, leſs elevated than the former, lying N. N. W. Between the S. S. E. point of this land and Cape l'Averdi, there remains a great gap, forming either a paſſage or a conſiderable gulph. At a great diſtance we ſaw ſome hillocks on it. Behind this new coaſt we perceived a much higher one, lying in the ſame direction. We ſtood as near as poſſible to come near the low lands. At noon we were about five leagues diſtant from it, and ſet its N. N. W. point bearing S. W. by W. In the afternoon three periaguas, in each of which were five or ſix negroes, came from the ſhore to view our ſhips. They ſtopped within muſket ſhot, and continued at that diſtance near an hour, when our repeated invitations at laſt determined them to come nearer. Some trifles which were thrown to them, faſtened on pieces of planks, inſpired them with ſome confidence. They came along-ſide of the ſhips, ſhewing cocoa-nuts, and crying *bouca, bouca, onelle!* They repeated theſe words inceſſantly, and we afterwards pronounced them as they did, which ſeemed to give them much pleaſure. They did not long keep along-ſide of the veſſel. They made ſigns that they were

Deſcriptio of ſome iſl ders who come nea the ſhip.

were going to fetch us cocoa-nuts. We applauded their resolution; but they were hardly gone twenty yards, when one of these perfidious fellows let fly an arrow, which happily hit nobody. After that, they fled as fast as they could row: our superior strength set us above punishing them.

These negroes are quite naked; they have curled short hair, and very long ears, which are bored through. Several had dyed their wool red, and had white spots on different parts of the body. It seems they chew *betel*, as their teeth are red. We found that the inhabitants of the Isle of Choiseul likewise make use of it; for in their periaguas we found little bags, containing the leaves, with areka and lime. From these negroes we got bows of six feet long, and arrows armed with points of a very hard wood. Their periaguas are less than those from the Warriors Creek; and we were surprised to find no resemblance in their construction. This last kind of periaguas had no great elevation at the head and stern; they were without any out-rigger, but broad enough for two men to work at the oar in one row. This isle, which we named *Bouka*, seems to be extremely well peopled, if we may judge so by the great number of huts upon it, and by the appearance of cultivation which it has. A fine plain, about the middle of the coast, all over planted with cocoa-nut trees, and other trees, offered a most agreeable prospect, and made me very desirous

of

of finding an anchorage on it; but the contrary wind, and a rapid current, which carried to the N. W. visibly brought us further from it. During night we stood as close as possible, steering S. by W. and S. S. W. and the next morning the Isle of Bouka was already very far from us to the east and S. E. The evening before, we had perceived a little isle, bearing N. W. and N. W. by W. We could not, upon the whole, be far from New Britain, where we hoped to take shelter at.

Anchorage on the coast of New Britain.

On the 5th, in the afternoon, we got sight of two little isles to the N. and N. N. W. ten or twelve leagues distant, and almost at the same instant another more considerable one between N. W. and W. Of this last, the nearest lands at half past five o'clock in the evening, bore N W. by W. about seven leagues distant. The coast was high, and seemed to form several bays. As we had neither water nor wood left, and our sick were growing worse, I resolved to stop here, and we made all night the most advantageous boards to keep this land under our lee. The 6th, at day-break, we were five or six leagues distant from it, and bore away for it, at the same moment when we discovered another new land, which was high, and in appearance very fine, bearing W. S. W. of the former, from eighteen to twelve, and to ten leagues distance. At eight o'clock, being about three leagues from the first land, I sent the chevalier du

Bou-

Bouchage with two armed boats to view it, and ſee whether there was an anchorage. At one o'clock in the afternoon he made ſignal of having found one; and I immediately gave order to fill the ſails, and bore down for a boat, which he ſent to meet us; at three o'clock we came to an anchor in 33 fathom, bottom of fine white ſand, and ooze. The Etoile anchored nearer the ſhore than we did, in 21 fathom, ſame bottom.

Qualities and marks of the anchorage.

In entering, you have a little iſle and a key to the weſtward, on the larboard ſide; they are about half a league off ſhore. A point, advancing oppoſite the key, forms within a true port, ſheltered againſt all the winds; the bottom being, in every part of it, a fine white ſand, from 35 to 15 fathom. On the eaſtern point there is a viſible ledge, which does not extend out to ſea. You likewiſe ſee, to the northward of the bay, two ſmall ledges, which appear at low water. Cloſe to the reefs, there is 12 fathom of water. The entrance to this port is very eaſy; the only precaution which muſt be taken, is to range the eaſtern point very near, and to carry much ſail; for as ſoon as you have doubled it, you are becalmed, and can enter only by the head-way, which the ſhip makes. Our bearings, when at an anchor, were as follows: The key, at the entrance, bore W. 9° 45′ S. the eaſtern point of the entrance, W. 10° S. the weſtern point, W. by N. the bottom of the harbour, S. E. by E.

We

We moored eaſt and weſt, ſpending the reſt of the day with thoſe manœuvres, and with ſtriking yards and top-maſts, hoiſting out our boats, and viſiting the whole circuit of the harbour.

Deſcription of the port and its environs.

It rained all the next night, and almoſt the whole day of the 7th. We ſent all our water-caſks on ſhore, pitched ſome tents, and began to fill water, take in wood, and make lies for waſhing, all which were abſolutely neceſſary occupations. The landing-place was handſome, on a fine ſand, without any rocks or ſurf; in the bottom of the port, in the ſpace of four hundred yards, we found four brooks. We took three for our uſe; the one for the Boudeuſe, and the other for the Etoile to water at, and the third for waſhing. The wood was near the ſea-ſide, and there were ſeveral ſorts of it, all very good fuel; ſome excellent for carpenters, joiners, and even for veneering. The two ſhips were within hail of each other, and of the ſhore. Beſides this, the harbour and its environs were not inhabited within a great diſtance, by which means we enjoyed a very precious and undiſturbed liberty. Thus we could not wiſh for a ſafer anchorage, a more convenient place for taking in water and wood, making thoſe repairs which the ſhips moſt urgently wanted, and letting our people, who were ſick of the ſcurvy, ramble about the woods at their eaſe.

Such

Such were the advantages of this harbour; but it likewise had its inconveniencies. Notwithstanding all our searches, we could neither find cocoa-nut trees and bananas, nor had we any other resources, which by good-will, or by force could have been obtained in an inhabited country. If the fishery should not happen to be abundant, we could expect nothing else here than safety and the mere necessaries. We had therefore great reason to fear, that our sick would not recover. It is true, we had none that were very ill, but many were infected; and if they did not mend, the progress of the disease must of course become more rapid.

Extraordinary adventure.

On the first day we found a periagua, as it were deposited, and two huts, on the banks of a rivulet, at a mile's distance from our camp. The periagua had an out-rigger, was very light, and in good order. Near it there were the remains of several fires, some great calcined shells, and some skeletons of the heads of animals, which M. de Commerçon said were wild boars. The savages had but lately been in this place; for some bananas were found quite fresh in the huts. Some of our people really thought they heard the cries of men towards the mountains; but we have since verified, that they have mistaken for such the plaintive notes of a large crested pigeon, of an azure plumage, and which

6 has

has the name of *crowned bird** in the Moluccas. We found ſomething ſtill more extraordinary on the banks of this river. A ſailor, belonging to my barge, being in ſearch of ſhells, found buried in the ſand, a piece of a plate of lead, on which we read theſe remains of Engliſh words,

HOR'D HERE
ICK MAJESTY's

There yet remained the mark of the nails, with which they had faſtened this inſcription, that did not ſeem to be of any ancient date. The ſavages had, doubtleſs, torn off the plate, and broke it in pieces.

This adventure engaged us carefully to examine all the neighbourhood of our anchorage. We therefore ran along the coaſt within the iſle which covers the bay; we followed it for about two leagues, and came to a deep bay of very little breadth, open to the S. W. at the bottom of which we landed, near a fine river. Some trees ſawed in pieces, or cut down with hatchets, im- Marks of an Engliſh camp.

* This bird is a native of the Iſle of Banda, one of the Moluccas, and is called by the Dutch *Kroon-Vogel*. Mr. Loten preſented one, ſome years ago, alive to the late princeſs royal of England and of Orange. Mr. Briſſon, in his Ornithology, vol i. p. 279. t. 26. f. 1. very improperly calls it a crowned Indian pheaſant (*Faiſan couronné des Indes*); and Mr. Buffon, in his Planches Enluminées, tab. 118. follows Briſſon, though every one will be convinced that it is a pigeon, at the very firſt examination of its bill. Mr. Edwards has deſcribed and figured it, p. 269. t. 338. of the third volume of his Gleanings. Its plumage is blue, or lead-colour-ed; the ſize, that of a turkey. In that noble repoſitory of natural hiſtory and learning, the Britiſh Muſeum, there is a fine ſpecimen of it. F.

mediately

mediately ſtruck our eyes, and ſhewed us that this was the place where the Engliſh put in at. We now had little trouble to find the ſpot where the inſcription had been placed. It was a very large, and very apparent tree, on the right hand ſhore of the river, in the middle of a great place, where we concluded that the Engliſh had pitched their tents; for we ſtill ſaw ſeveral ends of rope faſtened to the trees; the nails ſtuck in the tree; and the plate had been torn off but a few days before; for the marks of it appeared quite freſh. In the tree itſelf, there were notches cut, either by the Engliſh or the iſlanders. Some freſh ſhoots, coming up from one of the trees which was cut down, gave us an opportunity of concluding, that the Engliſh had anchored in this bay but about four months ago. The rope, which we found, likewiſe ſufficiently indicated it; for though it lay in a very wet place, it was not rotten. I make no doubt, but that the ſhip which touched here, was the Swallow; a veſſel of fourteen guns, commanded by captain Carteret, and which ſailed from Europe in Auguſt 1766, with the Dolphin, captain Wallace. We have ſince heard of this ſhip at Batavia, where I ſhall ſpeak of her; and where it will appear, that we from thence followed her track to Europe. This is a very ſtrange chance, by which we, among ſo many lands, come to the very ſpot where this rival nation had left a monument of an enterprize ſimilar to our's.

The rain was almoſt continual to the 11th. There ſeemed to be a very high wind out at ſea; but the port is ſheltered on all ſides, by the high mountains which ſurround it. We accelerated our works, as much as the bad weather would permit. I likewiſe ordered our long-boat to under-run the cables, and to weigh an anchor, in order to be better aſſured concerning the nature of the bottom; we could not wiſh for a better. One of our firſt cares had been to ſearch, (and certainly it was our intereſt to do ſo) whether the country could furniſh any refreſhments to our ſick, and ſome ſolid food to the healthy. Our ſearches were fruitleſs. The fiſhery was entirely unſucceſsful; and we only found in the woods a few thatch-palms, and cabbage-trees in very ſmall number; and even theſe we were obliged to diſpute with enormous ants, of which innumerable ſwarms forced us to abandon ſeveral of theſe trees, already cut down by us. It is true, we ſaw five or ſix wild boars; and, ſince that time, ſome huntſmen were always out in ſearch of them; but they never killed one. They were the only quadrupeds we ſaw here.

Productions of the country.

Some people likewiſe thought they had ſeen the footſteps of a tyger cat. We have killed ſome large pigeons of great beauty. Their plumage was green-gold; their neck and belly of a greyiſh-white; and they have a little creſt on the head. Here are likewiſe turtle-doves,

 ſome

ſome widow-birds, larger than thoſe of the Braſils, parrots, crown-birds; and another kind, whoſe cry ſo well reſembles the barking of a dog, that every one who hears it for the firſt time, muſt be deceived by it. We have likewiſe ſeen turtle in different parts of the channel; but this was not the ſeaſon when they lay eggs. In this bay are fine ſandy creeks, where I believe a good number of turtle could be caught at the proper time.

All the country is mountainous; the ſoil is very light, and the rocks are hardly covered with it. However, the trees are very tall, and there are ſeveral ſpecies of very fine wood. There we find the Betel, the Areca, and the fine Indian-reed, which we get from the Malays. It grows here in marſhy places; but whether it requires a peculiar culture, or whether the trees, which entirely overſhadow the earth, hinder its growth, and change its quality, or whether we were not here at the proper ſeaſon when it is in maturity, ſo much is certain, that we never found any fine ones here. The pepper-tree is likewiſe common to this country; but it had neither fruit nor flowers at this ſeaſon. The country, upon the whole, is not very rich for a botaniſt. There remain no marks in it of any fixed habitation: it is certain that the Indians come this way from time to time; we frequently found places upon the ſea-ſhore, where they had

ſtop-

ſtopped; the remnants of their meals eaſily betrayed them.

On the 10th, a ſailor died on board the Etoile, of a complication of diſorders, without any mixture of the ſcurvy. The three following days were fine, and we made good uſe of them. We refitted the heel of our mizen-maſt, which was worm-eaten in the ſtep; and the Etoile ſhortened hers, the head of it being ſprung. We likewiſe took in, from on board the ſtore-ſhip, the flour and biſcuit which ſtill belonged to us, in proportion to our number. There were fewer pulſe than we at firſt thought, and I was obliged to cut off above a third part of the allowance of the *(gourganes)* peaſe or caravanſes for our ſoup: I ſay ours, for every thing was equally diſtributed. The officers and the ſailors had the ſame nouriſhment; our ſituation, like death, rendered all ranks of men equal. We likewiſe profited of the fair weather, to make good obſervations.

Cruel fami[ne] which we ſuffer.

On the 11th, in the morning, M. Verron brought his quadrant and pendulum on ſhore, and employed them the ſame day, to take the ſun's altitude at noon. The motion of the pendulum was exactly determined by ſeveral correſponding altitudes, taken for two days conſecutively. On the 13th, there was an eclipſe of the ſun viſible to us, and we got every thing in readineſs to obſerve it, if the weather permitted. It was very fair; and

Obſervatio[n] of longitu[de]

 we

we ſaw both the moment of immerſion, and that of emerſion. M. Verron obſerved with a teleſcope of nine feet; the chevalier du Bouchage with one of Dollond's acromatic teleſcopes, four feet long; my place was at the pendulum. The beginning of the eclipſe was to us, on the 13th, at 10 h. 5′ 45″ in the morning, the end at 00 h. 28′ 16″ true time, and its magnitude 3′ 22″. We have buried an inſcription under the very ſpot where the pendulum had been; and we called this harbour *Port Praſlin*.

This obſervation is ſo much the more important, as it was now poſſible, by its means, and by the aſtronomical obſervations, made upon the coaſt of Peru, to determine, in a certain fixed manner, the extent of longitude of the vaſt Pacific Ocean, which, till now, had been ſo uncertain. Our good fortune, in having fair weather at the time of the eclipſe, was ſo much the greater, as from that day to our departure there was not a ſingle day but what was dreadful. The continued rains, together with the ſuffocating heat, rendered our ſtay here very pernicious to us. On the 16th, the frigate had completed her works, and we employed all our boats to finiſh thoſe of the Etoile. This ſtore-ſhip was quite light, and as there were no ſtones proper for ballaſt, we were obliged to make uſe of wood for that purpoſe; this was a long troubleſome labour, which in theſe foreſts,

where

where an eternal humidity prevails, is likewiſe unwholeſome.

Here we daily killed ſnakes, ſcorpions, and great numbers of inſects, of a ſingular ſort. They are three or four inches long, and covered over with a kind of armour; they have ſix legs, projecting points on the ſides, and a very long tail. Our people likewiſe brought me another creature, which appeared extraordinary to us all. It is an inſect about three inches long, and belongs to the Mantis genus. Almoſt every part of its body is of ſuch a texture, as one would take for a leaf, even when one looks cloſely at it. Each of its wings is one half of a leaf, which is entire when the two wings are cloſed together; the under ſide of its body reſembles a leaf, of a more dead colour than the upper one. The creature has two antennæ and ſix legs, of which the upper joints are likewiſe ſimilar to parts of leaves. M. de Commerçon has deſcribed this ſingular inſect; and I placed it in the king's cabinet, preſerved in ſpirits.

Deſcription of two inſects.

Here we found abundance of ſhells, many of them very fine. The ſhoals offered treaſures for the ſtudy of Conchology. We met with ten hammer-oyſters in one place, and they are ſaid to be a ſcarce ſpecies *. The curioſity

* They were found in a creek of the great iſle, which forms this bay; and which for that reaſon has been called Hammer Iſland, (*Iſle aux Marteaux*).

riosity of some of our people was accordingly raised to a great pitch; but an accident happening to one of our sailors abated their zeal. He was bit in the water by a kind of snake as he was hauling the seine. The poisonous effects of the bite appeared in half an hour's time. The sailor felt an excessive pain all over his body. The spot where he had been bit, which was on the left side, became livid, and swelled visibly. Four or five scarifications extracted a quantity of blood, which was already dissolved. Our people were obliged to lead the patient walking, to prevent his getting convulsions. He suffered greatly for five or six hours together. At last the treacle *(theriaque)* and flower de luce water which had been given him, brought on an abundant perspiration, and cured him.

Sailor bit by a water-snake.

This accident made every one more circumspect and careful in going into the water. Our Taiti-man curiously observed the patient during the whole course of his sickness. He let us know that in his country were snakes along the sea-shore, which bit the people in the sea, and that every one who was thus bit died of the wound. They have a kind of medicinal knowledge, but I do not believe it is extensive at all. The Taiti-man was surprised to see the sailor return to his work, four or five days after the accident had happened to him. When he examined the productions of our arts,

and

and the various means by which they augment our faculties, and multiply our forces, this islander would often fall into an extatic fit, and blush for his own country, saying with grief, *aouaou Taiti, fy upon Taiti.* However, he did not like to express that he felt our superiority over his nation. It is incredible how far his haughtiness went. We have observed that he was as supple as he was proud; and this character at once shews that he lives in a country where there is an inequality of ranks, and points out what rank he holds there.

Bad weather which persecutes us.

On the 19th in the evening we were ready to sail, but it seemed the weather always grew worse and worse. There was a high south wind, a deluge of rain, with thunder and tempestuous squalls, a great sea in the offing, and all the fishing birds retired into the bay. On the 22d in the morning, towards half an hour past ten o'clock, we sustained several shocks of an earthquake. They were very sensibly felt on board our ships, and lasted about two minutes. During this time the sea rose and fell several times consecutively, which greatly terrified those who were fishing on the rocks, and made them retreat to the boats. It seems upon the whole, that during this season the rains are uninterrupted here. One tempest comes on before the other is gone off, it thunders continually, and the nights are

Earthquake.

are fit to convey an idea of chaotic darkneſs. Notwithſtanding this, we daily went into the woods in ſearch of thatch palms and cabbage trees, and endeavouring to kill ſome turtle doves. We divided into ſeveral bodies, and the ordinary reſult of theſe fatiguing caravans, was, that we returned wet to the ſkin, and with empty hands. However, in theſe laſt days, we found ſome mangle-apples, and a kind of fruit called *Prunes de Monbin* *. Theſe would have been of ſome ſervice to us, had we diſcovered them ſooner. We likewiſe found a ſpecies of aromatic ivy, in which our ſurgeons believed they had diſcovered an antiſcorbutic quality; at leaſt, the patients who uſed an infuſion of it, and waſhed with it, found themſelves better.

Unſucceſsful endeavours to find proviſions.

Deſcriotion of a fine caſcade.

We all went to ſee a prodigious caſcade, which furniſhed the Etoile's brook with water. In vain would art endeavour to produce in the palaces of kings, what nature has here laviſhed upon an uninhabited ſpot. We admired the aſſemblage of rocks, of which the almoſt regular gradations precipitate and diverſify the fall of the waters; with admiration we viewed all theſe maſſes, of various figures, forming an hundred different baſons, which contain the limpid ſheets of water, coloured and

* It is not known to what genus this plant belongs; a general, but not ſyſtematical, deſcription of it may be found in Mr. *Valmont de Bomare's Dictionnaire d'Hiſtoire Naturelle*, article MONBAIN. F.

ſhaded

ſhaded by trees of immenſe height, ſome of which have their roots in the very reſervoirs themſelves. Let it ſuffice that ſome men exiſt, whoſe bold pencil can trace the image of theſe inimitable beauties: this caſcade deſerves to be drawn by the greateſt painter.

Mean while our ſituation grew worſe every moment of our ſtay here, and during all the time which we ſpent without advancing homeward. The number of thoſe who were ill of the ſcurvy, and their complaints encreaſed. The crew of the Etoile was in a ſtill worſe condition than ours. Every day I ſent boats out to ſea, in order to know what kind of weather there was. The wind was conſtantly at ſouth, blowing almoſt a ſtorm with a dreadful ſea. Under theſe circumſtances it was impoſſible to get under ſail, eſpecially as this could not be done without getting a ſpring upon an anchor that was to be ſlipped all at once; and in that caſe it would have been impoſſible in the offing to hoiſt in the boats that muſt have remained to weigh the anchor, which we could not afford to leave behind us. Theſe obſtacles determined me to go on the 23d to view a paſſage between Hammer iſland and the main land. I found one, through which we could go out with a ſouth wind, hoiſting in our boats in the channel. This paſſage had indeed great inconveniences, and happily we were not obliged to make uſe of it. It rained without intermiſſion

Our ſituation grows worſe every day.

We leaves Port Praſl

miſſion all the night between the 23d and 24th. At day-break the weather became fair and calm. We immediately weighed our ſmall bower, faſtened a warp to ſome trees, bent a hawſer to a ſtream-anchor, and hove a-peek on the off-anchor. During the whole day we waited for the moment of ſetting ſail; we already deſpaired of it, and the approach of night would have obliged us to moor again, when at half paſt five o'clock a breeze ſprung up from the bottom of the harbour. We immediately ſlipt our ſhore-faſt, veered out the hawſer of the ſtream-anchor, from which the Etoile was to ſet ſail after us, and in half an hour's time we were got under ſail. The boats towed us into the middle of the paſſage, where there was wind enough to enable us to proceed without their aſſiſtance. We immediately ſent them to the Etoile to bring her out. Being got two leagues out to ſea, we lay-to in order to wait for her, hoiſting in our long-boat and ſmall boats. At eight o'clock we began to ſee the Etoile which was come out of port; but the calm did not permit her to join us till two hours after midnight. Our barge returned at the ſame time, and we hoiſted her in.

During night we had ſqualls and rain. The fair weather returned at day break. The wind was at S. W. and we ſteered from E. by S. to N. N. E. turning to north-

northward with the land. It would not have been prudent to endeavour to paſs to windward of it: we ſuſpected that this land was New Britain, and all the appearances confirmed us in it. Indeed the lands which we had diſcovered more to the weſtward came very cloſe to this, and in the midſt of what one might have taken for a paſſage, we ſaw ſeparate hummocks, which doubtleſs joined to the other lands, by means of ſome low grounds. Such is the picture Dampier gives of the great bay, which he calls St. George's Bay, and we have been at anchor at the N. E point of it, as we verified on the firſt days after our leaving the port. Dampier was more ſucceſsful than we were. He took ſhelter near an inhabited diſtrict, which procured him refreſhments, and whereof the productions gave him room to conceive great hopes concerning this country; and we, who were as indigent as he was, fell in with a deſart, which, inſtead of ſupplying all our wants, has only afforded us wood and water.

When I left Port Praſlin, I corrected my longitude by that which we obtained from the calculation of the ſolar eclipſe, which we obſerved there; my difference was about 3°, which I was to the eaſtward. The thermometer during the ſtay which we made there, was conſtantly at 22° or 23°; but the heat was greater than it ſeemed to ſhew. I attribute the cauſe of this

 to

to the want of air, which is common here; this bason being closed in on all sides, and especially on the side of the reigning winds.

CHAP. VI.

Run from Port Praslin to the Moluccas; stay at Boero.

WE put to sea again after a stay of eight days, during which time, as we have before observed, the weather had been constantly bad, and the wind almost always southerly. The 25th it returned to S. E. veering round to E. and we followed the direction of the coast at about three leagues distance. It rounded insensibly, and we soon discovered in the offing a succession of islands, one after the other. We passed between them and the main, and I gave them the names of the principal officers. We now no longer doubted that we were coasting New Britain. This land is very high, and seemed to be intersected with fine bays, in which we perceived fires, and other marks of habitations.

Distribution of cloaths to the sailors.

The third day after our departure I caused our field-tents to be cut up, and distributed trowsers to the two ships companies. We had already, on several occasions, made

made the like diſtributions of cloathing of all kinds. Without that, how would it have been poſſible that theſe poor fellows ſhould be clad during ſo long a voyage, on which they were ſeveral times obliged to paſs alternately from cold to hot, and to endure frequent deluges of rain? I had, upon the whole, nothing more to give them, all was exhauſted, and I was even forced to cut off another ounce of the daily allowance of bread. Of the little proviſions that remained, part was ſpoiled, and in any other ſituation all our ſalt proviſions would have been thrown over-board; but we were under the neceſſity of eating the bad as well as the good, for it was impoſſible to tell when our ſituation would mend. Thus it was our caſe to ſuffer at once by what was paſt, which had weakened us; by our preſent ſituation, of which the melancholy circumſtances were every inſtant repeated before us; and laſtly, by what was to come, the indeterminate duration of which was the greateſt of all our calamities. My perſonal ſufferings encreaſed by thoſe of others. However, I muſt declare that not one ſuffered himſelf to be dejected, and that our patience under ſufferings has been ſuperior to the moſt critical ſituations. The officers ſet the example, and the ſeamen never ceaſed dancing in the evenings, as well in the time of ſcarcity, as in that of the greateſt plenty

Extreme want of victuals.

plenty. Nor has it been neceſſary to double their pay *.

Deſcription of the inhabitants of New Guinea.

We had New Britain conſtantly in ſight till the 3d of Auguſt, during which time we had little wind, frequent rain, the currents againſt us, and the ſhips went worſe than ever. The coaſt trenched more and more to the weſtward, and on the 29th in the morning, we found ourſelves nearer it than we had yet been: this approach procured us a viſit from ſome periaguas; two came within hail of the frigate, and five others went to the Etoile. They carried each of them five or ſix black men, with frizled woolly hair, and ſome of them had powdered it white. They had pretty long beards, and white ornaments round their arms, in form of bracelets. Their nudities were but indifferently covered with the leaves of trees. They are tall, and appeared active and robuſt. They ſhewed us a kind of bread, and invited us by ſigns to go aſhore. We deſired them to come on board; but our invitations, and even the gift

* M. de B. it ſeems can never ſufficiently elevate the courage and perſeverance of his countrymen; on all occaſions he praiſes their diſintereſtedneſs, and endeavours to depreciate the merits of the Britiſh ſailors, by balancing their ſufferings with the rewards which an equitable government diſtributed to them. I have already ſaid ſomething on this ſubject in a note to our author's Introduction (placed at the head of this work) and ſhall only add, that I ſhould be apt to ſuſpect M. de B. to envy the Britiſh circumnavigators thoſe very rewards which he ſeems ſo much to deſpiſe, if I could combine ſuch baſe ſentiments with his otherwiſe generous way of thinking. F.

of ſome pieces of ſtuff which we threw over-board, did not inſpire them with confidence ſufficient to make them venture along-ſide. They took up what was thrown into the water, and by way of thanks one of them with a ſling flung a ſtone, which did not quite reach on board; we would not return them evil for evil, ſo they retired, ſtriking all together on their canoes, and ſetting up loud ſhouts. They without doubt carried their hoſtilities farther on board the Etoile, for we ſaw our people fire ſeveral muſkets, which put them to flight. Their periaguas are long, narrow, and with out-riggers; they all have their heads and ſterns more or leſs ornamented with ſculptures, painted red, which does honour to their ſkill.

The next day there came a much greater number of them, who made no difficulty of coming along-ſide the ſhip. One of their conductors, who ſeemed to be the chief, carried a ſtaff about two or three feet long, painted red, with a knob at each end, which, in approaching us, he raiſed with both hands over his head, and continued ſome time in that attitude. All theſe negroes ſeemed to be dreſſed out in their beſt, ſome had their woolly hair painted red, others had plumes on their heads, certain ſeeds in their ears by way of ear-rings, or large white round plates hanging to their necks; ſome had rings paſſed through the cartilage of the noſe; but an ornament pretty common

mon to them all was bracelets, made of the mouth of a large ſhell, ſawed aſunder. We were deſirous of forming an intercourſe, in order to engage them to bring us ſome refreſhments, but their treachery ſoon convinced us that we could not ſucceed in that attempt. They ſtrove to ſeize what was offered them, and would give nothing in exchange. We could ſcarce get a few roots of yams from them; therefore we left off giving them, and they retired. Two canoes rowed towards the frigate at the beginning of night, but a rocket being fired for ſome ſignal, they fled precipitately.

They attack the Etoile.

Upon the whole, it ſeemed that the viſits they made us theſe two laſt days had been with no other view than to reconnoitre us, and to concert a plan of attack; for the 31ſt, at day-break, we ſaw a ſwarm of periaguas coming off ſhore, a part of them paſſed athwart us without ſtopping, and all directed their courſe for the Etoile, which they had no doubt obſerved to be the ſmalleſt veſſel of the two, and to keep aſtern. The negroes made their attacks with ſtones and arrows, but the action was ſhort, for one platoon diſconcerted their ſcheme, many threw themſelves into the ſea, and ſome periaguas were abandoned: from this time we did not ſee any more of them.

Deſcription of the northern part of New Britain.

The coaſt of New Britain now ran W. by N. and W. and in this part it became conſiderably lower. It was no no longer that high coaſt adorned with ſeveral rows of mountains; the northern point which we diſcovered was very low land, and covered with trees from ſpace to ſpace. The five firſt days of the month of Auguſt were rainy, the weather thick and unſettled, and the wind ſqually. We diſcovered the coaſt only by piece-meal, in the clear intervals, without being able to diſtinguiſh the particulars of it: however, we ſaw enough of it to be convinced that the tides continued to carry us a part of the moderate run we made each day. I then ſteered N. W. and N. W. by W. to avoid a cluſter of iſlands that ly off the northern extremity of New Britain. The 4th in the afternoon we diſcovered two iſlands, which I take to be thoſe that Dampier calls Matthias Iſland and Stormy or Squally Iſland. Matthias Iſland is high and mountainous, and extends to N. W. about eight or nine leagues. The other is not above three or four leagues long, and between the two lies a ſmall iſle. An iſland which we thought we perceived the 5th, at two o'clock in the morning, to the weſtward, cauſed us again to ſtand to the northward. We were not miſtaken; for at ten o'clock the fog, which till then had been thick, being diſſipated, we ſaw that iſland, which is ſmall and low, bearing S. E. by S. The tides then ceaſed to ſet to the ſouthward and eaſtward, which ſeemed to ariſe from our having got beyond the

1768 Auguſt.

northern point of New Britain, which the Dutch have called Cape Salomaſwer. We were then in no more than 00° 41 ſouth lat. We had ſounded almoſt every day without finding bottom.

Iſle of Anchorets.

We ſteered weſt till the 7th, with a pretty freſh gale and fair weather, without ſeeing land. The 7th in the evening, the ſky being very hazy, and appearing at ſun-ſet to be a horizon of land from W. to W. S. W. I determined to ſteer S. W. by S. for the night; at day-light we ſteered weſt again. In the morning we ſaw a low land, about five or ſix leagues a-head of us. We ſteered W. by S. and W. S. W. to paſs to the ſouthward of it, and we ranged along it at about a league and a half diſtance. It was a flat iſland, about three leagues long, covered with trees, and divided into ſeveral parts, connected together by breakers and ſand-banks. There are upon this iſland a great quantity of cocoa-nut trees, and the ſea-ſhore is covered with a great number of habitations, from which it may be ſuppoſed to be extremely populous. The huts were high, almoſt ſquare, and well covered. They ſeemed to us larger and handſomer than the huts built with reeds generally are, and we thought we again beheld the houſes of Taiti. We diſcovered a great number of periaguas employed in fiſhing all round the iſland; none of them ſeemed to be diſturbed at ſeeing us paſs, from which we judged that theſe

people,

people, who were not curious, were contented with their fate. We called this island the Isle of Hermits, or Anchorets. Three leagues to the westward of this, we saw another low island from the mast-head.

Archipelago by us called the *Echiquier*.

The night was very dark, and some fixed clouds to the southward made us suppose there was land; and, in fact, at day-light we discovered two small isles, bearing S. S. E. ¼ E. at eight or nine leagues distance. We had not yet lost sight of them, at half past eight o'clock, when we discovered another low island, bearing W. S. W. and a little after, an infinite number of little islands extending to W. N. W. and S. W. of this last, which might be about two leagues long; all the others, properly speaking, are nothing but a chain of little flat isles, or keys, covered with wood; which, indeed, was a very disagreeable discovery to us. There was, however, an island separated from the others, and more to the southward, which seemed to us more considerable. We shaped our course between that and the Archipelago of isles, which I called the Chess-board, *(l'Echiquier)* and which I wanted to leave to the northward. We were not yet near getting clear of it. This chain discovered, ever since the morning, extended much farther to the southwestward, than we were at that time able to determine.

Danger which we ran there.

We endeavoured, as I have observed before, to double it to the southward; but in the beginning of the

 night,

night, we were ftill engaged with it, without knowing precifely how far it extended. The weather being continually fqually, had never fhewn us at once, all that we had to fear; to add to our embarraffment, it became calm in the beginning of the night, and the calm fcarce ended at the return of day. We paffed the night under continual apprehenfions of being caft afhore by the currents. I ordered two anchors to be got clear, and the cables bitted with a range along the deck, which was almoft an unneceffary precaution; for we founded feveral times without finding bottom. This is one of the greateft dangers of thefe coafts; for you have not the refource of anchoring at twice the fhip's length from the ledges, by which they are bounded. The weather fortunately continued without fqualls; and about midnight a gentle breeze fprung up from the northward, which enabled us to get a little to the fouth-eaftward. The wind frefhened in proportion as the fun afcended, and carried us from thefe low iflands; which, I believe, are uninhabited; at leaft, during the time we were carried near enough to difcern them, we diftinguifhed neither fires, nor huts, nor periaguas. The Etoile had been, during the night, in ftill greater danger than us; for fhe was a very long time without fteerage-way, and the tide drew her infenfibly towards the fhore, when the wind fprung up to her relief. At two o'clock, in the after-

afternoon, we doubled the westermost of the islands, and steered W. S. W.

We get sight of New Guinea.

The 11th, at noon, being in 2° 17′ south latitude, we perceived, to the southward, a high coast, which seemed to us to be that of New Guinea. Some hours after, we saw it more distinctly. The land is high and mountainous, and in this part extends to the W. N. W. The 12th, at noon, we were about ten leagues from the nearest land; it was impossible to observe the coast minutely at that distance there: it appeared to us only a large bay, about 2° 25′ south latitude; in the bottom of which, the land was so low, that we only saw it from the mast-head. We also judged from the celerity with which we doubled the land, that the currents were become favourable to us; but in order to determine with any exactness, the difference they occasioned in our estimated run, it would have been necessary to sail at a less distance from the coast. We continued ranging along it, at ten or twelve leagues distance; its direction was constantly W. N. W. and its height immense. We remarked particularly two very high peaks, neighbours to each other, which surpassed all the other mountains in height. We called them the Two Cyclops. We had occasion to remark, that the tides set to the N. W. The next day we actually found ourselves further off from the coast of New Guinea; which here tended away west.

The

The 14th, at break of day, we discovered two islands and a little isle or key, which seemed to be between them, but more to the southward. Their corrected bearings are E. S. E. and W. N. W. They are at about two leagues distance from each other, of a middling height, and not above a league and a half in extent each.

Direction of the winds and currents.

We advanced but little each day. Since our arrival on the coast of New Guinea, we had pretty regularly a light breeze from east to N. E. which began about two or three o'clock in the afternoon, and lasted till about midnight; this breeze was succeeded with a longer or shorter interval of calm, which was followed by the land-breeze, varying from S. W. to S. S. W. and that terminated also towards noon, in two or three hours calm. The 15th, in the morning, we again saw the westmost of the two islands we had seen the preceding evening. We discovered at the same time other land, which seemed to us to be islands, extending from S. E. to W. S. W. very low, over which, in a distant point of view, we perceived the high mountains of the continent. The highest, which we set at eight o'clock in the morning, bearing S. S. E. by compass, detached from the others, we called the *Giant of Moulineau*, and we gave the name of *la Nymphe Alice* to the westmost of the low islands, to the N. W. of Moulineau. At ten in the morning we fell into a race of a tide, where the current seemed to

carry

carry us with violence to N. and N. N. E. It was so violent, that till noon it prevented our steering; and as it carried us much into the offing, it became impossible for us to fix a positive judgment of its true direction. The water, in the first tide-line, was covered with the trunks of drift trees, sundry fruits and rock-weeds; it was at the same time so agitated, that we dreaded being on a bank; but sounding, we had no bottom at 100 fathom. This race of a tide seems to indicate either a great river in the continent, or a passage which would here divide New Guinea; a passage whose entrance would be almost north and south. According to two distances, between the sun and moon, observed with an octant, by the chevalier du Bouchage and M. Verron, our longitude, the 15th at noon, was 136° 16′ 30″ east of Paris. My reckoning continued from the determined longitude of Port Praslin; differed from it 2° 47′. We observed the same day 1° 17′ south latitude.

Observatic compared with the r koning.

The 16th and 17th it was almost calm; the little wind that did blow, was variable. The 16th, we did not see the land till seven in the morning; and then only from the mast head, extremely high and rugged. We lost all that day in waiting for the Etoile, who, overcome by the current, could not keep her course; and the 17th, as she was very far from us, I was obliged to bear down to join her; but this we did not accom-

accomplish, till the approach of night, which proved very stormy, with a deluge of rain and frightful thunder. The six following days were all as unpropitious to us; we had rain and calms; and the little wind that did blow was right a-head. It is impossible to form an idea of this, without being in the situation we were then in. The 17th, in the afternoon, we had seen from S. by W. ½ W. to S. W. ½ W. by compass, at about sixteen leagues distance, a high coast, which we did not lose sight of till night came on. The 18th, at nine in the morning, we discovered a high island, bearing S. W. by W. distance about twelve leagues: we saw it again the next day; and at noon it bore from S. S. W. to S. W. at the distance of 15 or 20 leagues. During these three last days, the currents gave us ten leagues northing: we could not determine what they had helped us in longitude.

We cross the Equator.

The 20th we crossed the line, for the second time the voyage. The currents continued to set us from the land; and we saw nothing of it the 20th or 21st, although we had kept on those tacks by which we approached it most. It became, however, necessary to make the coast, and to range along it, near enough, so as not to commit any dangerous error, which might make us miss the passage into the Indian Sea, and carry us into one of the gulphs of Gilolo. The 22d, at break of day, we

7 had

had sight of a higher coast than any part of New Guinea that we had yet seen. We steered for it, and at noon we set it, when it bore from S. by E. ½ E. to S. W. where it did not seem to terminate. We passed the line for the third time. The land ran W. N. W. and we approached it, being determined not to quit it any more till we arrived at its extremity, which geographers call Cape Mabo. In the night we doubled a point, on the other side of which the land, still very high, trenched away W. by S. and W. S. W. The 23d at noon, we saw an extent of coast, of about twenty leagues; the westmost part of which bore from us S. W. thirteen or fourteen leagues. We were much nearer two low islands, covered with wood, distant from each other about four leagues. We stood within about half a league; and whilst we waited for the Etoile, who was a great distance from us, I sent the chevalier de Suzannet, with two of our boats armed, to the northermost of the two islands. We thought we saw some habitations there, and were in hopes of getting some refreshments. A bank, which lies the length of the island, and extends even pretty far to the eastward, obliged the boats to take a large circuit to double it. The chevalier de Suzannet found neither dwellings, inhabitants, nor refreshments. What had seemed to us at a distance to form a village, was nothing but

Cross the line again.

Unsuccessful attempt on shore.

but a heap of rocks, undermined and hollowed into caverns by the ſea. The trees that covered the iſland, bore no fruits proper to be eaten by man. We buried an inſcription here. The boats did not return on board till ten o'clock at night, when the Etoile had joined us. The conſtant ſight of the land ſhewed us that the currents ſet here to the N. W.

Continuation of New Guinea.

After hoiſting in our boats, we ſtrove to keep the ſhore on board, as well as the winds, which were conſtantly at S. and S. S. W. would permit us. We were obliged to make ſeveral boards, with an intent to paſs to windward of a large iſland, which we had ſeen at ſunſet, bearing W. and W. by N. The dawn of day ſurpriſed us, ſtill to leeward of this iſland. Its eaſtern ſide, which may be about five leagues long, runs nearly N. and S. and off the ſouth point lies a low iſland of ſmall extent. Between it and the coaſt of New Guinea, which runs here nearly S. W. by W. there appeared a large paſſage, the entrance of which, of about eight leagues, lay N. E. and S. W. The wind blew out of it, and the tide ſet to the N. W. it was not poſſible to gain in turning to windward againſt wind and ſea; but I ſtrove to do it till nine in the morning. I ſaw with concern that it was fruitleſs, and reſolved to bear away, in order to range the northern ſide of the iſland, abandoning with regret a paſſage, which I thought a fine one, to extricate me out of this everlaſting chain of iſlands.

We had two ſucceſſive alarms this morning. The firſt time they called from aloft, that they ſaw a long range of breakers a-head, and we immediately got the other tacks on board. Theſe breakers, at length, more attentively examined, turned out to be the ripling of a violent tide, and we returned to our former courſe. An hour after, ſeveral perſons called from the forecaſtle, that they ſaw the bottom under us; the affair was preſſing; but the alarm was fortunately as ſhort as it had been ſudden. We ſhould even have thought it falſe, if the Etoile, who was in our wake, had not perceived the ſame ſhoal for near two minutes. It appeared to them a coral-bank. Almoſt north and ſouth of this bank, which may have ſtill leſs water in ſome places, there is a ſandy creek, in which are built ſome huts, ſurrounded with cocoa-trees. This mark may ſo much the better ſerve for a direction, as hitherto we had not ſeen any traces of habitations on this coaſt. At one o'clock in the afternoon, we doubled the N. E. point of the great iſland; which from thence extended W. and W. by S. near 20 leagues. We were obliged to hug our wind to coaſt it; and it was not long ere we perceived other iſlands, bearing W. and W. by N. We ſaw one at ſun-ſet, which bore even N. E. by N. to which there joined a ledge, which ſeemed to extend as far as N. by W. thus were we once more hemmed in.

Hidden danger.

 This

Loſs of the maſter of the ſhip.

This day we loſt our firſt maſter, called Denys, who died of the ſcurvy. He was a native of St. Malo's, and aged about fifty years; moſt of them ſpent in the king's ſervice. The ſentiments of honour, and extenſive knowledge, that diſtinguiſhed him in his important charge, cauſed him to be univerſally regretted among us. Forty-five other perſons were afflicted with the ſcurvy; lemonade and wine only ſuſpended its fatal progreſs.

Difficult courſe.

We ſpent the night upon our tacks; and the 25th, at day-light, found ourſelves ſurrounded with land. Three paſſages preſented themſelves to us; one opened to the S. W. the ſecond to W. S. W. and the third almoſt eaſt and weſt. The wind was fair for none but the eaſt; and I did not approve of it, as I did not doubt that it would carry us into the midſt of the iſles of Papua. It was neceſſary to avoid falling any farther to the northward; for fear, as I have before obſerved, we ſhould be imbayed in one of the gulphs, on the eaſt ſide of Gilolo. The eſſential means for getting out of theſe critical parts, was therefore to get into a ſouthern latitude; for on the other ſide of the S. W. paſſage we obſerved to the ſouthward an open ſea, to the utmoſt extent of our view, therefore I reſolved to ply to windward, in order to gain that outlet. All theſe iſlands, which incloſed us, are very ſteep, of a moderate height, and covered with trees. We did not perceive the leaſt appearance of their being inhabited.

At eleven o'olock in the afternoon, we ſounded 45 fathom, a ſandy bottom; this was one reſource. At noon we obſerved in 00° 5′ N. latitude, having croſſed the line a fourth time. At ſix in the evening we were ſo far to windward, as to be able to fetch the W. S. W. paſſage, having gained about three leagues by working the whole day. The night was more favourable, thanks to the moon-ſhine, which enabled us to turn to windward between the rocks and iſlands. The current, which had been againſt us whilſt we were paſſing by the two firſt paſſages, likewiſe became favourable for us as ſoon as we opened the S. W. paſſage.

Fourth paſſage of the line.

The channel through which we at laſt paſſed out this night, may be about three leagues broad. It is bounded to the weſtward by a cluſter of pretty high iſlands and keys. Its eaſtern ſide, which at firſt ſight we took for the weſtmoſt point of the great iſland, is alſo nothing but a heap of ſmall iſlands and rocks, which, at a diſtance, ſeemed to form only one body; and the ſeparations between theſe iſlands ſhew at firſt the appearance of fine bays; this is what we diſcovered in each tack, that we made towards that ſhore. It was not till half paſt four o'clock in the morning, that we were able to double the ſouthmoſt of the little iſlands of the new paſſage, which we called the *French Paſſage*. We deepened our water in the midſt of this Archipelago of Iſlands,

Deſcriptio[n] of the cha[n]nel throug[h] which we paſs.

 in

in advancing to the ſouthward. Our ſoundings were from 55 to 75 and 80 fathom, grey ſand, ooze, and rotten ſhells. When we were entirely out of the channel, we ſounded and found no bottom. We then ſteered S. W.

The 26th, at break of day, we diſcovered an iſland, bearing S. S. W. and a little after another bearing W. N. W. At noon we ſaw no more of the labyrinth of iſlands we had left, and the meridian altitude gave us 00° 23′ ſouth latitude. This was the fifth time of our paſſing the line. We continued cloſe on a wind, with the larboard tacks on board, and in the afternoon we had ſight of a ſmall iſland in the S. E. The next day, at ſun-riſe, we ſaw it ſomewhat elevated, bearing N. E. about nine or ten leagues diſtance, ſeeming to extend N. E. and S. W. about two leagues. A large hummock, very ſteep, and of a remarkable height, which we named Big Thomas, *(Gros Thomas)* ſhewed itſelf at ten in the forenoon. At its ſouthern point there is a ſmall iſland, and there are two at the northern one. The currents ceaſed ſeting us to the northward; we had, on the contrary, a difference to the ſouthward. This circumſtance, together with our obſerved latitude, which made us to the ſouthward of Cape Mabo, totally convinced me that we were at length entered into the Archipelago of the Moluccas.

Paſs the line a fifth time.

Let

Let me now aſk, which this Cape Mabo is, and where it is ſituated? Some make it the Cape, which, to the northward, terminates the weſtern part of New Guinea. Dampier and Woods Rogers place it the former, in one of the gulphs of Gilolo in 30′ S. lat. The ſecond, eight leagues at fartheſt from this great iſland. But all this part is an extenſive Archipelago of little iſles; which, on account of their number, were called the Thouſand Iſles, by admiral Roggewein, who paſſed through them in 1722. Then in what manner does this Cape Mabo, which is in the neighbourhood of Gilolo, belong to New Guinea? Where ſhall we place it, if (as there is ſo much reaſon to believe) all New Guinea itſelf is a heap of great iſlands? the various channels between which are as yet unknown. It muſt certainly belong to the weſtmoſt of theſe conſiderable iſles.

Diſcuſſion concerning Cape Mabo.

On the 27th, in the afternoon, we diſcovered five or ſix iſlands, bearing from W. S. W. ½ W. to W. N. W. by compaſs. During night we kept the S. S. E. tack; ſo that we did not ſee them again the 28th in the morning. We then perceived five other little iſles, which we ſtood in for. At noon they bore from S. S. W. 1° W. to S. 10° W. at the diſtance of two, three, four, and five leagues. We ſtill ſaw Big Thomas bearing N. E. by E. ½ E. about five leagues. We likewiſe got ſight of another iſland, bearing W. S. W. ſeven or eight leagues diſtant. During

Entrance to the Ar[chi]pelago of Moluccas.

ing the laſt twenty-four hours we felt ſeveral ſtrong tides, which ſeemed to ſet from the weſtward. However, the difference between my reckoning, and the obſervation at noon, and at the ſetting of the bearings, gave us ten or eleven miles to S. W. by S. and S. S. W. At nine o'clock in the morning I ordered the Etoile to mount her guns, and ſent her cutter to the S. W. iſles, in order to ſee whether there was any anchorage, and whether theſe iſles had any intereſting productions.

Meeting with a negro.

It was almoſt a calm in the afternoon, and the boat did not return before nine o'clock in the evening. She had landed on two iſles, where our people had found no ſigns of habitation, or cultivation, and not even any kind of fruits. They were going to return, when, to their great ſurpriſe, they ſaw a negro, quite by himſelf, coming towards them in a periagua, with two outriggers. In one ear he had a golden ring, and his arms were two lances. He came up to our boat without ſhewing any marks of fear or ſurprize. Our people aſked him for ſomething to eat and to drink, and he offered them water, and a ſmall quantity of a ſort of flour, which ſeemed to be his ordinary food. Our men gave him a handkerchief, a looking-glaſs, and ſome other trifles of that ſort. He laughed when he received theſe preſents, and did not admire them. He ſeemed to know the Europeans, and we thought that he might poſſibly

be

be a run-away negro from one of the neighbouring iſlands where the Dutch have ſettlements; or that he had perhaps been ſent out a-fiſhing. The Dutch call theſe iſlands the Five Iſles, and ſend ſome people to viſit them from time to time. They told us that they were formerly ſeven in number, but that two have been ſunk by earthquakes, which happen frequently in theſe parts. Between theſe iſles there is a prodigious current, without any anchorage. The trees and plants are almoſt all the ſame here as upon New Britain. Our people took a turtle here of about two hundred weight.

From this time we continued to meet with violent tides, which ſet to the ſouthward, and we kept the courſe which came neareſt to their direction. We ſounded ſeveral times without finding bottom, and till the 30th in the afternoon, we got ſight of no other land than a ſingle iſle to the weſtward, ten or twelve leagues from us; but then we ſaw a conſiderable land bearing ſouth at a great diſtance. The current, which was of more ſervice to us than the wind, brought us nearer to it during night, and on the 31ſt at day-break we were about ſeven or eight leagues from it. This was the Iſle of Ceram. Its coaſt, which is partly woody and partly cleared, runs nearly eaſt and weſt, and we could not ſee it terminated. This iſle is very high;

Sight of Ceram.

high; prodigious mountains riſe on it from ſpace to ſpace, and the numerous fires which we ſaw on all ſides of it, indicate its being very populous. We paſſed the day and the next night in ranging the northern coaſt of this iſle, making our tacks in order to gain to the weſtward, and double its weſtermoſt point. The current was favourable to us, but the wind was ſcant.

Obſervations on the monſoons in theſe parts.

I ſhall here take an opportunity from the contrary winds we had now met with for a long time, to obſerve, that in the Moluccas, they call the weſterly monſoon the northern one, and the eaſterly monſoon the ſouthern one; becauſe, during the former, the winds blow more generally from N. N. W. than from W. and during the latter, they come moſt frequently from S. S. E. Theſe winds likewiſe prevail in the iſles of Papua, and on the coaſts of New Guinea; we got this information by fatal experience, having employed thirty-ſix days to make four hundred and fifty leagues in.

1768. September.

The firſt of September, at the dawn of day, we were at the entrance of a bay, in which we ſaw ſeveral fires Soon after we perceived two veſſels under ſail, built in form of the Malay boats. We hoiſted a Dutch enſign and pendent, and fired a gun, by which I committed a fault without knowing it. We have ſince learnt that the inhabitants of Ceram are at war with the Dutch, and that they have expelled the latter from almoſt every part

of

of their isle. Therefore we made a board into the bay without success, the boats retreated on shore, and we profited of the fresh breeze to proceed on our course. The shore at the bottom of the bay is low and level, surrounded by high mountains; and the bay itself contains several islands. We were obliged to steer W. N. W. in order to double a pretty large island, at the point of which you see a little isle or key, and a sand bank, with some breakers which seem to extend a league out to sea. This island is called *Bonao*; it is divided into two by a very narrow channel. When we had doubled it, we steered W. by S. till noon.

It blew very fresh from S. S. W. to S. S. E. and we plyed the remainder of the day between *Bonao*, *Kelang*, and *Manipa*, endeavouring to make way to the S. W. At ten o'clock in the evening we discovered the lands of the isle of Boero, by means of the fires which burnt on it; and as it was my intention to put in there, we passed the night on our tacks, in order to keep within reach, and if possible to the windward of it. I knew that the Dutch had a weak factory on this isle, which was however abundant in refreshments. As we were perfectly ignorant of the situation of affairs in Europe, it was not prudent to venture to learn the first intelligence concerning them among strangers, but at a place where we were almost the strongest.

Project fo[r] our safety

Sad condition of the ſhip's companies.

Exceſſive marks of joy accompanied our diſcovering the entrance of the gulph of Cajeli, at break of day There the Dutch have their ſettlement; there too was the place where our greateſt miſery was to have an end. The ſcurvy had made cruel havock amongſt us after we had left Port Praſlin; no one could ſay he was abſolutely free from it, and half of our ſhip's companies were not able to do any duty. If we had kept the ſea eight days longer, we muſt have loſt a great number of men, and we muſt all have fallen ſick. The proviſions which we had now left were ſo rotten, and had ſo cadaverous a ſmell, that the hardeſt moments of the ſad days we paſſed, were thoſe when the bell gave us notice to take in this diſguſting and unwholeſome food. I leave every one to judge how much this ſituation heightened in our eyes the beautiful aſpect of the coaſts of Boero. Ever ſince midnight, a pleaſant ſcent exhaled from the aromatic plants with which the Moluccas abound, had made an agreeable impreſſion upon our organs of ſmell, ſeveral leagues out at ſea, and ſeemed to be the fore-runner which announced the end of our calamities to us. The aſpect of a pretty large town, ſituated in the bottom of the gulph; of ſhips at anchor there, and of cattle rambling through the meadows, cauſed tranſports which I have doubtleſs felt, but which I cannot here deſcribe.

We were obliged to make ſeveral boards before we entered into this gulph, of which the northern point is called the point of *Liſſatetto*, and that on the S. E. ſide, point Rouba. It was ten o'clock before we could ſtand in for the town. Several boats were ſailing in the bay; we hoiſted Dutch colours, and fired a gun, but not one of them came along-ſide; I then ſent a boat to ſound a-head of the ſhip. I was afraid of a bank which lies on the S. E. ſide of the gulph. At half an hour paſt noon, a periagua conducted by Indians came near the ſhip; the chief perſon aſked us in Dutch who we were, but refuſed to come on board. However, we advanced, all ſails ſet, according to the ſignals of our boat, which ſounded a-head. Soon after we ſaw the bank of which we had dreaded the approach. It was low water, and the danger appeared very plain. It is a chain of rocks mixed with coral, ſtretching from the S. E. ſhore of the gulph to within a league of point Rouba, and its extent from S. E. to N. W. is half a league. About four times the length of a boat from its extremities, you have five or ſix fathoms of water, a foul coral bottom, and from thence you immediately come into ſeventeen fathoms, ſand and ooze. Our courſe was nearly S. W. three leagues, from ten o'clock to half paſt one, when we anchored oppoſite the factory, near ſeveral little Dutch veſſels, not quite a quarter of a league

Shoal of the gulph of Cajeli.

league off ſhore. We were in twenty-ſeven fathoms, ſand and ooze, and had the following bearings:

Point Liſſatetto, N. 4° E. two leagues.

Point Rouba, N. E. 2° E. half a league.

A peninſula, W. 10° N. three quarters of a league.

The point of a ſhoal, which extends above half a league to the offing from the peninſula, N. W. by W.

The flag of the Dutch factory, S. by W. ½ W.

We put in at Boero.

The Etoile anchored near us more to the W. N. W. We had hardly let go our anchor, when two Dutch ſoldiers, without arms, one of them ſpeaking French, came on board to aſk me on the part of the chief of the factory, what motives brought us to this port, when we could not be ignorant that the ſhips of the Dutch India company alone had the privilege of entering it. I ſent them back with an officer to declare to the chief, that the neceſſity of taking in proviſions forced us to enter into the firſt port we had met with, without permitting us to pay any regard to the treaties that exclude our ſhips from the ports in the Moluccas, and that we ſhould leave the harbour as ſoon as he ſhould have given us what help we ſtood moſt in need of. The two ſoldiers returned ſoon after, to communicate to me an order, ſigned by the governor of Amboina, upon whom the chief of Boero immediately depends, by which the latter is expreſsly forbid to receive foreign ſhips into his port.

Embarraſſment of the chief.

The

The chief at the ſame time begged me to give him a written declaration of my motives for putting in here, in order that he might thereby juſtify his conduct in receiving us here, before his ſuperior, to whom he would ſend the above declaration. His demand was reaſonable, and I ſatisfied it by giving him a ſigned depoſition, in which I declared, that having left the Malouines, and intending to go to India by the South Seas, the contrary monſoon, and the want of proviſions, had prevented our gaining the Philippinas, and obliged us to go in ſearch of the indiſpenſable ſupplies at the firſt port in the Moluccas, and that I deſired him to grant me theſe ſupplies in conſideration of humanity, the moſt reſpectable of obligations.

Good reception he gives us.

From this moment we found no difficulties; the chief having done his duty for his company, happily acted a very good natured character, and offered us all he had in as eaſy a manner as if he had every thing in his diſpoſal. Towards five o'clock I went on ſhore with ſeveral officers, in order to pay him a viſit. Notwithſtanding the embarraſſment which our arrival had cauſed him, he received us extremely well. He even offered us a ſupper, and we did not fail to accept of it. When he ſaw with what pleaſure and avidity we devoured it, he was better convinced than by our words, that we had reaſon to complain of being pinched by hunger.

hunger. All the Hollanders were ſtruck with the higheſt degree of ſurpriſe, and none of them durſt eat any thing for fear of wronging us. One muſt have been a ſailor, and reduced to the extremities which we had felt for ſeveral months together, in order to form an idea of the ſenſation which the ſight of greens and of a good ſupper produced in people in that condition. This ſupper was for me one of the moſt delicious moments of my life, eſpecially as I had ſent on board the veſſels what would afford as good a ſupper as ours to every one there.

We agreed that we ſhould have veniſon every day to ſupply our companies with freſh meat, during their ſtay; that at parting we were to receive eighteen oxen, ſome ſheep, and almoſt as much poultry as we ſhould require. We were obliged to ſupply the want of bread with rice, which the Dutch live upon. The iſlanders live upon ſago bread, which they get out of a palm of that name; this bread looks like the caſſava. We could not get great quantities of pulſe, which would have been extremely ſalutary to us. The people of this country do not cultivate them. The chief was ſo good as to give ſome to our ſick from the company's garden.

olice of the ompany.

Upon the whole, every thing here, directly or indirectly, belongs to the company; neat and ſmall cattle, grain,

grain, and victuals of all kinds. The company alone buys and ſells. The Moors indeed have ſold us fowls, goats, fiſh, eggs, and ſome fruit, but the money which they got for them will not long remain in their hands. The Dutch know how to get at it, by ſelling them very coarſe kinds of cloth, which however bear a very great price. Even ſtag-hunting is not allowed to every one, for the chief alone has a right to it. He gives his huntſmen three charges of powder and ſhot, in return they are obliged to bring him two deer, for which they are paid ſix-pence a-piece. If they bring home only one, he deducts from what is due to them the value of one charge of powder and ſhot.

On the 3d in the morning we brought our ſick on ſhore, to ly there during our ſtay. We likewiſe daily ſent the greateſt part of the crews on ſhore, to walk about and divert themſelves. I got the ſlaves of the company, whom the chief hired to us by the day, to fill the water of both ſhips, and to tranſport every thing from the ſhore to the ſhips, &c. The Etoile profited of this time to adjuſt the caps of her lower maſts, which had much play. We had moored at our arrival, but from what the Dutch told us of the goodneſs of the bottom, and of the regularity of the land and ſea breezes; we weighed our ſmall bower. Indeed, we ſaw all the Dutch veſſels riding at ſingle anchor.

 During

During our ſtay here we had exceeding fine weather. The thermometer generally roſe to 23° during the greateſt heat of the day; the breeze from N. E. and S. E. blowing in day time, changed in the evening; it then came from the ſhore, and the nights were very cool. We had an opportunity of ſeeing the interior parts of the iſle; we were allowed to go out a ſtag-hunting ſeveral times, in which we took a great deal of pleaſure. The country is charmingly interſperſed with woods, plains, and hillocks, between which the vallies are watered by fine rivulets. The Dutch have brought the firſt ſtags hither, which have multiplied prodigiouſly, and are delicious eating. Here are likewiſe wild boars in great plenty, and ſome ſpecies of wild fowls.

Particulars concerning the iſle of Boero.

The extent of the iſle of Boero or Burro from eaſt to weſt is reckoned at eighteen leagues, and from north to ſouth at thirteen. It was formerly ſubject to the king of Ternate, who got a tribute from thence. The principal place in it is Cajeli, ſituated at the bottom of the gulph of that name, in a marſhy plain, ſtretching about four miles between the rivers *Soweill* and *Abbo*. The latter is the greateſt river in the whole iſland, and its water is always very muddy. The landing is very inconvenient here, eſpecially at low water, during which, the boats are obliged to ſtop at a good diſtance from

6 the

the beach. The Dutch settlement, and fourteen Indian habitations, formerly dispersed in several parts of the isle, but now drawn together round the factory, form the village or town of Cajeli. At first, the Dutch had built a fort of stone here; it was blown up by accident in 1689, and since that time they have contented themselves with a simple enclosure of pallisadoes, mounted with six small cannon, forming a kind of battery; this is called Fort of Defence, and I took this name for a sort of ironical appellation. The garrison is commanded by the chief, and consists of a serjeant and twenty-five men; on the whole island are not above fifty white people. Some habitations of black people are dispersed on it, and they cultivate rice. Whilst we were here, the Dutch forces were encreased by three vessels, of which, the biggest was the Draak, a snow, mounting fourteen guns, commanded by a Saxon, whose name was Kop-le-Clerc; she was manned by fifty Europeans, and destined to cruise among the Moluccas, and especially to act against the people of Papua and Ceram.

Account of the natives of the country.

The natives of the country are of two classes, the Moors *(Maures)* and the Alfourians *(Alfouriens)*. The former live together under the factory, being entirely submitted to the Dutch, who inspire them with a great fear of all foreign nations. They are zealous observers of the Mahomedan religion, that is, they make frequent ablu-

ablutions, eat no pork, and take as many wives as they can ſupport, being very jealous of them, and keeping them ſhut up. Their food is ſago, ſome fruits, and fiſh. On holidays they feaſt upon rice, which the company ſells them. Their chiefs or *orencaies* are always about the Dutch chief, who ſeems to have ſome regard for them, and by their means keeps the people in order. The company have had the art of ſowing the ſeeds of a reciprocal jealouſy among theſe chiefs; this aſſures them of a general ſlavery, and the police which they obſerve here with regard to the natives, is the ſame in all their other factories. If one chief forms a plot, another diſcovers it, and immediately informs the Dutch of it.

Theſe moors are, upon the whole, ugly, lazy, and not at all warlike. They are greatly afraid of the Papous, or inhabitants of Papua; who come ſometimes in numbers of two or three hundred to burn their habitations, and to carry off all they can, and eſpecially ſlaves. The remembrance of their laſt viſit, made about three years ago, was ſtill recent. The Dutch do not make ſlaves of the natives of Boero; for the company gets thoſe, whom they employ that way, either from Celebes, or from Ceram, as the inhabitants of theſe two iſles ſell each other reciprocally.

Viſe people. The Alfourians are a free people, without being enemies of the company. They are ſatisfied with being in-

depen-

dependent, and covet not thoſe trifles, which the Europeans ſell or give them in exchange for their liberty. They live diſperſed in the inacceſſible mountains, which the interior parts of this iſle contain. There they ſubſiſt upon ſago, fruits, and hunting. Their religion is unknown; it is ſaid, that they are not Mahommedans; for they feed hogs, and likewiſe eat them. From time to time the chiefs of the Alfourians come to viſit the Dutch chief; they would do as well to ſtay at home.

Productions of the Boero

I do not know whether there were formerly any ſpice plantations on this iſle; but be this as it will, it is certain that there are none at preſent. The company get from this ſtation nothing but black and white ebony, and ſome other ſpecies of wood, which are much in requeſt with joiners. There is likewiſe a fine pepper plantation; the ſight of which has convinced us, that pepper is common on New Britain, as we conjectured before. Fruits are but ſcarce here; there are cocoa-nuts, bananas, ſhaddocks, ſome lemons, citrons, bitter-oranges, and a few pine-apples. There grows a very good ſort of barley, called *ottong*, and the *ſago-borneo*, of which they make ſoups, which ſeemed abominable to us. The woods are inhabited by a vaſt number of birds of various ſpecies, and beautiful plumage; and among them are parrots of the greateſt beauty. Here is likewiſe that ſpecies

ſpecies of wild cat *, which carries its young in a bag under its belly; the kind of bat, whoſe wings are of a monſtrous extent †; enormous ſerpents, which can ſwallow a whole ſheep at once, and another ſpecies of ſnakes, which is much more dangerous; becauſe it keeps upon trees, and darts into the eyes of thoſe who look into the air as they paſs by. No remedy is as yet found againſt the bite of this laſt kind; we killed two of them in one of our ſtag-hunts.

The river Abbo, of which the banks are almoſt every where covered with trees of a thick foliage, is infeſted by enormous crocodiles, which devour men and beaſts. They go out at night; and there are inſtances of their taking men out of their periaguas. The people keep them from coming near, by carrying lighted torches. The ſhores of Boero do not furniſh many fine ſhells. Thoſe precious ſhells, which are an article of commerce with the Dutch, are found on the coaſt of Ceram, at Amblaw, and at Banda, from whence they are ſent to Batavia. At Amblaw they likewiſe find the moſt beautiful kind of cockatoes.

* M. de Buffon has denied the exiſtence of the *Opoſſum* or *Didelphis*, Linn. in Eaſt India, though Piſo, Valentyn, and Le Brun have ſeen it in the Moluccas and in Java: M. de Buffon's own countryman, M. de Bougainville, now likewiſe aſſerts their being upon Boero, in a manner ſo little equivocal, that there can be no doubt of the Opoſſum genus inhabiting the Eaſt Indies, though the particular ſpecies is unknown. F.

† This is the great *Bat of Ternate*, Penn. Syn. Quad. p. 359. and Linnæus's *Veſpertilio Vampyrus*. F.

Henry

Henry Ouman, the chief at Boero, lives there like a sovereign. He has a hundred ſlaves for the ſervice of his houſe, and all the neceſſaries and conveniencies of life in abundance. He is an Under-Merchant *; and this degree is the third in the company's ſervice. This man was born at Batavia, and has married a Creole from Amboina. I cannot ſufficiently praiſe his good behaviour towards us. I make no doubt, but the moment when we entered this port, was a critical one for him; but he behaved like a man of ſenſe. After he had done what his duty to his ſuperiors required, he did what he could not be exempted from, with a good grace, and with the good manners of a frank and generous man. His houſe was ours; we found ſomething to eat and drink there at all times; and I think this kind of civility was as good as any other, eſpecially to people who ſtill felt the conſequences of famine. He gave us two repaſts of ceremony; the good order, elegance, and plenty of which, quite ſurpriſed us in ſo inconſiderable a place. The houſe of this honeſt Dutchman was very pretty, elegantly furniſhed, and built entirely in the Chineſe taſte. Every thing is ſo diſpoſed about it as to make it cool; it is ſurrounded by a garden, and a river runs acroſs it. You come to it from the ſea-ſhore, through an avenue of very great trees. His wife

Good proceedings of the reſident on our account.

* *Sous-Marchand.*

and

and daughter were dreſſed after the Chineſe faſhion, and performed the honours of the houſe very well. They paſs their time in preparing flowers for diſtillation, in making noſegays, and getting ſome betel ready. The air which you breathe in this agreeable houſe is moſt deliciouſly perfumed, and we ſhould all very willingly have made a long ſtay there: how great was the contraſt between this ſweet and peaceful ſituation, and the unnatural life we had now led for theſe ten months paſt?

Conduct of Aotourou at Boero.

I muſt mention what impreſſion the ſight of this European ſettlement made upon Aotourou. It will eaſily be conceived that his ſurpriſe muſt have been great at ſeeing men dreſſed like ourſelves, houſes, gardens, and various domeſtick animals in abundance, and great variety. He could not be tired with looking at theſe objects, which were new to him. He valued above all that hoſpitality, which was here exerciſed with an air of ſincerity and of acquaintance. As he did not ſee us make any exchanges, he apprehended that the people gave us every thing without being paid for it. Upon the whole, he behaved very ſenſibly towards the Dutch. He began with giving them to underſtand, that in his country he was a chief, and that he had undertaken this voyage with his friends for his own pleaſure. In the viſits, at table, and in our walks, he endeavoured to imitate

imitate us exactly. As I had not taken him with me on the first visit which we made, he imagined it was because his knees are distorted, and absolutely wanted some sailors to get upon them, to set them to rights. He often asked us, whether Paris was as fine as this factory?

Goodness of the provisions there.

On the 6th, in the afternoon, we had taken on board our rice, cattle, and all other refreshments. The good chief's bill was of a considerable amount; but we were assured, that all the prices were fixed by the company, and that he could not depart from their tariff. The provisions were indeed excellent; the beef and mutton are better by a great deal, than in any other hot country I know; and the fowls are most delicious there. The butter of Boero has a reputation in this country, which our sailors from Bretany found it had not lawfully acquired.

The 7th, in the morning, I took on board the sick people, and we made every thing ready, in order to set sail in the evening with the land-breeze. The fresh provisions, and the salubrious air of Boero, had done our sick much good. This stay on shore, though it lasted only six days, brought them so far, that they could be cured on board, or at least prevented from growing worse, by means of the refreshments which we could now give them.

Obſervations on the monſoons and currents.

It would doubtleſs have been very deſirable for them, and even for the healthy men, to have made a longer ſtay here; but the end of the eaſtern monſoon being at hand, preſſed us to ſet ſail for Batavia. If the other monſoon was once ſet in, it became impoſſible for us to go there; becauſe at that time, beſides having the winds contrary to us, we had likewiſe the currents againſt us, which follow the direction of the reigning monſoon. It is true, they keep the direction of the preceding monſoon for near a month after it; but the changing of the monſoon, which commonly happens in October, may come a month ſooner, as well as a month later. In September there is little wind: in October and November ſtill leſs; that being the ſeaſon of calms. The governor of Amboina chooſes at this ſeaſon to go his rounds to all the iſles which depend upon his government. June, July, and Auguſt, are very rainy. The eaſtern monſoon generally blows S. S. E. and S. S. W. to the north of Ceram and Boero; in the iſles of Amboina and Bandas it blows E. and S. E. The weſtern monſoon blows from W. S. W. and N. W. The month of April is the term when the weſtern winds ceaſe blowing; this is the ſtormy monſoon, as the eaſterly one is the rainy monſoon. Captain Clerk told us, that he had in vain cruized before Amboina, in order to enter it, during the whole month of July: he had there ſuffered continual rains, which had made all his people sick. It was at the same time that we were so well soaked in Port Praslin.

There had been three earthquakes this year at Boero, almost close after each other, on the 7th of June, the 12th and on the 17th of July. It was the 22d of the same month that we felt one on New Britain. These earthquakes have terrible consequences for navigation in this part of the world. Sometimes they sink known isles and sand-banks, and sometimes they raise some, where there were none before; and we gain nothing by such accidents. Navigation would be much safer, if every thing remained as it is.

Remarks on the earthquakes.

On the 7th after noon, all our people were on board, and we only waited for the land-breeze, in order to set sail. It was not felt till eight o'clock at night. I immediately sent a boat with a light to anchor at the point of the bank, which lies on the S. E. side, and we began to make every thing ready for setting sail. We had not been misled, when we were informed that the bottom was very good in this anchorage. We made fruitless efforts at the capstan for a long time; at last the voyal broke, and we could only by the help of our winding-tackle get our anchor out of this strong ooze, in which it was buried. We did not get under sail before eleven o'clock. Having doubled the point of the bank, we hoisted in our boats, as the Etoile did hers, and we

We leave Boero.

 steered

ſteered ſucceſſively N. E. N. E. by N. and N. N. E. in order to go out of the gulph of Cajeli.

Aſtronomical obſervations.

During our ſtay here, M. Verron had made ſeveral obſervations of diſtances on board; the mean reſult of which enabled him to determine the longitude of this gulph; and places it 2° 53′ more to the weſtward than our reckoning, which we had followed after determining the longitude on New Britain. Upon the whole, though we found the true European date current in the Moluccas, from which it was very natural, we had loſt a day by going round the world with the ſun's courſe, yet I ſhall continue the date of our journals, only mentioning, that inſtead of Wedneſday the 7th, they reckoned Thurſday the 8th in India. I ſhall not correct my date, till I come to the iſle of France.

CHAP.

Pl.V. to face page 381.

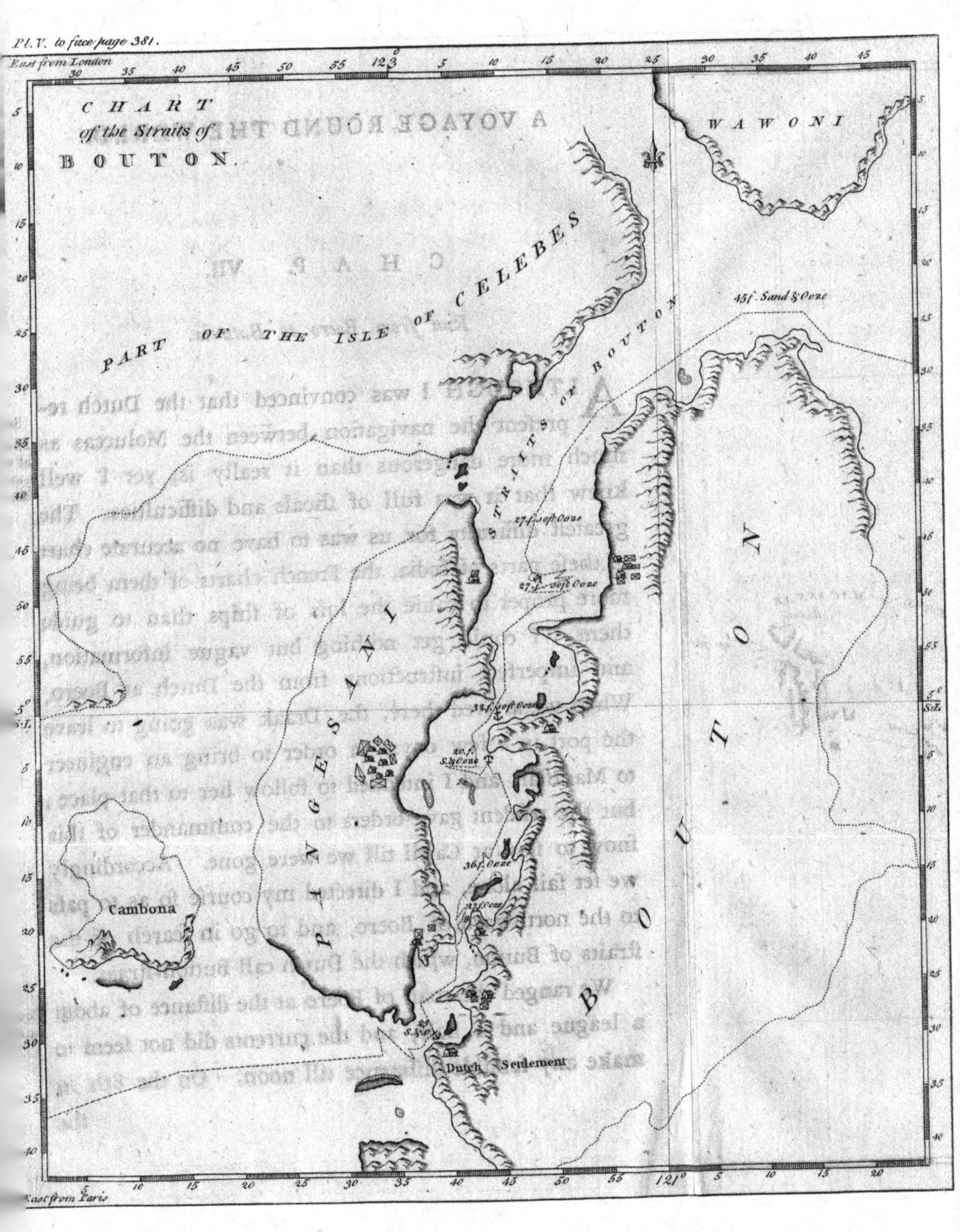

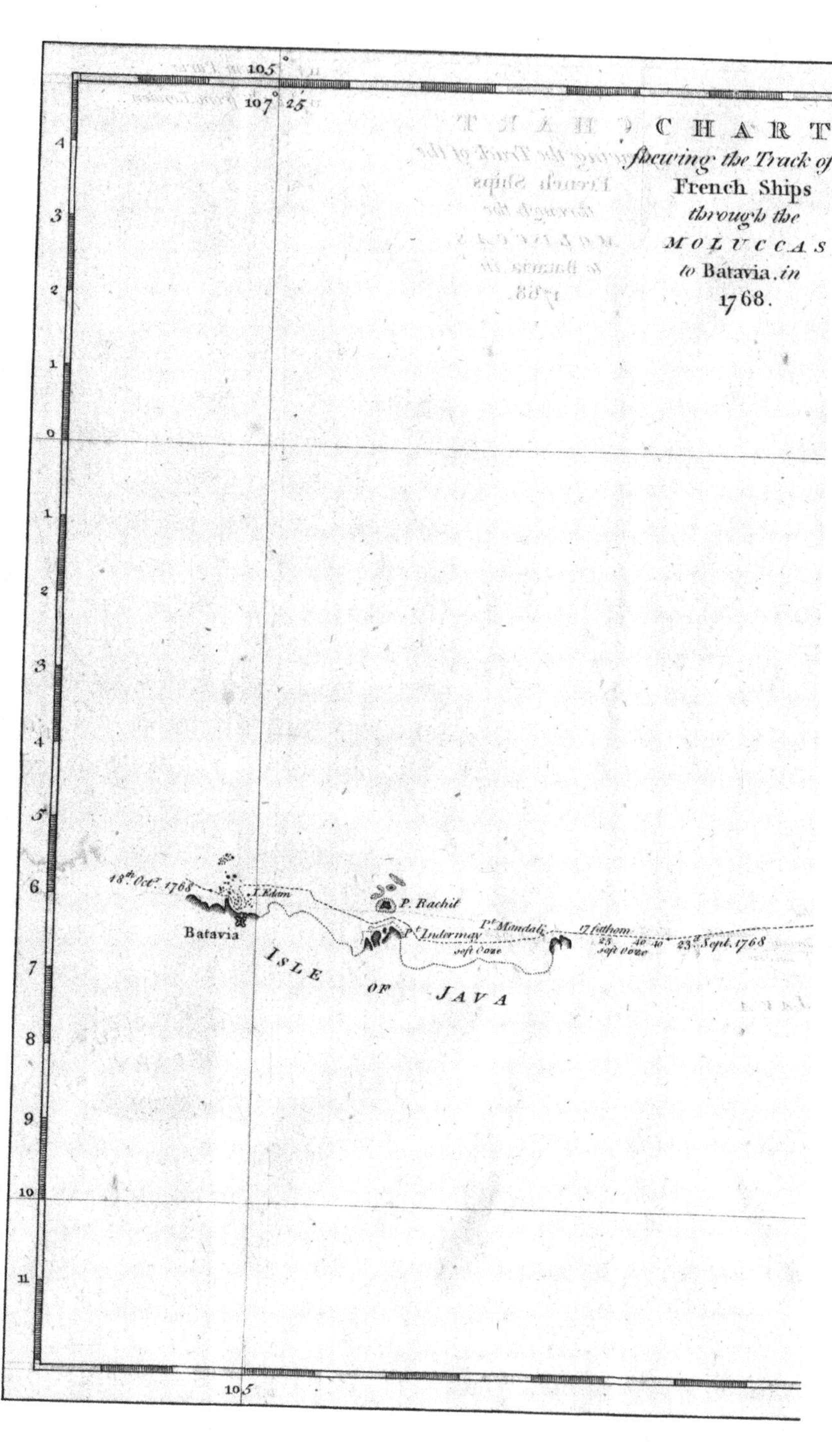
CHART
shewing the Track of
French Ships
through the
MOLUCCAS
to Batavia, in
1768.
105
107° 25
18th Octr. 1768
I. Edam
Batavia
P. Rachit
Pt. Indermay
Pt. Mondali
17 fathom
soft Ooze
25
40
40
soft Ooze
23d Sept. 1768
ISLE
OF
JAVA
4
3
2
1
0
1
2
3
4
5
6
7
8
9
10
11
105

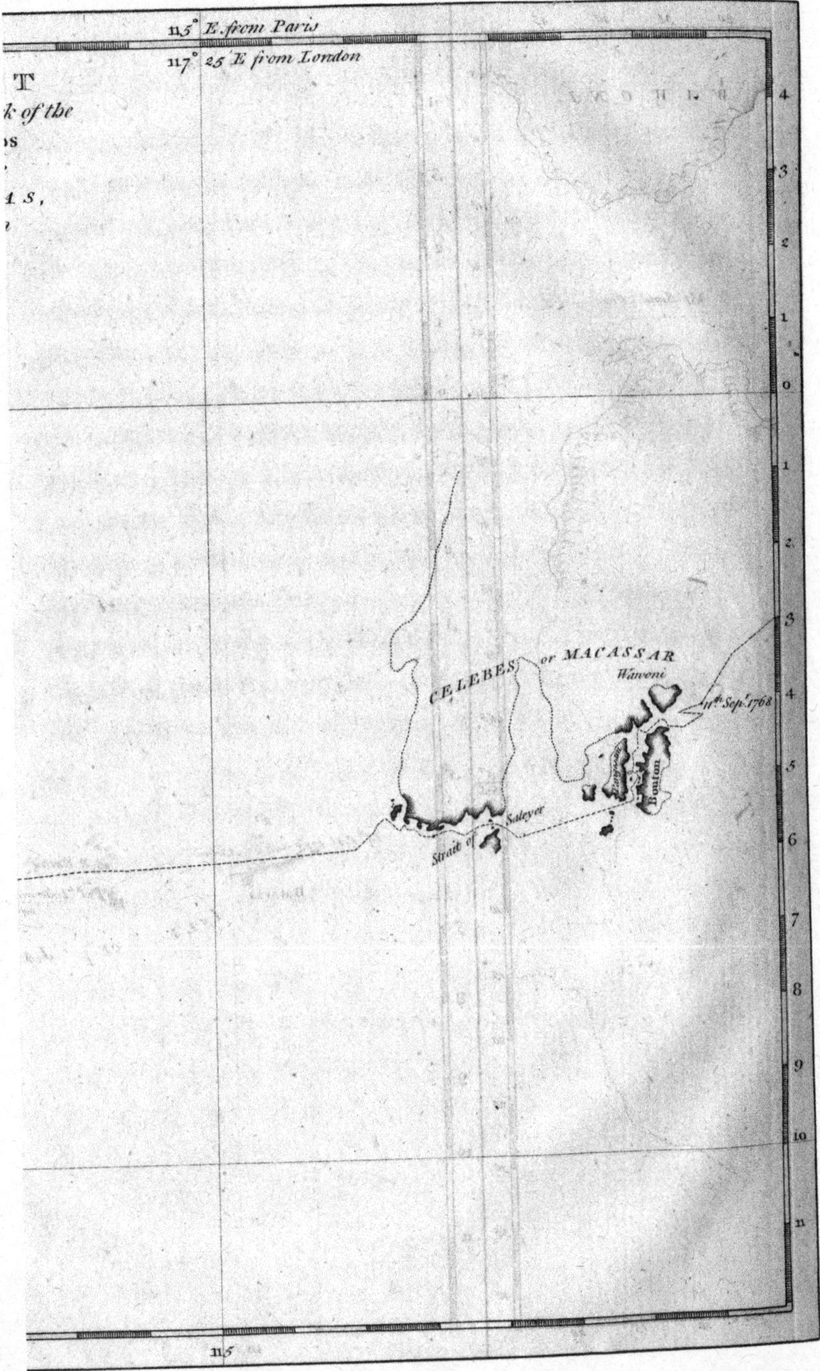

115° E. from Paris
117° 25' E. from London
k of the
os
AS,
CELEBES or MACASSAR
Waweni
11th Sep. 1768
Bouton
Saleyer
Strait of
4
3
2
1
0
1
2
3
4
5
6
7
8
9
10
11
115

CHAP. VII.

Run from Boero to Batavia.

1768. September. Difficulties of the navigation in the Moluccas.

ALTHOUGH I was convinced that the Dutch represent the navigation between the Moluccas as much more dangerous than it really is, yet I well knew that it was full of ſhoals and difficulties. The greateſt difficulty for us was to have no accurate chart of theſe parts of India, the French charts of them being more proper to cauſe the loſs of ſhips than to guide them. I could get nothing but vague information, and imperfect inſtructions from the Dutch at Boero. When we arrived there, the Draak was going to leave the port in a few days, in order to bring an engineer to Macaſſar, and I intended to follow her to that place; but the reſident gave orders to the commander of this ſnow to ſtay at Cajeli till we were gone. Accordingly we ſet ſail alone, and I directed my courſe ſo as to paſs to the northward of Boero, and to go in ſearch of the ſtraits of Button, which the Dutch call Button-ſtraat.

Courſe we take.

We ranged the coaſt of Boero at the diſtance of about a league and a half, and the currents did not ſeem to make any ſenſible difference till noon. On the 8th in the

the morning we perceived the isles of Kilang and Manipa. From the low land which you find after going out of the gulph of Cajeli, the coast is very high, and runs W. N. W. and W. by N. On the 9th in the morning we got sight of the isle of Xullabessie; it is a very inconsiderable one, and the Dutch have a factory there, in a redoubt, called *Cleverblad*, or the Clover leaf. The garrison consists of a serjeant and twenty five men, under the command of M. Arnoldus Holman, who is only book-keeper. This isle formerly was one of the dependences of the government of Amboina, at present it belongs to that of Ternate. Whilst we ran along Boero we had little wind, and the settled breezes almost the same as in the bay. The currents during these two days set us near eight leagues to the westward. We determined this difference with precision enough, on account of the frequent bearings which we took. On the last day they likewise set us a little to the southward, which was verified by the meridian altitude observed on the 10th.

We had seen the last lands of Boero on the 9th, at sun-setting; we found pretty fresh S. and S. S. E. winds out at sea, and we passed several very strong races of a tide. We steered S. W. whenever the winds permitted, in order to fall in with the land between Wawoni and Button, as I intended to pass through the straits of that

name.

name. It is pretended that during this season it is dangerous to keep to the eastward of Button, that one runs the risk of being thrown upon the coast by the winds and currents, and that then it is necessary, in order to lay it again, to wait for the western monsoon's being perfectly set in. This I have been told by a Dutch mariner, but I will not answer for the truth of it. I will however positively assert that the passage of the straits is infinitely preferable to the other course, either to the northward or to the southward of the shoal called Toukanbessie: this latter being full of visible and hidden dangers, which are dreaded even by those who know the coast.

Nautical advice.

On the 10th in the morning, one Julian Launai, taylor, died of the scurvy. He began already to grow better, but two excesses in drinking brandy carried him off.

Sight of the straits of Button.

The 11th, at eight o'clock in the morning, we saw the land, bearing from W. by S. to S. S. W. ½ W. At nine o'clock we found that it was the isle of Wawoni, which is high, especially in its middle: at eleven o'clock we discovered the northern part of Button. At noon we observed in 4° 6′ of south lat. The northermost point of the isle of Wawoni then bore W. ½ N. its southermost point S. W. by W. 4° W. eight or nine leagues distant, and the N. E. point of Button, S. W. ½ W. about nine

nine leagues diſtant. In the afternoon we ſtood within two leagues of Wawoni, then ſtood out into the offing, and kept plying all night, in order to keep to windward of the ſtraits of Button, and be ready to enter them at day-break. The 12th, at ſix o'clock in the morning, it bore between N. W. by W. and W. N. W. and we ſtood in for the north point of Button. At the ſame time we hoiſted out our boats, and kept them in tow. At nine o'clock we opened the ſtraits, with a fine breeze, which laſted till half paſt ten o'clock, and freſhened again a little before noon.

Deſcription of the entrance.

When you enter theſe ſtraits, it is neceſſary to range the land of Button, of which the north point is of a middling height, and divided into ſeveral hummocks. The cape on the larboard ſide of the entrance is ſteep and bold-to. Several white rocks ly before it, pretty high above the water, and to the eaſtward is a fine bay, in which we ſaw a ſmall veſſel under ſail. The oppoſite point of Wawoni is low, tolerably level, and projects to the weſtward. The land of Celebes then appears before you, and a paſſage opens to the north, between this great iſle and Wawoni; this is a falſe paſſage: the ſouthern one indeed appears almoſt entirely ſhut up; there you ſee at a great diſtance a low land, divided as it were into little iſles or keys. As you advance in the ſtraits, you diſcover upon the coaſt of Button, great round capes, and fine creeks. Off one of these capes are two rocks, which one must absolutely take at a distance for two ships under sail; the one pretty large, and the other a small one. About a league to the eastward of them, and a quarter of a league off the coast, we sounded in forty-five fathoms, sand and ooze. The straits from the entrance run successively S. W. and south.

At noon we observed in 4° 29′ south lat. and were then somewhat beyond the rocks. They ly off a little isle, behind which there appears to be a fine inlet. There we saw a kind of vessel in form of a square chest, having a periagua in tow. She made way both by sailing and rowing, and ranged the shore. A French sailor, whom we took in at Boero, and who for these four years past had sailed with the Dutch in the Moluccas, told us that it was a boat of piratical Indians, who endeavour to make prisoners in order to sell them. They seemed to be rather troubled at meeting with us. They furled their sail, and set their vessel with setting poles close under the shore, behind the little isle.

Aspect of country.

We continued our course in the straits, the winds turning round with the channel, and permitting us to come by degrees from S. W. to south. Towards two o'clock in the afternoon we thought the tide began to set against us; the sea then washed the lower parts of

 the

the trees upon the coaſt, which ſeems to prove that the flood-tide comes here from the northward, at leaſt during this ſeaſon. At half an hour after two o'clock we paſſed a very fine port upon the coaſt of Celebes. This land offers a charming proſpect, on account of the variety of low lands, hills, and mountains. The landſcape is adorned with a fine verdure, and every thing announces a rich country. Soon after, the iſle of Pangeſani, and the keys to the northward of it, appear ſeparated, and we diſtinguiſhed the ſeveral channels which they form. The high mountains of Celebes appeared above, and to the northward of theſe lands. The ſtraits are afterwards formed by this long iſle of Pangaſani, and by that of Button. At half paſt five o'clock we were locked in ſo that we could not ſee either the entrance or the out-let, and we ſounded in twenty-ſeven fathoms of water, and an excellent oozy bottom.

First anchorage.

The breeze which then came from E. S. E. obliged us to ſail cloſe upon it, in order to keep the coaſt of Button on board. At half paſt ſix o'clock, the wind coming more contrary, and the tide ſetting pretty ſtrong againſt us, we let go a ſtream-anchor almoſt in the midſt of the channel, in the ſame ſoundings which we had before, twenty-ſeven fathoms, ſoft ooze; which is a mark of an equal depth in all this part. The breadth of the ſtraits from the entrance to this firſt anchorage,

6 varies

varies from ſeven to eight, nine and ten miles. The night was very fine. We ſuppoſed there were habitations on this part of Button, becauſe we ſaw ſeveral fires there. Pangaſani appeared much better peopled to us, if we judge by the great number of fires on every part of it. This iſle is here low, level, and covered with fine trees, and I ſhould not wonder if it contained ſpices.

Traffic with the inhabitants.

On the 13th, a great many periaguas, with outriggers, ſurrounded the ſhips. The Indians brought us fowls, eggs, bananas, perrokeets and cockatoes. They deſired to be paid in Dutch money, and eſpecially in a plated coin, which is of the value of two French ſous and a half. They likewiſe willingly took knives with red handles. Theſe iſlanders came from a conſiderable plantation on the heights of Button, oppoſite our anchorage, occupying the ſkirts of five or ſix mountains. The land is there entirely cleared, interſected with ditches, and well planted. The habitations lay together in villages, or ſolitary in the midſt of fields, ſurrounded by hedges. They cultivate rice, maize, potatoes, yams, and other roots. We have no where eaten better bananas than we got at this place. Here are likewiſe abundance of cocoa-nuts, citrons, mangle-apples, and ananas or pine-apples. All the people are very tawny, of a ſhort ſtature, and ugly. Their language, the ſame

 as

as that of the Molucca iſles, is the Malays, and their religion the Mahometan. They ſeem to have a great experience in their trade, but are gentle and honeſt. They offered us for ſale ſome pieces of coloured but very coarſe cotton. I ſhewed them ſome nutmegs and cloves, and aſked them to give me ſome. They anſwered that they had ſome dried in their houſes, and that whenever they wanted any, they went to get it upon Ceram, and in the neighbourhood of Banda, where the Dutch certainly are not the people to provide them with it. They told me that a great ſhip belonging to the company had paſſed through the ſtraits about ten days ago.

From ſun-riſing the wind was weak and contrary, varying from ſouth to S. W. I ſet ſail at half paſt ten, with the firſt of the flood, and we made many boards without gaining much way. At half paſt four o'clock in the afternoon we entered a paſſage, which is only four miles broad. It is formed on the ſide of Button, by a low, but much projecting point, and leaves to the northward a great bay, in which are three iſles. On the ſide of Pangaſani it is formed by ſeven or eight little iſles or keys, covered with wood, and lying at moſt half a quarter of a league from the coaſt. In one of our boards we ranged theſe keys almoſt within piſtol ſhot, ſounding cloſe to them with fifteen fathoms with-

out

out finding bottom. In the channel our soundings were in thirty-five, thirty, and twenty-seven fathoms, oozy bottom. We passed without, that is, on the west side of the three isles, upon the coast of Button. They are of a considerable size, and inhabited.

Second anchorage.

The coast of Pangasani here rises like an amphitheatre, with a low land at bottom, which I believe is often overflowed. I conclude it from seeing the islanders always fix their habitations upon the sides of the mountains. Perhaps too, as they are almost always at war with their neighbours, they choose to leave an interval of wood between their huts and the enemies who should attempt the landing. It seems even that they are dreaded by the inhabitants of Button, who consider them as pirates, upon whom no reliance can be had. Both parties are likewise used to wear the *criss* or dagger constantly in their girdle. At eight o'clock in the evening, the wind dying away entirely, we let go our stream-anchor in thirty-six fathoms, bottom of soft ooze. The Etoile anchored to the northward, nearer the land. Thus we had passed the first narrow gut or gullet.

Third and fourth anchorage.

The 14th, at eight o'clock in the morning, we weighed and made all the sail possible, the breeze being faint, and we plied till noon; when, upon seeing a bank to the S. S. W. we anchored in twenty fathoms, sand and ooze,

ooze, and I ſent a boat to ſound round the bank. In the morning ſeveral periaguas came alongſide, one among them diſplaying Dutch colours at her poop. At her approach, all the others retired to make way for her. She had on board one of their *orencaies* or chiefs. The company allow them their colours, and the right to carry them. At one o'clock in the afternoon we ſet ſail again, with a view to gain ſome leagues farther; but this was impoſſible, the wind being too light and ſcant; we loſt about half a league, and at half paſt three o'clock we let go our anchor again, in thirteen fathom bottom of ſand, ooze, ſhells, and coral.

Nautical advice.

Mean while M. de la Corre, whom I had ſent in the boat, to ſound between the bank and the ſhore, returned and made the following report: Near the bank there is eight or nine fathom of water; and as you go nearer the coaſt of Button, which is high and ſteep, oppoſite a fine bay, you always deepen your water, till you find no bottom with eighty fathom of line, almoſt mid-channel between the bank and the land. Conſequently, if one was becalmed in this part, there would be no anchoring, except near the bank. The bottom is, upon the whole, of a good quality hereabouts. Several other banks ly between this and the coaſt of Pangaſani. We cannot therefore ſufficiently recommend it, to keep as cloſe as poſſible to the land of Button in all this ſtrait.

The

The good anchorages are along this coast; it hides no danger; and, besides this, the winds most frequently blow from thence. From hence, almost to the out-let of the strait, it seems to be nothing but a chain of isles; but the reason of this is, its being intersected by many bays, which must form excellent ports.

Continuatio and descrip- tion of the straits.

The night was very fair and calm. The 15th, at five o'clock in the morning, we set sail with a breeze at E. S. E. and we steered so as to come close to the east of Button. At half past seven o'clock we doubled the bank, and the breeze dying away, I hoisted out the long-boat and barge, and made signal for the Etoile to do the same. The tide was favourable, and our boats towed us till three o'clock in the afternoon. We passed by two excellent bays, where I believe an anchorage might be found; but all along, and very near the high-shores, there is no bottom. At half after three o'clock the wind blew very fresh at E. S. E. and we made sail to find an anchorage near the narrow pass, by which one must go out of these straits. We did not yet discover any appearances of it. On the contrary, the farther we advanced, the less issue did we perceive. The lands of both shores, which over-lap here, appear as one continued coast, and do not so much as let one suspect any out-let.

At

At half paſt four o'clock we were oppoſite, and to the weſtward of a very open bay, and ſaw a boat of the country-people's, which ſeemed to advance into it, to the ſouthward. I ſent my barge after her, with orders to bring her to me, as I intended to get a pilot by this means. During this time our other boats were employed in ſounding. Somewhat off ſhore, and almoſt oppoſite the north point of the bay, they found twenty-five fathom, ſand and coral bottom; and after that they were out of ſoundings. I put about, then lay-to under top-ſails, in order to give the boats time to ſound. After paſſing by the entrance of the bay, you find bottom again, all along the land which joins to its ſoutherly point. Our boats made ſignal of 45, 40, 35, 29, and 28 fathom, oozy bottom; and we worked to gain this anchorage with the help of our long-boats. At half paſt five, we let go one of our bower-anchors there, in thirty-five fathom of water, bottom of ſoft ooze. The Etoile anchored to the ſouthward of us.

Fifth anchorage.

As we were juſt come to an anchor, my barge returned with the Malayo boat. He had not found it difficult to determine the latter to follow her; and we took an Indian, who aſked four ducatoons (about thirteen ſhillings ſterling) for conducting us; this bargain was ſoon concluded. The pilot came to ly on board, and his periagua went to wait for him on the other ſide of the paſſage. He told us, ſhe was going thither through the bottom of a neighbouring bay, from whence there was but a ſhort portage, or carrying-place, for the periagua. We were, upon the whole, enabled to do without the aſſiſtance of this pilot; for ſome moments before we anchored, the ſun ſhining very favourably upon the entrance of the gut, was the occaſion of our diſcovering the larboard point of the out-let, bearing S. S. W. 4° W. but one muſt gueſs which it is; for it laps over a double rock, which forms the ſtarboard point. Some of our gentlemen employed the reſt of the day in walking about on ſhore; they found no habitations near our anchorage. They likewiſe ſearched the woods, with which all this part is entirely covered, but found no intereſting production in it. They only met with a little bag near the ſhore, containing ſome dried nutmegs.

The next morning we began to heave a-head at half paſt two o'clock in the morning, and it was four before we got under ſail. We could hardly perceive any wind; however, being towed by our boats, we got to the entrance of the paſſage.

The water was then quite low on both ſhores; and as we had hitherto found that the flood-tide ſet from the northward, we expected the favourable return of it every inſtant; but we were much deceived in our hopes; for here the flood ſets from the ſouthward, at leaſt during this ſeaſon, and I know not which are the limits of the two powers. The wind had freſhened conſiderably, and was right aft. In vain did we with its aſſiſtance endeavour to ſtem the tide for an hour and a half; the Etoile, which firſt began to fall aſtern, anchored near the entrance of the paſſage, on the ſide of Button, in a kind of elbow, where the tide forms a ſort of eddy, and is not very ſenſibly felt. With the help of the wind I ſtill ſtruggled near an hour without loſing ground; but the wind having left me, I ſoon loſt a good mile, and anchored at one o'clock in the afternoon, in thirty fathom, bottom of ſand and coral. I kept all the ſails ſet, and ſteering the ſhip, in order to eaſe my anchor, which was only a light ſtream-anchor.

Sixth anchorage.

Leaving the Straits of Button; deſcription of the paſſage.

All this day our ſhips were ſurrounded with periaguas. They went to and fro as at a fair, being laden with refreſhments, curioſities, and pieces of cotton. This commerce was carried on without hindering our manœuvres. At four o'clock in the afternoon, the wind having freſhened, and it being almoſt high-water, we weighed our anchor, and with all our boats a-head of the frigate we entered the paſſage, and were followed by the Etoile, who was towed in the ſame manner by her boats. At half paſt five o'clock, the narroweſt paſs was happily cleared; and at half an hour after ſix we anchored without, in the bay called Bay of Bouton, under the Dutch ſettlement.

Let us now return to the deſcription of the paſſage. When you come from the northward, it does not begin to open till you are within a mile of it. The firſt object which ſtrikes one, on the ſide of Button, is a detached rock, hollow below, repreſenting exactly the figure of a tented galley *, half of whoſe cut water is carried away: the buſhes which cover it ſeem to form the tent; at low water, this galley joins to the bay; at high water, it is a little iſle. The land of Button, which is tolerably high in this part, is covered with houſes, and the ſea-ſhore full of encloſures, for catching fiſh in. The other ſhore of the paſſage is perpendicular; its point is diſtinguiſhable by two ſections, which form as it were two ſtories in the rock. After paſſing the galley, the lands on both ſides are quite ſteep, and in ſome parts even hang over the channel. One would think, that the god of the ſea had opened a paſſage here for his ſwelled waters, by a ſtroke of his trident. However, the aſpect of the coaſt is charming; that of Button is cultivated, riſes like an amphitheatre, and every where full of habitations, unleſs in ſuch places, which by their ſteepneſs exclude men from coming at them. The coaſt of Pangaſani, which is ſcarce any thing but one ſolid

* *Galere tentée:* we ſuppoſe M. de Bougainville means a galley, with her awnings ſpread. F.

 rock.

rock, is however covered with trees; but there appear only two or three habitations on it.

About a mile and a half to the northward of the passage, nearer Button than Pangasani, we find 20, 18, 15, 12, and 10 fathom, oozy bottom; as we advance to the southward in the channel, the bottom changes; there is sand and coral at different depths, from thirty-five to twelve fathom, and after that you are out of soundings.

Advice on this navigation.

The passage is about half a league long; its breadth varies from about 150 to 400 toises *, as we judged from appearance. The channel goes winding, and on the side of Pangasani; for at about two-thirds of its length, there is a fishery, which must be considered as a mark to avoid this shore, and range that of Button. In general it is necessary, as much as possible, to keep the middle of the gut. It is likewise fit, unless you have a brisk and favourable wind, to have your boats out a head, in order to steer well in the sinuosities of the channel. The current, upon the whole, is strong enough there to carry you past in a calm, and even when there is a light contrary wind; but it is not sufficient to overcome a brisk head-wind, and to permit your passing the channel, making short boards under top-sails. When you come out of the gullet,

* Of six feet French measure each.

the land of Button, ſeveral iſles to the S. W. of it, and the lands of Pangaſani look as the entrance of a great gulph. The beſt anchorage there, is oppoſite the Dutch ſettlement, about a mile off ſhore.

Our pilot from Button, had aſſiſted us with his knowledge, as far as was poſſible for a man who knows the particular ſituation of theſe parts, but underſtands nothing of the manœuvres of our ſhips. He took the greateſt care to inform us of all dangers, banks, and anchorages; only he always required, that we ſhould ſteer right in for the place where we wanted to go, making no allowance for our manner of hugging the wind, in order to be to windward, and to ſecure our point. He likewiſe believed, that we drew eight or ten fathom of water. In the morning another Indian came on board; he was an experienced old man, and we took him to be the father of our pilot. They ſtayed with us till the evening, and I ſent them back in one of my boats. Their habitation is near the Dutch factory. They would abſolutely eat none of our proviſions, not even bread; ſome bananas and betel were their only food. They were not ſo religious about drinking. Both the pilot and his father drank great quantities of brandy; being, doubtleſs, aſſured that Mahomed had only forbid them wine.

The

Great visit of the islanders.

The 17th, at five o'clock in the morning, we were under sail. The wind was on-end; at first faint, then pretty fresh, and we continued plying. At day-break we saw a whole swarm of periaguas come out from all parts; they soon surrounded the ships, and a commerce was established, with which all parties were pleased. The Indians, without doubt, disposed of their provisions to us, to much greater advantage than they could have done to the Dutch; however they sold them at a low rate, and all our sailors could get poultry, eggs, and fruit. Both ships were full of fowls, up as high as the tops. I must here advise those that pass this way, to provide themselves, if they can, with the coin which the Dutch make use of in the Moluccas; and especially with the plated pieces; the value of which is $2\frac{1}{2}$ sous. As the Indians did not know the coin which we had, they did not value the Spanish reals, nor our pieces of 12 and 24 sous; and often refused to take them. These Indians likewise offered to sell some finer and handsomer cottons than we had hitherto seen, and a prodigious quantity of cockatoes and parroquets, of the finest plumage.

Towards nine o'clock in the morning, we were visited by five *orencaies* of Button. They came in a boat, which looked like a European one, except its being managed with paddles instead of oars. They had a great

Dutch

Dutch flag at their poop. These *orencaies* are well dressed; they have long breeches, jackets with metal buttons, and turbans; whereas the other Indians are naked. They have likewise the distinctive mark which the company gives them; and which is a cane with a silver head, and this mark VOC on it. The oldest amongst them had above this mark an M, in the following manner, $\overset{\text{M}}{\text{VOC}}$. They came, as they said, to be obedient to the company, and when they heard that we were French, they were not disconcerted; and said, that they very willingly did homage to France. They accompanied their first compliments of welcoming us, with the gift of a roe-buck; I presented them in the king's name with some silk stuffs, which they divided into five lots; and I taught them how to distinguish the colours of our nation. I offered them some liquor; this was what they expected, and Mahomed permitted them to drink some to the health of the sovereign of Button, and to that of France; to the prosperity of the Dutch company, and to our happy voyage. They then offered me all the assistance they were able to give; and told me, that within three years there had passed at different times, three English ships, which they had furnished with water, wood, fowls, and fruit; that they were their friends, and that they conceived, we should be their friends also. That instant their glasses were filled, and they had already drank off ſeveral bumpers. They further informed me, that the king of Button reſided in this diſtrict; and I ſaw plainly that they were uſed to the more civilized manners of the capital. They call him Sultan *; and have certainly received that name from the Arabians, together with their religion. The Sultan is deſpotic and powerful, if power can be ſaid to conſiſt in the number of ſubjects; for his iſle is large and well peopled. The *orencaies*, after taking leave of us, made a viſit on board the Etoile. There they likewiſe drank to the health of their new friends, who were obliged to hand them down into their periaguas.

Situation of the Dutch at Button.

I aſked them when they were drinking, whether their iſle produced ſpices? and they anſwered in the negative; and I readily believe they ſpoke the truth, conſidering the weak ſettlement which the Dutch have here. This ſtation is compoſed of ſeven or eight bamboo huts, with a kind of palliſadoes, decorated by the pole of a tent. There a ſerjeant and three men reſide for the company. This coaſt, upon the whole, offers a moſt pleaſing proſpect; it is every where cultivated and covered with huts. The plantations of cocoa-nut trees are very frequent on it. The land riſes with a gentle ſlope, and every where offers cultivated and encloſed fields. The ſea-ſhore is all full of fiſheries. The coaſt, which is oppoſite Button, is no leſs pleaſing, nor leſs peopled.

* The word Sultan is not of Arabic, but of Tartarian origin; but early introduced into the Arabian language by the Turks that were in the ſervice of the Caliphs. F.

Our pilot likewiſe returned to ſee us in the morning, and brought me ſome cocoa-nuts, which were the beſt I had as yet taſted. He told me, that when the ſun ſhould be at its greateſt height, the S. E. breeze would be very freſh, and I gave him a good draught of brandy for ſuch good news. We actually ſaw all the periaguas retire towards eleven o'clock; they would not venture out to ſea at the approach of the briſk wind, which did not fail to blow as the Indian had foretold. A freſh and pretty ſtrong breeze at S. E. took us as we made a board upon an iſle to the weſt of Button; it permitted us to ſteer W. S. W. and made us gain a good way againſt the tide. I muſt here obſerve, that one muſt take heed of a bank, which runs pretty far out to ſea from the iſle of which I have juſt ſpoken. As we plied in the morning, we ſounded ſeveral times without finding bottom, with fifty fathoms of line.

Nautical advice.

At noon we obſerved in 5° 31′ 30″ ſouth lat. and this obſervation, together with that which we had made at the entrance of the ſtrait, ſerved to determine its length with preciſion. At three o'clock we perceived the ſouthern extremity of Pangaſani. We had ever ſince this morning ſeen the high mountains of the iſle of Cambona, on which there is a peak, whoſe ſummit riſes up above the clouds. About half an hour after four we diſcovered a part of the land of Celebes. We hoisted in our boats at sun-setting, and set all sails, steering W. S. W. till ten o'clock in the evening, when we stood W. by S. and we continued this course all night, with studding-sails set alow and aloft.

Remarks on this navigation.

My intention was to fall in with the isle of Saleyer, about three or four leagues from its northern point, that is, in 5° 55′ or 6° of latitude, in order afterwards to go in search of the strait of the same name, between this isle and that of Celebes, along which you sail without seeing it, as its coast almost from Pangasani forms a gulph of immense depth. It is likewise necessary to return in search of the strait of Saleyer, when you pass through the Toukan-bessie; and from the above details it must certainly be concluded, that the course through the strait of Button is in every respect preferable. It is one of the safest and most agreeable navigations that can be made. It joins all the advantages of the best harbour to excellent anchorage, and to the pleasure of making way at one's ease. We had now as great an abundance of fresh provisions on board our ships as there had been want before. The scurvy disappeared visibly; a great many fluxes were indeed complained of, occasioned by the change of food; this inconvenience, which is dangerous in the hot countries, where it commonly is converted into a bloody-flux, still more frequently becomes a severe sickness in the Moluccas.

Advantages of the preceding track.

Both

Both on ſhore and at ſea it is deadly there to ſleep in the open air, eſpecially when the dew falls.

Paſſing the ſtraits of Saleyer.

The 18th in the morning we did not ſee land, and I believe, that, during night, we loſt three leagues by the currents; we ſtill continued our courſe to W. by S. At half paſt nine o'clock we had a fair view of the high lands of Saleyer, bearing from W. S. W. to W. by N. and as we advanced, we diſcovered a leſs elevated point, which ſeems to terminate this iſle to the northward. I then ſteered from W. by N. ſucceſſively to N. W. by N. in order to view the ſtraits well. This paſſage, which is formed by the lands of Celebes and thoſe of Saleyer, is likewiſe made more narrow by three iſles which ſeem to ſhut it up. The Dutch call them Bougerones; and the paſſage, the Bout-ſaron. They have a ſettlement upon Saleyer, commanded at preſent by Jan Hendrik Voll, book-keeper.

Deſcription of this paſſage.

At noon we obſerved in 5° 55′ ſouth lat. At firſt, we thought we ſaw an iſland to the northward of the middle land, which we had taken for the point of Saleyer; but this is a pretty high land, terminated by a point which is connected with Saleyer, by an exceeding low neck of land. Afterwards we diſcovered at once two pretty long iſles, of a middle height, about four or five leagues aſunder. And laſtly, between thoſe two we perceived a third, which is very little and very low. The

The good paſſage is near this little iſle, either to the northward or ſouthward of it. I determined upon the latter, which appeared to me to be the largeſt. In order to facilitate the deſcription, we ſhall call the little iſle, *Iſle of the Paſſage*; and the two others, the one *South Iſland*, and the other *North Iſland*.

When we had ſufficiently viewed them, I lay-to at the beginning of night, to wait for the Etoile. She did not come up with us till eight o'clock in the evening, and we entered the paſſage, keeping in the middle of the channel, which is about ſix or ſeven miles broad. At half paſt nine o'clock we bore north and ſouth with the Iſle of the Paſſage, and the middle of South Iſland bore from ſouth to S. by E. I then ſteered W. by S. at one o'clock in the morning, then lay-to with the larboard-tack till four o'clock in the morning. Before, and in the paſſage, we ſounded ſeveral times with the hand-lead, finding no bottom with twenty and twenty-five fathoms of line. On the 19th at day-break we came near, and ranged the coaſt of Celebes at the diſtance of three or four miles. It is really difficult to ſee a finer country in the world. In the back-ground there appear high mountains, at the foot of which extends an immenſe plain, every where cultivated, and covered with houſes. The ſea-ſhore forms a continued plantation of cocoa-nut trees, and the eye of a ſailor, who

Deſcription of this part of Celebes.

who has but juſt left off ſalt proviſions, ſees with rapture great herds of cattle grazing in theſe agreeable plains, embelliſhed with groves at various diſtances. The population ſeems to be conſiderable in this part. At half an hour after noon we were oppoſite a great village, of which, the habitations, ſituated amidſt the cocoa-nut trees, for a conſiderable ſpace, followed the direction of the coaſt, along which you find eighteen and twenty fathoms of water, bottom of grey ſand; but this depth decreaſes as you approach the ſhore.

This ſouthern part of Celebes is terminated by three long points, which are level and low, and between which there are two pretty deep bays. Towards two o'clock we chaſed a Malayo boat, hoping to find ſomebody in it who might have practical knowledge of theſe ſhores. The boat immediately fled towards the ſhore, and when we joined her within reach of muſket-ſhot, ſhe was between the land and us, and we were in no more than ſeven fathoms of water. I fired three or four guns at her, which ſhe did not attend to. She certainly took us for a Dutch ſhip, and was afraid of ſlavery. Almoſt all the people of this coaſt are pirates, and the Dutch make ſlaves of them whenever they take any. Being obliged to abandon the purſuit of this boat, I ordered the Etoile's canoe to ſound a-head of us.

We

Difficulty of the navigation in this part.

We were at this time almoſt oppoſite the third point of Celebes, named Tanakeka, after which, the coaſt tends to N. N. W. Almoſt to the N. W. of this point are four iſles, of which the moſt conſiderable named Tanakeka, like the S. W. point of Celebes, is low, level, and about three leagues long. The three others, more northerly than theſe, are very ſmall. It was not neceſſary to double the dangerous ſhoal of *Brill* or the Spectacles, which I take to be north and ſouth of Tanakeka, at the diſtance of four or five leagues to the utmoſt. Two paſſages lay before us, one between point Tanakeka and the iſles, (and it is pretended that this is followed by the Dutch) the other between the iſle of Tanakeka and the Spectacles; I preferred the latter, through which the courſe is more ſimple, and which I took to be the wideſt.

I ordered the Etoile's boat to direct her courſe in ſuch a manner as to paſs within a league and a half of the iſle of Tanakeka, and I followed her under topſails, the Etoile keeping in our wake. We paſſed over eight, nine, ten, eleven and twelve fathoms of water, ſteering from W. N. W. to W. by N. and then weſt, when we came into thirteen, fourteen, fifteen, and ſixteen fathoms, the northermoſt iſle bearing N. N. E. I then recalled the Etoile's boat, and ſtood S. W. by S. ſounding every half hour, and always finding fifteen or ſixteen

teen fathoms, bottom of coarſe grey ſand and gravel. At ten o'clock in the evening, the depth encreaſed; at half paſt ten o'clock we ſounded in ſeventy fathoms, ſand and coral; then we found none with 120 fathoms of line. At midnight I made ſignal for the Etoile to hoiſt in her boat, and carry as much ſail as ſhe could, and I ſteered S. W. in order to paſs mid-channel, between the Spectacles and a bank called Saras, ſounding every hour without finding bottom. Whenever the wind is not briſk or favourable for doubling the Spectacles, it is neceſſary to anchor on the coaſt of Celebes, in one of the bays, and to wait for ſettled weather there; otherwiſe you run the riſk of being thrown upon this dangerous ſhoal by the currents, without your being able to prevent it.

The next day we ſaw no land; at ten o'clock we ſtood to W. S. W. and at noon had an obſervation in 6° 10′ ſouth latitude. Then reckoning that we had doubled the bank of Saras, at leaſt being ſure, by obſervation, of being to the ſouthward of it, I ſteered weſt, and after making five or ſix leagues by this courſe, I ſtood W. by N. ſounding every hour without finding bottom. Thus we kept in the channel between the Seſtenbank and the Hen (*Poule)*, to the northward, and the Pater-noſter and Tangayang to the ſouthward, carrying all ſails ſet, both night and day, in order to get

Continuatic of the direc tion of our courſe.

get time to ſound, by gaining upon the Etoile. I was told, that the currents here ſet towards the iſles and bank of Tangayang. By the obſervation at noon, which was in 5° 44′, we had, on the contrary, at leaſt nine minutes of difference north. The beſt advice I can give, is to keep ſuch a courſe as to be out of ſoundings; you are then ſure of being in the channel; if you approach too near the ſouthern iſles, you would begin to find only thirty fathom of water.

We made ſail all the day of the 21ſt, in order to view the iſles of Alambaï. The French charts mark three of them together, and a much larger one to the S. E. of them, ſeven leagues diſtant. This laſt does not exiſt where they place it; and the iſles of Alambaï are all the four iſles together. I reckoned myſelf in their latitude at ſun-ſet, and ſteered W. by S. till we had run the length of them. During day-time we had diſpenſed with ſounding. At eight o'clock in the evening we had forty fathom of water, bottom of ſand and ooze. We then ſtood S. W. by W. and W. S. W. till ſix in the morning; then reckoning that we had paſſed the iſles of Alambaï, we ſtood W. by S. till noon. During night we always found forty fathom, bottom of ſoft ooze, till four o'clock, when we found only thirty eight. At mid-night we ſaw a boat coming towards us; as ſoon as ſhe perceived us, ſhe hauled her wind, and would not bear down to us, though we twice fired a gun. These people are more afraid of the Dutch, than of the firing of guns. Another boat, which we saw in the morning, was not more curious to come near us. At noon we observed in 6° 8′ of latitude, and this observation further gave us a distance of 8′ north of our reckoning.

General remarks on th navigation.

We were now past all the dangers which are so much dreaded in the navigation from the Moluccas to Batavia. The Dutch take the greatest precautions to keep those charts secret by which they sail in these parts. It is probable that they magnify the dangers; at least I have seen very few in the straits of Button, Saleyer, and in the last passage we had now left, though all these three parts had been described to us at Boero, as perilous beyond measure. I own that this navigation would be much more difficult from east to west. The points of landfall to the eastward are not fine, and can easily be missed, whereas those to the west are fine and safe. However, in both courses, it is essential to have good observations of latitude every day. The want of this help might lead one into dangerous mistakes. We could not, in these last days, compute whether the currents set us eastward or westward, as we had had no bearings.

Inexactne of the kn charts of part.

I must here mention, that all the French charts of these parts cannot be depended upon. They are inac-

 curate,

curate, not only in regard to the position of the coasts and isles, but even in the essential latitudes. The straits of Button and Saleyer are extremely faulty; our charts even have omitted the three isles which make this last passage narrower, and those which ly to the N. N. W. of the isle of Tanakeka. M. d'Aprės, at least mentions, that he does not answer for the exactness of his chart of the Moluccas and Philippinas, because he had not been able to obtain satisfactory memoirs concerning that part. For the safety of navigators, I wish that all those, who compile charts, would display the same candour. The map which gave me the greatest assistance, is that of Asia, by M. d' Anville, published in 1752. It is very good from Ceram to the isles of Alambaï. On the whole course I have verified, by my observations, the exactness of his positions, and of the bearings which he gives to the most interesting parts of this difficult navigation. I shall add, that New Guinea and the isles of Papua come nearer the truth in this map, than in any other which I had in my hands. I do this justice to M. d' Anville's work with pleasure. I have known him particularly; and he seemed to me to be as good a citizen as he was a good critic, and a man of great erudition.

From the 22d in the morning, we continued our course W. by S. till the 23d, at eight o'clock in the morning, when we steered W. S. W. We found 47,

45, 42, and 41 fathom; and the bottom, I ſhall ſay it once for all, is here, and upon the whole coaſt of Java, an excellent bottom of ſoft ooze. We ſtill found ſeven minutes difference north by the altitude at noon, which we obſerved in 6° 24′. The Etoile had made ſignal of ſeeing the land by ſix o'clock in the morning; but the weather becoming ſqually, we did not then perceive it. After noon I ſhaped our courſe more to the ſouthward, and at two o'clock we diſcovered at maſt-head the north coaſt of the iſle of Maduré. At ſix o'clock we ſet it, bearing from S. E. by S. to W. ½ S. The horizon was too thick to enable us to compute at what diſtance it was. The ſoundings in the afternoon conſtantly gave forty fathom. We ſaw a great many fiſhing-boats, ſome of which were at anchor, and had thrown out their nets.

Sight of the iſle of Java.

The winds, during night, varied from S.E. to S.W. We ran cloſe-hauled, with the larboard tacks on board; and from ten o'clock in the evening had ſoundings in 28, 25, and 20 fathom. At nine o'clock in the morning, when we had approached the land, we found 17 fathom, and at noon only ten. The great lands of point Alang upon Java, then bore S. E. by S. of us, about two leagues; the iſle of Mandali S. W. 9° W. two miles; and the moſt weſterly lands, W. S. W. four leagues. Having theſe bearings, we obſerved in 6° 22′ 30″, which was pretty con-

conformable to our eſtimated latitude, pricking off our point at noon upon the chart of M. d' Après, according to the bearings I found.

Geographical obſervations.

1ſt, That the coaſt of Java is there placed nine or twelve minutes more to the ſouthward than it ought to be, by the mean reſult of our meridian obſervation.

2d, That the poſition of point Alang is not exact in it, as he makes it run W. S. W. and S. W. by W. whereas it really runs from the iſle of Mandali W. by S. for about 15 miles; after which it turns to the ſouthward, and forms a great gulph.

3d, That he gives too little extent to this part of the coaſt; and that if we had followed the bearings on his chart, we muſt, from noon to noon, have made thirteen miles leſs to the weſtward; either becauſe the coaſt had really ſo much more extent, or becauſe the currents ſet us to the eaſtward.

Meeting ſome Dutch ſhips.

Beſides a great number of fiſhing-boats, we ſaw in the morning four ſhips, of which two ſtood the ſame courſe as ourſelves, and diſplayed Dutch colours. Towards three o'clock we joined one of them, and ſpoke with her; ſhe was a ſnow from Malacca, bound for Japara. Her conſort, a three-maſted ſhip, likewiſe coming from Malacca, was bound for Saramang. They ſoon came to an anchor upon the coaſt. We ranged it, at the diſtance of about three quarters of a league, till

four

four o'clock in the evening. We then ſteered W. by N. in order not to get deeper into this gulph, and to paſs on the off ſide of a coral-bank, which is about five or ſix leagues off ſhore. As far as this part, the coaſt of Java is not much elevated near the ſea-ſhore, but in the interior parts we perceived high mountains. At half paſt five o'clock, the middle of the iſles of Carimon-Java bore N. 2° W. about eight leagues.

Courſe alon[g] Java.

We ſtood W. by N. till four o'clock in the morning, then weſt till noon. The day before we had ſounded in nine and ten fathom near the ſhore; we deepened our water by ſeven o'clock in the evening; when we found 30, and in the night 32, 34, and 35 fathom. At ſun-riſe we ſaw no lands, only ſome ſhips; and, as uſual, an infinite number of fiſhing-boats. Unluckily it was a calm almoſt the whole 25th, till five o'clock in the evening. I ſay unluckily, by ſo much the more, as it was neceſſary we ſhould have ſight of the land before night, in order to direct our courſe in conſequence thereof, between Point Indermay, and the Iſles of Rachit, and afterwards to keep towards the offing of ſome rocks under water, which are to the weſtward of them. From noon, when we had obſerved in 6° 26′ of latitude, we ſteered W. and W. by S. but the ſun ſet before we could ſee the land. Some of our people thought, but without any certainty, that they perceived the blue mountains,

tains, which are forty leagues off Batavia. From ſix o'clock in the evening to midnight, we ſteered W. and W. by N. ſounding every hour in twenty-five, twenty-four, twenty-one, twenty, and nineteen fathoms. At one o'clock in the morning we ran W. by N. from two o'clock to four, N. W. then N. W. by W. till ſix o'clock. My intention, expecting to be in the middle of the channel between the iſles of Rachit and the land of Java at one o'clock in the morning, was to get to the northward of the rocks. We ſounded thrice in twenty fathoms, than twenty two, next twenty-three, and I then reckoned myſelf three or four leagues to the N. N. W. of the iſles of Rachit.

Error in the reckoning of our courſe.

I was very much out in my reckoning. On the 26th, the rays of the riſing ſun ſhewed us the coaſt of Java, bearing from S. by W. to weſt, ſome degrees north, and at half paſt ſeven o'clock we ſaw from maſt-head the iſles of Rachit, about ſeven leagues diſtant, bearing N. N. W. and N. W. by N. Theſe bearings gave me a prodigious and dangerous difference with the chart of M. d'Après. But I ſuſpended my judgment till the obſervation at noon ſhould determine whether this difference was to be attributed to the currents, or whether the chart ought to be charged with it. I ſteered W. by N. and W. N. W. in order to view the coaſt well, it being in this part extremely low, and without any

moun-

mountains in the interior parts. The wind was at S. S. E. S. E. and E. pretty fresh.

Causes of this error.

At noon the southermost point of Indermay bore E. by S. 2° S. about four leagues distant; the middle of the isles of Rachit, N. E. five leagues distant, and the mean result of the altitude observed on board, placed us in 6° 12′ of latitude. By this observation, and the bearings, it seemed to me that the gulph between the isle of Mandali and point Indermay, is in the chart laid down less broad from E. to W. by twenty-two minutes than it really is, and that the coast is therein laid down 16′ more southerly than our observations place it. The same correction must take place in regard to the isles of Rachit, by adding, that the distance between these isles and the coast of Java, is at least two leagues greater than that which is expressed in the chart. In regard to the bearings of the several parts of the coast from each other, they appeared to me to be exact enough, as much as we could judge of it by our successive estimations made by sight, and as we ran along. Upon the whole, the differences above-mentioned are very dangerous for one who sails in this part in night-time.

All this morning we had found twenty-one, twenty-three, nineteen, and eighteen fathoms. The E. S. E. breeze continued, and we ranged the coast at three or four miles distance, in order to pass to the southward of the

the hidden rocks, of which I have already ſpoken, and which are laid down five or ſix leagues to the weſtward of the iſles of Rachit. At one o'clock in the afternoon, a boat which lay at anchor a head of us, made ſail upon the ſtarboard-tack, which made me think that the current then changed, and became contrary to us. We ſpoke with her at two o'clock; a Dutchman who commanded her, and who ſeemed to be the only white man on board, having ſome mulattoes with him, ſaid, he was bound for Amboina and Ternate; and that he came from Batavia, from whence he was twenty-ſix leagues by his reckoning. After coming out of the paſſage of Rachit, and paſſing within the rocks which are under water, I wanted to ſtand N. W. in order to double two ſand-banks, named Perilous Banks, which run pretty far out to ſea, between the points Indermay and Sidari. The wind would not admit of it, and as I could only ſtand W. N. W. I let go a ſtream anchor, at ſeven o'clock in the evening, in thirteen fathoms, oozy bottom, about a league off ſhore. We could only ply with very ſhort and unſafe tacks between the rocks under water on one ſide, and the perilous banks on the other. We had ſounded ſince noon in nineteen, fifteen, fourteen, and ten fathoms. Before we anchored, we made a ſhort board to the offing, which brought us into thirteen fathoms.

We

We weighed on the 27th, at two o'clock in the morning, with the land-breeze, which this night came from the weſt, whereas on the preceding nights they had veered all round from north to ſouth by the eaſt. Having ſteered N. W. we did not ſee the land again till eight o'clock in the morning, it being then very low, and almoſt overflowed; we kept the ſame courſe till noon, and from our ſetting ſail to that hour, our ſoundings varied from thirteen to ſixteen, twenty, twenty-two, twenty-three, and twenty-four fathoms. At half paſt ten o'clock we found a coral bottom; I ſounded again the moment after, and the bottom was oozy as uſual.

At noon we obſerved 5° 48′ of latitude; we could not ſee the land from the deck, as it is ſo very low. We ſet it from maſt-head, bearing from ſouth to S. W. by W. at the computed diſtance of five or ſix leagues. This day's obſervation, compared with the bearings, did not differ above two or three minutes, which this part of Java is placed too much ſouth in the chart of M. d'Après; but this difference is equal to nothing, becauſe, to make it real, we muſt ſuppoſe the computation of the diſtances of the bearings perfectly exact. The currents had ſtill ſet us to the northward, and I believe likewiſe to the weſtward.

The

New error in our reckoning.

The weather was very fine all day, and the wind favourable; in the afternoon I ſhaped our courſe a little more to the northward, in order to avoid the ſhallows of the point of Sidari. At midnight, thinking to have paſt them, we ſtood W. by S. and W. S W. then S. W. ſeeing that the water, which was nineteen fathoms at one o'clock, was ſucceſſively encreaſed to twenty-ſeven fathoms. At three o'clock in the morning we perceived an iſle, bearing N. W. ½ N. about three leagues. Being then convinced that I was more advanced than I at firſt thought, and even being afraid of paſſing by Batavia, I came to an anchor, in order to wait for day-light. At ſun-riſe we diſcerned all the iſles of the bay of Batavia; the iſle of Edam, on which there is a flag, bore S. E. by S. about four leagues, and the iſle of Onruſt, or of Careening, S. by W. ½ W. near five leagues: thus we were ten leagues more to the weſtward than we had thought; a difference which may have been occaſioned both by the currents, and by the inaccuracy with which the coaſt is laid down in the chart.

At half paſt ten o'clock in the morning I attempted to ſet ſail, but the wind dying away immediately, and the tide being contrary, I let go a ſtream-anchor under ſail. We weighed again at half an hour after noon, ſtanding in for the middle of the iſle of Edam, till we were within three quarters of a league of it. The cu-

 pola

pola of the great church at Batavia then bearing ſouth, we ſteered for it, paſſing between the beacons which indicate the channel. At ſix o'clock we anchored in the road in ſix fathoms, oozy bottom, without mooring, as it is uſual here to be content only with having another anchor ready to let go. An hour after, the Etoile anchored to the E. N. E. at two cable's lengths from us. Thus, after keeping the ſea for ten months and a half, we arrived on the 28th of September, 1768, at one of the fineſt colonies in the univerſe, where we all looked upon each other as having completed our voyage. Anchorage at Batavia.

Batavia, by my reckoning is in 6° 11′ ſouth latitude, and 104° 52′ eaſt longitude, from the meridian of Paris.

 CHAP.

C H A P. VIII.

Stay at Batavia, and particulars concerning the Moluccas.

THE unhealthy feafon, which here generally begins at the end of the monfoon, and the approach of the rainy weftern monfoon, determined us to make our ftay at Batavia as fhort as poffible. However, notwithftanding our impatient defire of leaving it, our wants forced us to pafs a certain number of days there, and the neceffity of getting fome bifcuit baked, which we did not find ready, detained us longer than we had imagined. On our arrival, there were thirteen or fourteen of the Dutch company's fhips in the road, one of which was a flag fhip. This is an old fhip which is left at this ftation; it has the jurifdiction of the road, and returns the falute of all the merchant fhips. I had already fent an officer to inform the governor-general of our arrival, when a barge from this flag fhip came on board, with a paper written in Dutch, which I knew nothing of. There was no officer in the barge, and the cockfwain, who doubtlefs acted in his ftead, afked me who we were, and required a certificate, written and figned by me. I anfwered him, that I had fent my decla-

Ceremonies at our arrival.

declaration on ſhore, and ſo I put him off. He returned ſoon after, inſiſting upon his firſt demand; I ſent him away once more with the ſame anſwer, and he put up with it. The officer who had been ſent to the general, did not return till nine o'clock in the evening. He had not ſeen his excellency, who was in the country, and he was brought before the *Sabandar*, or introducer of ſtrangers, who appointed him to return the next morning, and told him, that if I would come on ſhore, he would conduct me to the general.

Viſit to the general in t country.

Viſits are made very early in this country, on account of the exceſſive heat. We ſet out at ſix o'clock in the morning, conducted by the ſabandar M. Vanderluys, and we went to M. Vander Para, general of the Eaſt-Indies, who was at one of his country-houſes, about three leagues from Batavia. We found him a plain, but civil man, who received us perfectly well, and offered us all the aſſiſtance we could be in need of. He appeared neither ſurpriſed nor diſpleaſed at our having touched at the Moluccas; he even approved of the conduct of the chief reſident at Boero, and of his good proceedings in our behalf. He conſented to have our ſick put into the hoſpital of the company, and immediately ſent orders thither for their reception. As to the ſupplies which the king's ſhips were in want of, it was agreed, that we ſhould give in an account of our demands to the ſabandar, who ſhould be charged with providing us with every thing. One of the perquiſites of his place was to gain ſomething by us, and ſomething by the undertakers. When all was ſettled, the general aſked me, whether I would not ſalute the flag; I anſwered I would, on condition that the ſalute was returned gun for gun from the place. Nothing, ſays he, is more equitable, and the citadel has got the proper orders. As ſoon as I was returned on board, I ſaluted with fifteen guns, and the town anſwered with the ſame number.

I immediately ſent our ſick to the hoſpital, from both ſhips, being in number twenty-eight, ſome ſtill afflicted with the ſcurvy, but the greater part ill of a bloody-flux. We likewiſe prepared to give in to the ſabandar an account of our wants, in biſcuit, wine, flour, freſh meat, and pulſe; and I begged him to let us have our proviſion of water by the company's people. We at the ſame time thought of getting a lodging in town, during our ſtay: this we got in a great and fine houſe, here called the *inner logement*, where you are lodged and boarded for two rix dollars a day, ſervants not included, which amounts to about a piſtole of our money. This houſe belongs to the company, who let it to a private perſon, and by that means give him the excluſive privilege of lodging all ſtrangers. However,

men

men of war are not subject to this law, and therefore the officers of the Etoile went to lodge in a private family. We likewise hired several carriages, which one cannot absolutely do without in this great town, especially as we intended to visit its environs, which are infinitely more beautiful than the town itself. These hired carriages have room for two persons, are drawn by horses, and their price every day is somewhat more than ten francs (between eight and nine shillings sterling).

On the third day of our arrival, we went in a body to pay a visit of ceremony to the general, the sabandar having previously given him notice of it. He received us in another country-seat, named Jacatra, of which the distance from Batavia is only about a third of that of the house where I had been on the first day. The road which leads to it cannot be better compared than to the place called Boulevards, at Paris, supposing them to be embellished with a canal of running water on the right and the left. We ought to have made several other visits of ceremony; likewise introduced by the sabandar, namely, to the director-general, the president of justice, and the chief of the marine. M. Vanderluys told us nothing of it, and we only visited the latter. His title is Scopenhagen *. Though this

* This name is wretchedly disfigured from the Dutch, *Schout-by-Nacht*, which signifies *Rear Admiral*. F.

officer

officer only ranks as rear-admiral in the company's ſervice, the preſent is however vice-admiral of the ſtates general, by a particular favour of the ſtadtholder. This prince was willing thus to diſtinguiſh a man of quality, who, on account of his broken fortune, has been obliged to leave the ſervice of the ſtates, whom he has well ſerved, and to take the place which he now occupies.

The *Schout-by-Nacht* is a member of the high regency, in whoſe aſſembly he has a ſeat, and a vote in their debates for the marine affairs; he likewiſe enjoys all the honours which are due to the Edel-heers. He keeps a great retinue, lives very high, and makes himſelf amends for the bad moments he has often paſſed at ſea, by ſpending his time in a delicious villa.

Amuſements which are to be found at Batavia.

The principal inhabitants of Batavia endeavoured to make our ſtay agreeable to us. Great feaſts in the town and country, concerts, charming walks, the variety of objects united here, and moſt of them new to us, the ſight of the emporium of the richeſt commerce in the world; and, more than this, the appearance of ſeveral people who, though of entirely oppoſite manners, cuſtoms, and religion, however form one ſociety; every thing in fine concurred to charm the eye, inſtruct the navigator, and intereſt even the philoſopher. Here is likewiſe a pretty good play-houſe; we could judge only of of the theatre itſelf, which ſeemed handſome to us; as we did not underſtand the language, we had ſeen enough of it by going there once. We were much more curious to ſee the Chineſe comedies, though we did not underſtand any more what was ſaid there; it would not be very agreeable to ſee them every day, but one ought to ſee one of each kind. Independent of the great pieces which are repreſented on a theatre, there are every day ſmall pieces and pantomimes, repreſented on ſcaffolds, built at every corner in the Chineſe quarter. The Roman people called for bread and ſhows; the Chineſe muſt have commerce and farces. God forbid I ſhould ever again hear the declamations of their actors and actreſſes, which is always accompanied with ſome inſtruments. It is an overſtrained recitative accompanied, and I know of nothing that can be more ridiculous, except their geſtures. I muſt likewiſe obſerve, that I cannot properly ſpeak of their actors, becauſe the parts of men are always acted by women. I ſhall add too, and allow the reader to make what inferences he pleaſes, that I have ſeen blows as frequent on the Chineſe ſtage, and gain as much applauſe there, as at the Italian comedy, or at Nicolet's.

Beauty of environs.

We could never be tired with walking in the environs of Batavia. Every European, though he be uſed to live in the greateſt capitals, muſt be ſtruck with the mag-

 nificence

nificence of the country around it. This is adorned with houses and elegant gardens, which are kept in order, in that taste and with that neatness which is peculiarly observable in all the Dutch possessions. I can venture to assert that these environs surpass those of the greatest cities in France, and approach the magnificence of those of Paris. I ought not to omit mentioning a monument, which a private person has there erected to the Muses. Mr. Mohr, the first clergyman at Batavia, a man of immense riches, but more valuable on account of his knowledge and taste for the sciences, has built an observatory, in a garden belonging to one of his country-houses, which would be an ornament to any royal palace. This building, which is scarce completed, has cost prodigious sums. Its owner now does something still better, he makes observations in it. He has got the best instruments of all kinds from Europe, necessary for the nicest observations, and he is capable of making use of them. This astronomer, who is doubtless the richest of all the children of Urania, was charmed to see M. Verron. He desired he should pass the nights in his observatory; unluckily, not a single one has been favourable to their purposes. M. Mohr has observed the last transit of Venus, and has communicated his observations to the academy of Harlem; they will serve to determine the longitude of Batavia with precision.

2

Interior part of the town.

Though this city is really very fine, it is however far from anſwering what one may expect, after ſeeing its environs. We ſee few great buildings in it; but it is well laid out: the houſes are convenient and pleaſant; the ſtreets large, and adorned with a well embanked canal, and bordered with trees; the firſt to promote cleanlineſs, and the latter to procure conveniency by their ſhade. It is true, theſe canals keep up an unwholſome humidity, which renders the ſtay at Batavia pernicious to Europeans. The inſalubrity of this climate is likewiſe in part attributed to the bad quality of the water; therefore the rich people at Batavia drink nothing but Seltzer water, which they get from Holland at a vaſt expence. The ſtreets are not paved; but on each ſide there is a broad and fine foot-pavement of free-ſtone, or of bricks; and Dutch cleanlineſs conſtantly keeps it in the beſt repair. I do not pretend to give an exact and particular deſcription of Batavia; that ſubject has often been exhauſted. One may form an idea of that famous place, by knowing that it is built in the taſte of the fineſt towns in Holland, with this difference, that on account of the frequent earthquakes, the people cannot raiſe their houſes very high, and conſequently they have only one ſtory. I ſhall likewiſe not deſcribe the Chineſe camp, which is out of town, nor the police it is ſubject to, to, nor their cuſtoms, nor a number of other things, which have already been repeatedly ſaid by others.

Riches and luxury of the inhabitants.

The luxury which prevails at Batavia is very ſtriking; the magnificence and taſte, with which the interior parts of the houſes are decorated, are proofs of the riches of their inhabitants. We have however been told, that Batavia was not near ſo great as it had been. For ſome years paſt, the company have forbid private perſons to carry on the commerce between the two Indies, which was to them the ſource of an immenſe circulation of riches. I do not cenſure this new regulation of the company, as I do not know what advantages they may have in view in this prohibition. I only know, that the perſons in their ſervice ſtill know the ſecret of making thirty, forty, an hundred, and up to two hundred thouſand livres, of yearly revenues, of their places, to which the ſalaries of fifteen hundred, three thouſand, and at moſt, ſix thouſand livres are annexed. But almoſt all the inhabitants of Batavia are employed by the company. However it is certain, that the price of houſes, both in the town and country, is more than two thirds below their ancient value; yet Batavia will always remain more or leſs rich; both by means of the ſecret I have juſt ſpoken of, and becauſe thoſe who make a fortune here, find it difficult to bring it over to Europe. There are no other means of conveying it to Holland than through the

hands

hands of the company, who take charge of it at the rate of eight per cent. diſcount; but they take but a very little at a from each perſon. Beſides this, it is impoſſible to ſend over ſuch caſh by ſtealth; the ſpecie, which is current here, loſing twenty-eight per cent. in Europe. The company employs the emperor of Java to ſtrike a particular coin, which is the currency throughout India.

Particulars concerning the adminiſtration of the company

In no place in the world the different claſſes of people are leſs confounded together, than at Batavia; every one has his rank aſſigned to him; this is fixed unalterably by ſome exterior marks; and the ſtiff *etiquette* is more rigidly obſerved here than it ever was at any congreſs. The ranks of the different ſtates are the high regency, the court of juſtice, the clergy, the ſervants of the company, the officers of the marine, and, laſt of all, the military.

The high regency conſiſts of the general, who preſides there; of the counſellors of the Indies, whoſe title is *Edele-heeren*, of the preſident of the court of juſtice, and of the Schout-by-Nacht. They meet at the caſtle twice a week. The counſellors of the Indies are now ſixteen in number; but they are not all at Batavia. Some of them have the important governments of the Cape of Good Hope, of Ceylon, of the coaſt of Coromandel, of the eaſtern part of Java, of Macaſſar, and of Amboina, and they reſide there. Theſe *Edele-heercn* have the prerogative

rogative of gilding their carriages all over, and having two running-footmen before them; whereas every private person can only keep one. It is further settled, that all coaches must stop, when those of the *Edele-heeren* pass by; and the people within, either men or women, are obliged to rise up.

The general, besides this distinction, is alone permitted to go with six horses; he is always followed by a guard on horseback, or at least by the officers of that guard, and some of the private men; when he passes by, both men and women must step out of their carriages; and the coaches of none but those of the *Edele-heeren* can drive to the flight of steps before his door. I have seen some of them, who had good sense enough to laugh with us in private at all these pompous prerogatives.

The court of Justice decides without appeal in all civil and criminal causes. About twenty years ago, they condemned a governor of Ceylon to death. That Edeleheer was convicted of exercising horrible oppressions in his government, and was executed at Batavia, on the place opposite the citadel. The appointment of the general of the Indies, of the Edele-heeren, and of the members of the court of Justice, is made out in Europe. The general, and the high regency of Batavia, propose persons for the other employments, and their choice must always

ways be confirmed in Holland. However, the general has the right of giving away all the military preferments. One of the most considerable and best places, in point of emolument after the governments, is that of commissary of the country. This officer has the inspection over every thing, which forms the company's demesnes upon the Isle of Java, even over the possessions and conduct of the several sovereigns of the island; he has likewise an absolute jurisdiction over those Javanese, who are the company's subjects. The regulations of the police concerning them are very severe, and every considerable offence is rigorously punished. The constancy of the Javanese, in suffering the most barbarous torments, is incredible; but when they are executed, they must have white drawers on, and never be beheaded. If the company should refuse to have this complaisance for them, their authority would be in danger, and the Javanese would revolt. The reason of this is obvious: as, according to their tenets, they believe that they would meet in the other world with a bad reception, if they should arrive there without their heads, and without white drawers; they likewise dare to believe, that despotism has a power over them only in this world.

Another employment, which is much sought after, of which the functions are agreeable, and the revenues considerable

ſiderable, is that of Sabandar, or miniſter for foreigners. There are two of them, the ſabander of the Chriſtians, and that of the Pagans. The former is charged with every thing that regards the European foreigners. The latter is veſted with the affairs relative to all the divers nations of India, compriſing the Chineſe. Theſe laſt are the brokers of all the interior commerce of Batavia, where their numbers at preſent exceeds a hundred thouſand. The abundance which has reigned for ſome years paſt in the markets of this great city, is likewiſe owing to their labour and care. In general, the order of employments in the company's ſervice is as follows: aſſiſtant, book-keeper, under merchant, merchant, great-merchant, governor. All theſe civil degrees have a uniform, and the military ranks have a kind of correſpondence with them. Thus for example; the major ranks as great-merchant, the captain as under-merchant, &c. but the military can never come to any places in the adminiſtration, without changing their condition. It is very natural, that in a trading company, the military body ſhould have no influence at all; they are there looked upon merely as a body who are kept in pay; and this idea is here ſo much the more applicable, as it conſiſts entirely of ſtrangers.

Order of the places in the ſervice of the company.

The company poſſeſſes, in their own right, a conſiderable part of the iſle of Java. All the north coaſt, to

[D]emeſnes of [th]e company [u]pon the iſle [o]f Java.

6 the

the eastward of Batavia, belongs to them. They have added, several years ago, to their possessions, the isle of Maduré, of which the sovereign had revolted; and the son is at present the governor of that isle, where his father had been king. The company have likewise profited of the revolt of the king of Balimbuan, in order to appropriate to themselves that fine province, which forms the eastermost point of Java. That prince, who was the brother of the emperor, ashamed of being subject to merchants, and by the advice, as it is said, of the English, (who furnished him with arms and gunpowder, and even built him a fort) attempted to throw off the yoke. The company spent two years, and great sums, in conquering him, and had concluded the war but two months before we came to Batavia. The Dutch had been worsted in the first battle; but in the second, the Indian prince had been taken with all his family, and conducted to the citadel of Batavia, where he died a few days after. His son, and the other persons of that unhappy family, were to be put aboard the first vessels, and brought to the Cape of Good Hope, where they will end their days upon the Isle of Roben.

Number o[f] principalit[ies] into whic[h] the isle o[f] Java is di[-]vided.

The remaining part of the isle of Java is divided into several kingdoms. The emperor of Java, whose residence lies in the southern part of the island, has the first rank; next to him is the sultan of Mataran, and the

 king

king of Bantam. Tſeribon is governed by three kings, vaſſals of the company, whoſe conſent is likewiſe neceſſary to all the other ſovereigns on the iſle, for mounting their tottering thrones. They place a European guard round every one of theſe kings, and oblige them to anſwer for their perſons. The company have likewiſe four fortified factories in the emperor's dominion; one in the ſultan's, four in Bantam, and two in Tſeribon. Theſe ſovereigns are obliged to furniſh the company with proviſions, at a certain rate fixed by the latter. The company receives rice, ſugar, coffee, tin, and arrack from them; and again have the excluſive right of furniſhing them with opium, of which the Javaneſe conſume great quantities, and the ſale of which brings in conſiderable profits.

Commerce of Batavia.

Batavia is the emporium or ſtaple of all the productions of the Moluccas. The whole crop of ſpices is carried thither: the ſhips are annually laden with as much as is neceſſary for the conſumption in Europe, and what remains is burnt. This commerce alone forms the riches, and I may ſay aſſures the exiſtence of the Dutch Eaſt India Company; it enables them to bear, not only the immenſe expences, which they muſt incur, but likewiſe the depredations of the people whom they employ, and which often come to as much as the expences themſelves. They accordingly direct their principal

cares

cares to this exclusive commerce, and that of Ceylon. I shall say nothing of Ceylon, because I do not know that isle; the company have just put an end there to an expensive war, with more success than to another in the Persian gulph, where all their factories have been destroyed. But as we are almost the only ships of the king that penetrated into the Moluccas, I must beg leave to give some particulars concerning the present state of that important part of the world, which is kept from the knowledge of other nations by the silence of the Dutch, and its great distance.

Particulars concerning the Moluccas.

Formerly they only comprised under the name of Moluccas, the little isles situated almost under the line between 15′ S. lat. and 50 N. lat. along the western coast of Gilolo; of which the most considerable were Ternate, Tidor, Mothier or Mothir, Machian, and Bachian. By degrees that name became common to all the isles which produced spices. Banda, Amboina, Ceram, Boero, and all the adjacent isles are ranged under the same denomination, under which some have unsuccessfully attempted to bring Bouton and Celebes. The Dutch now divide these countries, which they call the *Countries of the East*, into four principal governments, from which the other factories depend, and which again stand under the high regency of Batavia. These four governments are Amboina, Banda, Ternate, and Macassar.

 Am-

Government of Amboina.

Amboina, of which an Edel-heer is governor, has six factories dependent upon it, viz. on Amboina itself, Hila, and Larique; of which the chiefs rank, the one as merchant, and the other as under-merchant; to the westward of Amboina, the isles of Manipa and Boero, on the former of which is only a book-keeper, on the latter our benefactor Hendrick Ouman, under-merchant; Haroeko, a little isle, nearly to the E. S. E. of Amboina, where an under-merchant resides; and, lastly, Saparoea, an isle likewise to the S. E. and about 15 leagues off Amboina. There resides a merchant, in whose dependency is the little isle of Neeslaw, whither he sends a serjeant and fifteen men: there is a little fort built upon a rock at Saparoea, and a good anchorage in a fine bay. This isle, and that of Neeslaw, could furnish a whole ship's lading of cloves. All the forces of the government of Amboina actually consist of no more than 150 men, under the command of a captain, a lieutenant, and five ensigns: they have likewise two artillery officers, and an engineer.

Government of Banda.

The government of Banda is more considerable, as to its fortifications, and its garrison is likewise more numerous; it consists of three hundred men, commanded by a captain, captain-lieutenant, two lieutenants, four ensigns, and one artillery officer. This garrison, the same as that of Amboina, and of the other

chief

chief places, ſupplies all the detached ſtations. The entrance to Banda is very difficult to thoſe who are unacquainted with it. It is neceſſary to range cloſe along the mountain of Gunongapi, on which there is a fort, taking care to avoid a bank of rocks, which muſt be left on the larboard ſide. The paſs is only a mile broad, and there are no ſoundings in it. You muſt then range along the bank, in order to get to the anchorge in eight or ten fathom under the fort London, where five or ſix ſhips can ly at anchor.

Three ſtations depend upon the government of Banda; Ouriën, where a book-keeper reſides; Wayer, where an under-merchant is ſtationed; and the iſle of Pulo Ry en Rhun, which is nearer Banda, and covered with nutmegs. A great-merchant commands upon this iſle, where the Dutch have a fort; none but ſloops can anchor in the harbour; and they muſt ly upon a bank, which prevents their approaching the fort. It would even be neceſſary (in caſe of an attack) to canonade it under ſail; for cloſe to the bank there is no bottom to be found. There is no freſh-water upon the iſle; the garriſon is obliged to get it from Banda. I believe that the Iſle of Arrow is likewiſe in the diſtrict of this government. There is a factory on it, with a ſerjeant and fifteen men, and the company get pearls from thence. Timor and Solor, though they are pretty near

it,

it, depend immediately upon Batavia. These isles furnish sandal-wood. It is singular enough, that the Portuguese should keep a station upon Timor; and still more singular, that they make but little advantage of it.

Government of Ternate.

Ternate has four principal factories in its dependency, viz. Gorontalo, Manado, Limbotto, and Xullabessie. The chiefs of the two first rank as under-merchants; the latter are only book-keepers; several little stations, commanded by serjeants, likewise depend upon it. Two hundred and fifty men are garrisoned in the government of Ternate, under the command of a captain, a lieutenant, nine ensigns, and one artillery officer.

Government of Macassar.

The government of Macassar, upon the isle of Celebes, which is occupied by an Edel-heer, has four factories in its department; Boelacomba en Bonthain, and Bima, where two under-merchants reside; Saleyer and Maros, of which the chiefs are only book-keepers. Macassar, or Jonpandam, is the strongest place in the Moluccas; however, the natives are careful to confine the Dutch there within the limits of their station. The garrison there consists of three hundred men, commanded by a captain, captain-lieutenant, two lieutenants, and seven ensigns; there is likewise an artillery officer.

There are no spices to be found within the district of this government, unless it is true that Button produces some, which I have not been able to ascertain. The intention

tention in eſtabliſhing it, was to make ſure of a paſſage, which is one of the keys to the Moluccas, and to open an advantageous trade with Celebes and Borneo. Theſe two great iſles furniſh the Dutch with gold, ſilk, cotton, precious ſorts of wood, and even diamonds, in return for iron, cloths, and other European or Indian merchandizes.

Dutch politics in the Moluccas.

This account of the different ſtations which the Dutch occupy in the Moluccas, is pretty exact. The police which they have there eſtabliſhed does honour to the underſtanding of thoſe who were then at the head of the company. When they had driven the Spaniards and Portugueſe from thence, by the moſt ſenſible combination of courage with patience, they well gueſſed that the expulſion of the Europeans from the Moluccas would not ſecure them the excluſive ſpice-trade. The great number of theſe iſles made it almoſt impoſſible for them to guard them all; and it was not leſs difficult to prevent an illicit intercourſe of theſe iſlanders with China, the Phillippinas, Macaſſar, and all ſmuggling veſſels or interlopers that ſhould attempt it. The company had ſtill more to fear, that ſome of the trees might be carried off, and that people might ſucceed in planting them elſewhere. They reſolved therefore to deſtroy as far as they could the ſpice trees in all the iſlands, only leaving them on ſome ſmall iſlands, which

which might eaſily be kept; then nothing remained, but to fortify well theſe precious depoſitories. They were obliged to keep thoſe ſovereigns in pay, whoſe revenues conſiſted chiefly of this drug, in order to engage them to conſent, that the fountain thereof ſhould be annihilated. Such is the ſubſidy of 20,000 rix-dollars, which the Dutch company pays annually the king of Ternate, and ſome other princes of the Moluccas. When they could not prevail on any one of theſe ſovereigns to burn his ſpice-plants, they burnt them in ſpite of him, if they were the ſtrongeſt; or elſe they annually bought up the green leaves of the trees, well knowing that they would periſh, after being for three years thus robbed of their foliage, which the Indians were doubtleſs ignorant of.

By this means, whilſt cinnamon is gathered upon Ceylon only, Banda alone has been conſecrated to the culture of nutmegs; Amboina, and Uleaſter, adjoining to it, to that of cloves, without its being allowed to cultivate either cloves at Banda, or nutmegs at Amboina. Theſe places furniſh more than the whole world can conſume. The other ſtations of the Dutch, in the Moluccas, are intended to prevent other nations from ſettling there, to make continual ſearches for diſcovering and burning all the ſpice-trees, and to furniſh ſubſiſtence for thoſe iſles where they are cultivated. Upon the whole, all the engineers and mariners employed in this part, are obliged when they leave the ſervice, to give up all their charts and plans, and to make oath that they keep none. It is not long ſince that an inhabitant of Batavia has been whipped, branded, and baniſhed to a diſtant iſle, for having ſhewed a plan of the Moluccas to an Engliſhman.

The ſpice-harveſt begins in December, and the ſhips which are deſtined to take in ladings of it, arrive at Amboina and Banda in the courſe of January, and go from thence for Batavia in April and May. Two ſhips likewiſe go annually to Ternate, and their voyages are regulated by the monſoons. There are likewiſe ſome ſnows of twelve, or fourteen guns, deſtined to cruize in theſe parts.

Every year the governors of Amboina and Banda aſſemble, towards the middle of September, all the orencaies or chiefs in their department. They at firſt give them feaſts and entertainments for ſeveral days; and then they ſet out with them in a kind of large boats, called *coracores*, in order to viſit their governments, and burn all the ſuperfluous ſpice-plants. The chiefs of every particular factory are obliged to come to their governors-general, and to accompany them on this viſitation, which generally ends with the end of October, or at the beginning of November; and the return from this tour is celebrated by new feſtivals. When we were

 at

at Boero, M. Ouman was preparing to ſet out for Amboina, with the Orencaies of his iſland.

The Dutch are now at war with the inhabitants of Ceram; an iſland that is very rich in cloves. Its inhabitants would not ſuffer their plants to be extirpated, and have driven the company from the principal ſtations which they occupied on their ground; they have only kept the little factory of Savaï, ſituated in the northern part of the iſle, where they keep a ſerjeant and fifteen men. The Cerameſe have fire-arms and gun-powder, and they all ſpeak the Malayo pretty well, beſides their national jargon. The inhabitants of Papua are likewiſe conſtantly at war with the company and their vaſſals. They have been ſeen in veſſels armed with pedereroes, and containing two hundred men. The king of Salviati *, which is one of their greateſt iſlands, has been taken by ſurprize, as he was going to do homage to the king of Ternate, whoſe vaſſal he was, and the Dutch keep him priſoner.

Nothing can be better contrived than the above plan and no meaſures could be better concerted for eſtabliſhing and keeping up an excluſive commerce. Accordingly the company have long enjoyed it; and owe that ſplendour to it, which makes

* *Salawati.* F.

them

them more like a powerful republic, than a ſociety of merchants. But I am much miſtaken, or the time is nigh at hand, when this commerce will receive a mortal ſtroke. I may venture to ſay, that to deſire the deſtruction of this excluſive trade, would be enough to effect it. The greateſt ſafety of the Dutch conſiſts in the ignorance of the reſt of Europe concerning the true ſtate of theſe iſles, and in the myſterious clouds which wrap this garden of the Heſperides in darkneſs. But there are difficulties which the force of man cannot overcome, and inconveniencies for which all his wiſdom cannot find a remedy. The Dutch may conſtruct reſpectable fortifications at Amboina and Banda; they may ſupply them with numerous garriſons; but when ſome years have elapſed, an almoſt periodical earthquake ruins theſe works to the very foundations; and every year the malignity of the climate carries off two thirds of the ſoldiers, mariners, and workmen which are ſent thither. Theſe are evils without remedy; the forts of Banda, which have thus been overthrown three years ago, are but juſt rebuilt; and thoſe of Amboina are ſtill in ruins. The company may likewiſe have been able to deſtroy in ſome iſles, a part of the known ſpices; but there are iſles which they do not know, and others too, which they are acquainted with, but which defend themſelves againſt their efforts.

 The

The English now frequent the Moluccas very much; and this is doubtless not done without some design. Several years ago, some small vessels sailed from Bencoolen, and came to examine the passages, and pick up the necessary intelligence concerning this difficult navigation. We have seen above, that the natives of Bouton told us of three English ships lately passing through those straits; we have likewise made mention of the assistance they gave to the unfortunate sovereign of Balimbuan; and it seems to be certain, that they likewise furnish the Ceramese with powder and arms; they had even built them a fort, which captain le Clerc told us he had destroyed, and in which he had found two pieces of cannon. In 1764, M. Watson, who commanded the Kingsberg, a frigate of twenty-six guns, came to the entrance of Savaï, obliged the people, by firing muskets at them, to give him a pilot, who could bring him to the anchorage, and committed many outrages in that weak factory. He likewise made some attempt against the people of Papua; but it did not succeed. His long-boat was seized by the Indians, and all the Europeans in it, among the rest, a son of lord Sandwich's, who was a midshipman, and commanded the boat, were fastened to posts, circumcised, and then cruelly murdered *.

* Lord Sandwich's son never was in any of these expeditions; it therefore is evident, that M. de Bougainville has been misinformed in regard to this particular. F.

It

It ſeems, upon the whole, as if the Engliſh do not mean to hide their projects from the Dutch company. About four years ago they formed a ſtation in one of the iſles of Papua, called *Soloo* or *Tafara*. M. Dalrymple, who founded it, was its firſt governor; but the Engliſh kept it only for three years. They have now abandoned it, and M. Dalrymple came to Batavia in 1768, on board the Patty, captain Dodwell, from whence he went to Bencoolen, where the Patty ſunk in the road *. This ſtation furniſhed bird's neſts, mother of pearl, ivory, pearls, and *tripans* or *ſwallops*, a kind of glue or froth, of which the Chineſe are very fond. What I find extraordinary is, their coming to ſell their cargoes at Batavia; which I know from the merchant who bought them. The ſame man aſſured me that the Engliſh likewiſe got ſpices by means of this ſtation; perhaps they obtained them from the Cerameſe. I cannot ſay why they have abandoned it. It is poſſible, that they may already have got a great number of ſpice-trees tranſplanted in one of their poſſeſſions in India, and that believing they were ſure of their ſucceſs, they have abandoned an expenſive ſtation, which is but too ſufficient to alarm one nation, and give information to another. At Batavia we had the firſt ac-

* Mr. Dalrymple never was at Batavia, nor Bencoolen; he left China in January 1765, and arrived in England in July 1765, ſince when he has never been out of the kingdom From whence it muſt be obvious, M. de Bougainville is entirely miſtaken in what he ſays concerning M. Dalrymple.

count of the ſhips, of which we had met with the traces ſeveral times on our voyage. Mr. Wallace arrived at Batavia in January 1768, and ſailed from thence again almoſt immediately. Mr. Carteret, who was involuntarily ſeparated from his chief, ſoon after leaving the ſtraits of Magalhaens, has made a much longer voyage, and his adventures I believe muſt have been far more complicated. He came to Macaſſar at the end of March of the ſame year, having loſt almoſt all his crew, and his ſhip being in a rotten condition. The Dutch would not bear him at Jonpandam, and ſent him back to Bontain, hardly conſenting to his taking Moors to replace the loſs of his people; after ſtaying two months in the iſle of Celebes, he came to Batavia on the 3d of June, careened there, and ſailed from thence the 15th of September, that is, only twelve days before we arrived there. M. Carteret has ſaid very little about his voyage here; however, he has mentioned enough to let the people know, that in a paſſage by him called St. George's ſtrait, he had had a fight with the Indians, whoſe arrows he ſhewed, with which they have wounded ſeveral of his people, and among the reſt, the next in command after him, who even left Batavia without being cured.

1768. October. Diſeaſes contracted at Batavia.

We had ſcarce been above eight or ten days at Batavia, when the diſeaſes began to make their appearance.

ance. From the beſt ſtate of health, in all appearance, people were in three days brought to the grave. Several of us fell ill of violent fevers, and our ſick found no relief at the hoſpital. I accelerated as much as I could the diſpatch of our affairs; but our ſabandar likewiſe falling ſick, and not being able to do any buſineſs, we met with difficulties and delays. I was not ready before the 16th of October to go out, and I weighed, in order to anchor without the road. The Etoile was to get her biſcuit on board that day. She completed the ſtowing of it in the night, and as ſoon as the wind permitted, ſhe came to anchor near us. Almoſt every officer on board my ſhip was already ſick, or felt a diſpoſition towards it. The number of fluxes had not decreaſed among the crews, and if we had made a longer ſtay at Batavia, it would certainly have made greater havock among us than the whole voyage. Our man from Taiti, who had doubtleſs been ſheltered from the influence of the climate by the extaſy into which every thing that he ſaw threw him, fell ſick during the laſt days, and his illneſs has been of a long duration, though his docility in taking phyſick was quite equal to that of a man born at Paris: however, when he ſpeaks of Batavia, he always calls it the land which kills, *enoua maté*.

CHAP.

CHAP. IX.

Departure from Batavia; touching at the isle of France; return to France.

THE 16th of October, I set sail alone from the road of Batavia, in order to anchor in seven fathom and a half, bottom of soft ooze, about a league in the offing. I was thus half a mile W. N. W. of the beacon, which is left on the starboard side, when you enter the road of Batavia. The isle of Edam bore N. N. E. ½ E. three leagues. Onrust, N. W. by W. two leagues and a third. Rotterdam, N. 2° W. a league and a half. The Etoile, who could not get her bread before it was late, weighed at three o'clock in the morning; and steering for the lights, which I kept lighted all night, she came to an anchor near me.

Particulars concerning the course which must be taken in going out from Batavia.

As the course for leaving Batavia is interesting, I hope I shall be allowed to mention the particulars of that which I have taken. On the 17th we were under sail, by five o'clock in the morning, and we steered N. by E. in order to pass to the eastward of the isle of Rotterdam, about half a league; then N. W. by N. in order to pass to the southward of Horn and Harlem; then W. by by N. and W. by S. to range to the northward of the isles of Amsterdam and Middleburg, upon the last of which there is a flag; then west, leaving on the starboard side a beacon, placed south of the Small Cambuis. At noon we observed in 5° 55′ of south latitude, and we were then north and south with the S. E. point of the Great Cambuis, about one mile. From thence I steered between two beacons, placed, the one to the southward of the N. W. point of the Great Cambuis, the other east and west of the isle of Anthropophagi, or Canibals, otherwise called Pulo Laki. Then you range the coast at what distance you will or can. At half past five o'clock, the currents setting us towards the shore, I let go a stream-anchor in eleven fathoms, oozy bottom, the N. W. point of the bay of Bantam bearing W. 9° N. about five leagues, and the middle of Pulo Baby, N. W. ½ W. three leagues.

In order to sail out of Batavia, there is another passage besides that which I have taken. When you leave the road, range the coast of Java, leaving on the larboard side a buoy, which serves as a beacon, about two leagues and a half from the town; then you range the isle of Kepert to the southward; you follow the direction of the coast, and pass between two beacons, situated, the one to the southward of Middelburg island, the other opposite this, on a bank which joins to the point of the main land; you then find the beacon, which lies to the ſouthward of the ſmall Cambuis, and then the two routes unite. The particular chart which I give of the run from Batavia, exactly points out both tracks.

Clearing the ſtraits of Sonda.

The 18th, at two o'clock in the morning we were under ſail; but we were forced to anchor again in the evening: it was not till the 19th in the afternoon that we cleared the ſtraits of Sonda, paſſing to the northward of Prince's iſland. At noon we obſerved in 6° 30′ ſouth latitude, and at four o'clock in the afternoon, being about four leagues off the N. W. point of Prince's iſland, I took my departure upon the chart of M. d'Après, in 6° 21′ ſouth lat. and 102° eaſt longitude, from the meridian of Paris. In general, you can anchor every where along the coaſt of Java. The Dutch keep ſome ſmall ſtations on it, at ſhort diſtances from each other, and every ſtation has orders to ſend a ſoldier on board the ſhips which paſs, with a regiſter, on which he begs that the ſhip's name, from whence ſhe come, and whither ſhe is bound, may be inſcribed. You put into this regiſter what you pleaſe; but I am far from blaming the cuſtom of keeping it, as it may be the means of getting news of a ſhip, concerning which, one is often in great anxiety, and as the ſoldier who carries it on board always brings along with him fowls, turtle, and other refreſhments, which he turns to

good

good account. There was now no longer any ſcorbutic complaint, at leaſt, no apparent one on board my ſhips; but ſeveral of the crew were ill of a bloody-flux. I therefore reſolved to ſhape my courſe for the Iſle of France, without waiting for the Etoile, and on the 20th I made her the ſignal for that purpoſe.

Run to the iſle of France

In this run we found nothing remarkable, except the fine weather, which has much ſhortened the voyage. We had conſtantly a very freſh wind at S. E. Indeed we ſtood in need of it, for the number of the diſeaſed encreaſed daily, they recovered but ſlowly, and beſides the bloody-flux, ſome were likewiſe afflicted with hot fevers, of which one of my carpenters died in the night between the 30th and the 31ſt. My maſts likewiſe gave me much concern; I had reaſon to fear that the main-maſt would break five or ſix feet below the cat-harpings; we fiſhed it, and to eaſe it, we got down the main-top-gallant-maſt, and always kept two reefs in the main-top-ſail. Theſe precautions conſiderably retarded our run; yet notwithſtanding this, on the 18th day after leaving Batavia, we got ſight of the Iſle of Rodrigue *, and the ſecond day after that, of the Iſle of France.

1768. November.

Sight of the iſle of Rodrigue.

The 5th of November, at four o'clock in the evening, we were north and ſouth of the north eaſt point of the

* Diego Rays. F.

Isle of Rodrigue, whence I concluded the following difference in our reckoning from Prince's island to Rodrigue. M. Pingré has there observed 60° 52′ east longitude from Paris, and at four o'clock, I was, by my reckoning, in 61° 26′. These supposing, that the observation made upon the isle at the habitation, had been taken two minutes to the westward of the point with which I bore north and south at four o'clock, my difference in a run of twelve hundred leagues, was thirty-four minutes a-stern of the ship; the difference of the observations made on the 3d, by M. Verron, gave for the same time 1° 12′ a head of the ship.

Land-fall at the Isle of France.

We had sight of Round Island the 7th at noon; at five o'clock in the evening we bore north and south with its middle. We fired some guns at the beginning of night, hoping that the fire on the Cannoniers Point would be lighted; but this fire, which M. d'Aprés mentions in his instructions, is now never lighted; so that, after doubling the Coin de Mire, which you may range as close as you please, I was much embarrassed in order to avoid a dangerous shoal, which runs above half a league out into the sea off the Cannoniers Point. I kept plying, in order to keep to windward of the port, firing a gun from time to time; at last, between eleven and twelve o'clock at night, one of the pilots of the harbour, who are paid by the king, came on board. I then

6 thought

thought I was out of danger, and had given him the charge of the ſhip, when at half paſt three o'clock he run us a-ground, near the Bay of Tombs. Luckily there was no ſwell; and the manœuvre which we quickly made, in order to endeavour to caſt the ſhip off ſhore, ſucceeded; but it may eaſily be conceived, how great our grief would have been, if after happily avoiding ſo many dangers, we had been caſt away cloſe to our port, through the fault of an ignorant fellow, to whom we were obliged to leave the management of the ſhip, by the regulation of the ſervice. We got off with the loſs of only forty five feet of our falſe keel, which was carried away.

Danger which the frigate runs.

Nautical advice.

This accident, of which we had like to have been the victims, gives me an opportunity of making the following reflection: When you are bound for the Iſle of France, and ſee that it is impoſſible to reach the entrance of the port in day-time, prudence requires, that you muſt take care in time, not to be too much entangled with the land. It is neceſſary to keep all night on the off ſide, and to windward of Round-iſland, not lying-to, but plying to windward, under a good deal of ſail, on account of the currents. Beſides, there is anchorage between the little iſles; we have found from thirty to twenty-five fathom there, and a ſandy bottom; but one muſt only anchor there in an extreme caſe of neceſſity.

On

Anchorage of the Isle of France.

On the 8th, in the morning, we entered the port, where we moored that day. The Etoile appeared at six o'clock in the evening, but could not come in till the next morning. Here we found our reckoning was a day too late, and we again followed the date of the whole world.

Particulars of our proceedings there.

The first day of my arrival, I sent all my sick people to the hospital, I gave in an account of what I wanted in provisions and stores, and we immediately fell to work in preparing the frigate for heaving down. I took all the workmen in the port, that could be spared, and those of the Etoile, being determined to depart as soon as I should be ready. The 16th and 18th we breamed the frigate. We found her sheathing worm-eaten, but her bottom was as sound as when she came off the stocks.

We were obliged to change some of our masts here. Our main-mast had a defect in the heel, and therefore might give way there, as well as in the head, where the main-piece was broken. I got a main-mast all off one piece, two top-masts, anchors, cables, and some twine, which we were in absolute want of. I returned my old provisions into the king's stores, and took others for five months. I likewise delivered to M. Poivre, the *intendant* of the Isle of France, all the iron and nails embarked on board the Etoile; my alembic and recipient,

many

many medicines, and a number of merchandiſes, which now became uſeleſs to us, and were wanted in this colony. I likewiſe gave three and twenty ſoldiers to the legion, as they aſked my leave to be incorporated in it. Meſſieurs Commerçon and Verron, both conſented to defer their return to France; the former, in order to enquire into the natural hiſtory of theſe iſles, and of Madagaſcar; the latter, in order to be more ready to go and obſerve the tranſit of Venus in India; I was likewiſe deſired to leave behind M. de Romainville, an engineer, ſome young volunteers, and ſome under-pilots, for the navigation in the ſeveral parts of India.

Loſs of two officers.

We were happy, after ſo long a voyage, to be ſtill in a condition to enrich this colony with men and neceſſary goods. The joy which I felt on this occaſion, was cruelly converted into grief, by the loſs which we here ſuffered, by the death of the chevalier du Bouchage, enſign of the king's ſhips, and a man of diſtinguiſhed merit, who joined all the qualities of the heart and mind which endear a man to his friends, to that knowledge which forms a complete ſea-officer. The friendly care and all the ſkill of M. de la Porte, our ſurgeon, could not ſave him. He expired in my arms, the 19th of November, of a flux, which had begun at Batavia. A few days after, a young ſon of M. le Moyne, *commiſſaire ordonnateur* of the marine, who embarked as a volun-

volunteer with me, and had lately been made a *garde de la marine* *, died of a pectoral diſeaſe.

In the Iſle of France I admired the forges, which have been eſtabliſhed there by Meſſrs. Roſting and Hermans There are few ſo fine ones in Europe, and the iron which they make is of the beſt kind. It is inconceiveable how much perſeverance, and how great abilities have been neceſſary to make this undertaking more complete, and what ſums it has coſt. He has now nine hundred negroes, from which M. Hermans has drawn out and exerciſed a battalion of two hundred men, who are animated by a kind of ambition. They are very nice in the choice of their comrades, and refuſe to admit all thoſe who have been guilty of the leaſt roguery. Thus we ſee ſentiments of honour combined with ſlavery †.

* Equal to our midſhipman. F.

† We are very ready to do juſtice to Mr. Bougainville, when he preſents us with a new and intereſting obſervation; but when he, without the leaſt neceſſity, becomes the advocate of tyranny and oppreſſion, we cannot let theſe ſentiments paſs unnoticed. It would have appeared to us impoſſible, that ſuch an idea as this could enter into any man's head who is in his right ſenſes: he wrote down this ſtrange aſſertion, either being carried away by the itch to ſay ſomething extraordinary and paradoxical, or in order to make ſlavery more tolerable to his fellow Frenchmen— Slavery endeavours to extirpate and to ſmother all ſentiments of honour, which only can operate in the breaſt of a really free man; true honour, therefore, and ſlavery, are in direct oppoſition, and can be combined as little as fire and water. If Mr. B. threw this ſentence out, in order to alleviate the yoke of tyranny his country groans under, we think we could excuſe it in ſome meaſure, as he would then act from principles of humanity. But if the irreſiſtible deſire of ſaying ſomething new was the prevalent motive with him, it has much the appearance as if he were willing to inſult the poor victims of deſpotiſm. The generous and amiable character which M. B. from other inſtances appears in, prompts us to wiſh, that this ſentence had been omitted by him. F.

During our ſtay here, we conſtantly enjoyed the faireſt weather imaginable. The 5th of December, the ſky began to be covered with thick clouds, the mountains were wrapt in fogs; and every thing announced the approaching ſeaſon of rain, and the hurricane which is felt in theſe iſles almoſt every year. The 10th I was ready to ſet ſail. The rain and the wind right on end did not allow it. I could not ſail till the 12th in the morning, leaving the Etoile juſt when ſhe was going to be careened. This veſſel could not be fit for going out before the end of the month, and our junction was now no longer neceſſary. This ſtore-ſhip left the iſle of France towards the end of December, and arrived in France a month after me. I took my departure at noon, in the obſerved S. lat. of 20° 22′ and 54° 40′ eaſt longitude from Paris.

1768. December.

Departure from the Iſle of France.

The weather was at firſt very cloudy, with ſqualls and rain. We could not ſee the iſle of Bourbon. As we got further from the land, the weather cleared up by degrees. The wind was fair and blew freſh; but our new main-maſt ſoon cauſed us as much anxiety as the firſt. It deſcribed ſo conſiderable an arch at the head, that I durſt not make uſe of the top gallant-ſail, nor carry the top-ſails hoiſted up.

Run to the Cape of Good Hop[e]

From the 22d of December to the 8th of January, we had a conſtant head-wind, bad weather, or calms. I was

Bad weat[her] which we meet wit[h]

I was told, that theſe weſt winds were quite without example at this ſeaſon. They however retarded us for a fortnight ſucceſſively, during which we kept trying or beating to windward with a very great ſea. We got ſight of the coaſt of Africa before we had any ſoundings. When we firſt ſaw this land, which we took to be the Cape of Shoals, *(Cabo dos Baxos)* we had no bottom. On the 30th we ſounded 78 fathom, and from that day we kept on Bank Aguilhas, being almoſt conſtantly in ſight of the land. We ſoon fell in with ſeveral Dutch ſhips, of the Batavia fleet; their fore runner ſet ſail from thence on the 20th of October, and the fleet the 6th of November; the Dutch were ſtill more ſurpriſed than we were at the weſterly winds blowing ſo much out of ſeaſon.

1768. January.

At laſt, on the 18th of January in the morning, we had ſight of Cape Falſe, and ſoon after of the land of the Cape of Good Hope. I muſt here obſerve, that five leagues E. S. E. of Cape Falſe, there is a very dangerous rock under the water; that to the eaſtward of the Cape of Good Hope, there is a reef extending about one third of a league to the offing, and that at the foot of the Cape itſelf there is a rock running out to ſea to the ſame diſtance. I was come up with a Dutch ſhip, which I had perceived in the morning, and I had ſhortened ſail, in order not to paſs by her, but to follow her if ſhe intended to enter in the night-time. At ſeven o'clock in the evening,

Nautical advice.

evening, she took in her top-gallant-sails, studding-sails, and even top-sails; I then stood out to sea, and plyed all night, with a very fresh southerly wind, varying from S. S. E. to S. S. W.

At day-break, the currents had set us near nine leagues to the W. N. W. the Dutch ship was above four leagues to the leeward of us, and we were obliged to croud sail, in order to make good again what we had lost. Therefore those who must pass the night on their boards, with the intention of entering the bay of the Cape in the morning, would do well to bring-to at the eastern point of the Cape of Good Hope, keeping about three leagues off shore; being in this position, the currents will set them in a good situation for entering early in the morning. At nine o'clock in the morning we anchored in Table-bay, at the Cape, at the head of the road, and we moored N. N. E. and S. S. W. Here were fourteen ships of several nations, and several others arrived during our stay. Captain Carteret had sailed from hence on Epiphany-day. We saluted the town with fifteen guns, and they returned the salute with an equal number.

We touc the Cape Good H

We had all possible reasons to be content with the governor and inhabitants of the Cape of Good Hope; they were desirous of procuring us all that is useful and agreeable. I shall not stop to describe this place, which every body knows. The Cape immediately depends upon Europe, and not upon Batavia, neither with regard to its civil and military adminiſtration, nor to the appointment of perſons to places. It is even ſufficient to have had an employment at the Cape, to exclude one from obtaining one at Batavia. However, the council of the Cape correſponds with that of Batavia, with regard to commercial affairs. It conſiſts of eight perſons, among which is the governor, who is the preſident. The governor does not belong to the court of juſtice, where the ſecond in command preſides; he only ſigns the ſentences of death.

There is a military ſtation at Falſe Bay, and one at the bay of Saldagna. The latter, which forms an excellent harbour, ſheltered from all winds, could not be made the chief place, becauſe it has no water. They are now working to encreaſe the ſettlement at Falſe Bay; there the ſhips anchor in winter, when they are forbid lying in the bay of the Cape. There you find the ſame aſſiſtance, and every thing as cheap as at the Cape itſelf. The diſtance over land of theſe two places, is eight leagues, and the road very bad.

Particulars concerning the vineyards at Conſtantia.

Nearly half way between them both is the diſtrict of Conſtantia, which produces the famous wine of that name. This vineyard, where they cultivate the Spaniſh muſcade vines, is very ſmall, but it is not true that it belongs

belongs to the company, or that it is ſurrounded, as people believe here, by walls, and watched. It is diſtinguiſhed into High and Little Conſtantia, ſeparated by a hedge, and belonging to two different proprietors. The wine which is made there is nearly alike in quality, though each of the two Conſtantias has its partiſans. In common years they make a hundred and twenty or a hundred and thirty *barriques* of this wine, of which the company takes a third at a ſtated price, and the reſt is ſold to every buyer that offers. The price at preſent is thirty piaſtres or dollars the barrel of ſeventy bottles of white wine, and thirty-five piaſtres for the ſame quantity of red wine. My officers and myſelf went to dine with M. Vanderſpie, the proprietor of High Conſtantia. He treated us in the beſt manner poſſible, and we there drank a good deal of his wine, both at dinner, and in taſting the different ſorts, in order to make our proviſion of them.

The ſoil of Conſtantia is a ſandy gravel, lying on a gentle ſlope. They cultivate the vines without props, and leave only a ſmall number of buds when they cut them. They make the wine by putting the grapes without their grains into the veſſel. The full caſks are kept in a cellar level with the ground, in which the air has a free circulation. As we returned from Conſtantia, we viſited two country-houſes belonging to the governor.

The

The largest, named Newland, has a garden which is much larger than the company's, at the Cape. This last we have found much inferior to the reputation it has acquired. Some long walks of very high horn-beams, give it the appearance of a garden for fryars, and it is planted with oaks, which thrive very ill there.

Situation of the Dutch at the Cape.

The Dutch plantations have spread very much on the whole coast, and plenty is every where the consequence of cultivation, because the cultivator is free, subject to the laws only, and sure of his property. There are inhabitants almost a hundred and fifty leagues off the capital; they have no other enemies to fear than the wild beasts; for the Hottentots do not molest them. One of the finest parts of the Cape is the colony, which has been called Little Rochelle. This is a settlement of French, driven out of France by the repeal of the edict of Nantes. It surpasses all the rest in the fertility of the soil, and the industry of the colonists. They have given this adopted mother the name of their old country, which they still love, though it has treated them so hardly.

The government sends caravans out from time to time to search the interior parts of the country. One was out for eight months in 1763. This detachment advanced to the northward, and made, as I was told, some important discoveries; however, this journey had

not

not the ſucceſs which one might have expected; diſcontent and diſcord got amongſt them, and forced the chief to return home, leaving his diſcoveries imperfect. The Dutch got ſight of a yellow nation, with long hair, and ſeeming very ferocious to them.

On this journey they found a quadruped of ſeventeen feet high, of which I have given the drawing to M. de Buffon; it was a female ſuckling a young one, (fawn) which was only ſeven feet high. They killed the mother, and took the fawn alive, but it died after a few days march. M. de Buffon aſſured me that this is the animal which naturaliſts call the *giraffe*. None of them had been ſeen after that which was brought to Rome in the time of Cæſar, and ſhewn there in the amphitheatre. About three years ago they have likewiſe found and brought to the Cape, a quadruped of great beauty, which is related to the ox, horſe, and ſtag, and of which the genus is entirely new. It only lived two months at the Cape; I have likewiſe given M. de Buffon an exact drawing of this animal, whoſe ſtrength and fleetneſs equal its beauty. Is is not without reaſon that Africa has been named the mother of monſters.

Being provided with good proviſions, wines, and refreſhments of all ſorts, we ſet ſail from the road of the Cape the 17th in the afternoon. We paſſed between the iſle of Roben and the coaſt; at ſix o'clock in the evening, Departure from the Cape.

ing, the middle of that iſle bore S. by E. ½ E. about four leagues diſtant, from whence I took my departure in 33° 40′ ſouth latitude, and 15° 48′ eaſt longitude from Paris. I wanted to join M. Carteret, over whom I had certainly a great advantage in ſailing; but he was ſtill eleven days before me.

I directed my courſe ſo as to get ſight of St. Helena, in order to make ſure of putting in at Aſcenſion iſland, an anchorage which I intended to make beneficial to my

Sight of St. Helena.

crew. Indeed we got ſight of it the 29th, at two o'clock after noon, and the bearings which we ſet of it gave us no more than eight or ten leagues difference in our

1769. February.

reckoning. In the night between the 3d to the 4th of February, being in the latitude of Aſcenſion iſland, and being about eighteen leagues from it by my reckoning, I went only under the two top-ſails. At day-break we ſaw the iſle nearly nine leagues diſtant, and at eleven o'clock we anchored in the north weſt creek, or Creek of the Mountain of the Croſs, in twelve fathoms, bottom of ſand and coral. According to the Abbé la Caille's obſervations, this anchorage is in 7° 54′ ſouth latitude, and 16° 19′ weſt longitude from Paris.

topping at ſcenſion.

We had hardly caſt anchor, when I hoiſted out the boats, and ſent out three detachments to catch turtle; the firſt in the N. E. creek, the ſecond in the N. W. creek, oppoſite which we were; and the third in the Engliſh

Engliſh creek, which is in the S. W. of the iſland. Every thing promiſed a favourable capture; there was no other ſhip than ours, the ſeaſon was advantageous, and we entered with the new moon. As ſoon as the detachments were ſet off, I made every thing ready for fiſhing my two greater maſts under the rigging, viz. the main maſt with a fore top-maſt, the heel upwards; and the fore-maſt which was ſplit horizontally between the cheeks, with an oak fiſh.

In the afternoon the bottle was brought to me which contains the paper whereon the ſhips of every nation generally write their name, when they touch at Aſcenſion iſland. This bottle is depoſited in a cavity of the rocks of this bay, where it is equally ſheltered from rain and the ſpray of the ſea. In it I found written the Swallow, that Engliſh ſhip which captain Carteret commanded, and which I was deſirous of joining. He arrived here the 31ſt of January, and ſet ſail again on the firſt of February; thus we had already gained ſix days upon him, after leaving the cape of Good Hope. I inſcribed the Boudeuſe, and ſent back the bottle.

The 5th was ſpent in fiſhing our maſts under the rigging, which is a very nice operation in a road where the ſea is rough; in over-hauling our rigging, and embarking the turtle. The fiſhery was abundant; ſeventy turtle had been turned in the night, but we could only take on board fifty-six, the others were set at liberty again. We observed at our anchorage 9° 45′, variation N. W. The 6th, at three o'clock in the morning, the turtle being got on board, and the boats hoisted in, we began to weigh our anchors; at five o'clock we were under sail, happy on account of our capture, and of the hope that our next anchorage would be in our own country. Indeed, we had had a great many since our departure from Brest.

Departure from Ascension.

In leaving Ascension isle, I kept my wind in order to range the Cape Verd isles as close as possible. The 11th in the morning we passed the line for the sixth time on this voyage, in 20° of estimated longitude. Some days after, when, notwithstanding the fish with which we had strengthened our fore-mast, it cut a very bad figure, we were obliged to support it by preventer-shrouds, getting down the fore-top-gallant-mast, and almost always keeping the fore-top-sail close reefed, and sometimes handed.

Passing of the line.

The 25th in the evening we perceived a ship to windward, and a-head of us; we kept sight of her during the night, and joined her the next morning; it was the Swallow. I offered captain Carteret all the services that one may render to another at sea. He wanted nothing, but upon his telling me that they had given him letters for France at the Cape, I sent on board for

Meeting with the Swallow.

6 them.

them. He presented me with an arrow which he had got in one of the isles he had found on his voyage round the world, a voyage that he was far from suspecting we had likewise made. His ship was very small, went very ill, and when we took leave of him, he remained as it were at anchor. How much he must have suffered in so bad a vessel, may well be conceived. There were eight leagues difference between his estimated longitude and ours; he reckoned himself so much more to the westward.

Error in the reckoning of our course.

1769. March.

We expected to pass to the eastward of the Acores, when the 4th of March in the morning we had sight of the Isle of Tercera, which we doubled in day-time, ranging very close along it. The sight of this isle, supposing it well placed on M. Bellin's great chart, would give us about sixty-seven leagues of error to the westward, in the reckoning of our run; which indeed is a considerable error on so short a track as that from Ascension to the Acores. It is true that the position of these isles in longitude, is still uncertain. But I believe, that in the neighbourhood of the Cape Verd islands, there are very strong currents. However, it was essential to us to determine the longitude of the Acores by good astronomical observations, and to settle their distances and bearings among themselves. Nothing of all this is accurate on the charts of any nation. They only differ by by a greater or leſſer degree of error. This important taſk has juſt been executed by M. de Fleurieu, enſign of the king's ſhips.

I corrected my longitude in leaving the Iſle of Tercera, by that which M. de Bellin's great chart aſſigns to it. We had ſoundings the 13th in the afternoon, and the 14th in the morning we had ſight of Uſhant. As the wind was ſcant, and the tide contrary to double this iſland, we were forced to ſtand off, the wind blowing very freſh at weſt, and a very great ſea. About ten o'clock in the morning, in a violent ſquall, the fore-yard broke between the two jear-blocks, and the main-ſail at the ſame inſtant was blown out of the bolt-rope from clue to ear-ring. We immediately brought to under our main, fore, and mizen-ſtay-ſails, and we ſet about repairing the damage; we bent a new main-ſail, made a fore-yard with a mizen-yard, a main-top-ſail-yard, and a ſtudding-ſail-boom, and at four o'clock we were again enabled to make ſail. We had loſt ſight of Uſhant, and whilſt we lay-to, the wind and ſea drove us into the channel.

Sight of Uſhant.

Squall which damaged our rigging.

Being determined to put into Breſt, I reſolved to ply with variable winds, from S. W. to N. W. when the 15th in the morning our people came to inform me, that our fore-maſt was near being carried away under the rigging. The ſhock it had received when its yard

Arrival at St. Maloes.

broke, had made it worse; and though we had eased its head by lowering the yard, taking in the reefs in the fore-sail, and keeping the fore-top-sail upon the cap close reefed, yet we found, after an attentive examination, that this mast could not long resist the pitching caused by the great sea, we being close-hauled; besides this, all our rigging and blocks were rotten, and we had none to replace them; then how was it possible in such a condition to combat the bad weather of the equinoxes between two coasts? I therefore resolved to bear away, and conduct the frigate to St. Maloes. That was then the nearest port, which could serve us as an asylum. I entered it on the 16th in the afternoon, having lost only seven men, during two years and four months, which were expired since we had left Nantes.

Puppibus & læti Nautæ imposuere Coronas. VIRG. Æneid. Lib. iv.

VOCA-

VOCABULARY

OF THE LANGUAGE OF

TAITI ISLAND.

A.

ABOBO*	*Tomorrow.*
Aibou	*Come.*
Ainé	*Girl, (fille)*
Aiouta	*There is some*
Aipa	*The term of negation, there is none.*
Aneania	*Importune, tedious.*
Aouaou	*Fy; term of contempt, and of displeasure.*
Aouereré	*Black.*
Aouero	*Egg.*
Aouri	*Iron, gold, silver, every metal, or instrument of metal.*
Aoutti	*Flying fish.*
Aouira	*Lightning.*
Apalari	*To break or destroy,*
Ari	*Cocoa-nut.*
Arioi	*Bachelor, and a man without children.*
Ateatea	*White.*

I know of no word that begins with these consonants of ours, B, C, D.

E.

Ea	*Root.*
Eaï	*Fire.*
Eaia	*Parroquet.*
Eaiabou	*Vase.*
Eaiabou-maa	*Vase which is used to put their victuals in.*
Eame	*Drink made of cocoa nuts.*
Eani	*All manner of fighting.*
Eao	*Clouds. also a flower in bud, before it opens.*
Eatoua	*Divinity. The same word likewise expresses his ministers, and also the subordinate good or evil genii.*
Eeva	*Mourning.*
Eie	*Sail of a periagua.*
Eiva-eoura	*Dance or festival of the Taitians.*
Eivi	*Little.*

* I must here observe, that I have not altered the spelling of the words at all; and the reader will therefore take notice, that they should be pronounced according to the rules of the French language. F.

Eite	*To understand.*
Elao	*A fly.*
Emaa	*A sling.*
Emao	*A shark; it likewise signifies to bite.*
Emeitai	*To give.*
Emoé	*To sleep.*
Enapo	*Yesterday.*
Enene	*To discharge.*
Enia	*In, upon.*
Enninnito	*To stretch one's self yawning.*
Enoanoa	*To smell well.*
Enomoi	*Term to call, come hither.*
Enoo-te-papa	*Sit down.*
Enoua	*The earth and its different parts (a country).*
Enoua-Taiti	*The country of Taiti.*
Enoua-Paris	*The country of Paris.*
Eo	*To sweat.*
Eoe-tea	*An arrow.*
Eoe-pai	*A paddle or oar.*
Emoure-papa	*The tree from which they get the cotton, or substance for their stuffs, the cloth-tree.*
Eone	*Sand, dust.*
Eonou	*Turtle.*
Eote	*To kiss (baiser).*
Eouai	*Rain.*
Eonao	*To steal or rob.*
Eououa	*Pimples in the face.*
Eoui	*To belch or eruct.*
Eounoa	*Daughter-in-law.*
Eouramaï	*Light (not darkness).*
Eouri	*A dancer.*
Eouriaye	*A dancing girl.*
Epao	*Luminous vapour in the atmosphere, called a shooting-star. At Taiti they are looked upon as evil genii.*
Epata	*Exclamation to call one's wife.*
Epepe	*Butterfly.*
Epija	*Onion.*
Epoumaa	*Whistle; they make use of it to call the people to their meals.*
Epouponi	*To blow the fire.*
Epouré	*To pray.*
Epouta	*A wound; this word likewise signifies the scar.*
Era	*The sun.*
Era-ouao	*Rising sun.*
Era-ouopo	*Setting sun.*
Era-ouavatea	*Noon sun.*
Eraï	*Heaven.*
Erepo	*Dirty, unclean.*
Ero	*Ant.*
Eri	*King.*
Erie	*Royal.*
Eroï	*To wash, to cleanse.*
Eroleva	*Slate.*
Eroua	*A hole.*
Erouai	*To vomit.*
Eroupe	*Very large species of blue pigeon, like those which are in the possession of marshal Soubise.*
Etai	*Sea.*
Etao	*To dart, or throw.*
Etaye	*To weep.*
Eteina	*Elder brother or sister.*

Etouana	*Younger brother or sister.*
Etere	*To go.*
Etere-maine	*To come back.*
Etio	*Oyster.*
Etipi	*To cut, cut (particip.)*
Etoi	*A hatchet.*
Etoumou	*A turtle dove.*
Etouna	*An eel.*
Etooua	*To grate.*
Evaï	*The water.*
Evaie	*Moist.*
Evaine	*A woman.*
Evana	*A bow.*
Evare	*A house.*
Evaroua-t-eatoua	*A wish to persons when they sneeze, meaning that the evil genius may not lull thee asleep, or that the good genius may awaken thee.*
Evero	*A lance.*
Evetou	*A star.*
Evetou-eave	*A comet.*
Evi	*An acid fruit, like a pear, and peculiar to Taiti.*
Evuvo	*A flute.*

The following words are pronounced with a long e, *like the Greek* η.

ηti	*Wooden figures representing subordinate genii, and called* ηti-tane, *or* ηti-aine *according as they are of the masculine or feminine gender. These figures are employed in religious ceremonies, and the people of Taiti have several of them in their houses.*
ηieie	*Basket.*
ηou	*A fart. They detest it, and burn every thing in a house where one has farted.*
ηouou	*A muscle-shell (moule.)*
ηreou-tataou	*Colour for marking the body; with it they make indelible impressions on different parts of the body.*
ηriri, *and likewise* ouariri	*To be vexed, to be angry.*

I know of no word beginning with the consonants F, G.

H.

Horreo	*A kind of instrument for sounding, made of the heaviest shells.*

I.

Ióre	*A rat.*
Ioiroi	*To fatigue.*
Iroto	*In.*
Ivera	*Hot.*

I do not know any of their words beginning with the consonant L.

M.

Maa	*Eating.*
Maea	*Twin children.*
Maeo	*To scratch one's self, to itch.*
Maï	*More, is likewise said* maine; *it is an adverb of repetition* etere, *to go,* etere-maï *or* etere-maine, *to go once*

once more, to go and come again.

Maglli	*Cold.*
Mala	*More.*
Malama	*The moon.*
Malou	*Considerable, great.*
Mama	*Light, not heavy.*
Mamaï	*Sick.*
Manoa	*Good-day, your servant; expression of politeness or friendship.*
Manou	*A bird, swift (leger.)*
Mao	*Hawk for fishing.*
Mataï	*Wind.*
Mataï-malac	*East or S. E. wind.*
Mataï-aoueraï	*West or S.W. wind.*
Matao	*Fish-hook.*
Matapo	*One-eyed, squinting.*
Matari	*The* Pleïades.
Matïe	*Grass-herbage.*
Mato	*Mountain.*
Mate	*To kill.*
Mea	*A thing (chose.)*
Meïa	*Banana-tree, bananas.*
Metoua	*Parents.* Metouatane, *or* eoure, *father;* Metouaaine, *or* erao, *mother.*
Mimi	*To make water, to piss.*
Móa	*Cock, hen.*
Moea	*Mat.*
Mona	*Fine, good.*
Moreou	*A calm.*
Motoua	*Grand son.*

N.

Nate	*To give.*
Nie	*A sail of a boat.*
Niouniou	*A jonquil.*

O.

Oaï	*Walls and stones.*
Oaite	*To open.*
Oorah	*The piece of cloth which they wrap themselves in.*
Ooróa	*Generous, he that gives.*
Opoupoui	*To drink.*
Oualilo	*To steal, to rob.*
Ouaouara,	*Aigret of feathers.*
Ouaora	*To cure, or cured.*
Ouanao	*To ly in.*
Ouare	*To spit.*
Ouatere	*The helm's-man.*
Ouera	*Hot.*
Oueneo	*That does not smell well, it infects.*
Ouetopa	*To lose, lost.*
Ouhi	*Ho! ah! (hé.)*
Ouopé	*Ripe.*
Oupani	*Window.*
Oura	*Red.*
Ouri	*Dog and quadrupeds.*

P.

Pai	*Periagua.*
Paia	*Enough.*
Papa	*Wood, chair, and every piece of furniture of wood.*
Papanit	*To shut, to stop up.*
Paoro	*A shell, mother-of-pearl.*
Parouai	*Dress, cloth.*
Patara	*Grandfather.*
Patiri	*Thunder*
Picha	*Coffer, trunk.*
Pirara	*Fish.*
Piropiro	*Stink of a fart, or of excrements.*
Pirioi	*Lame.*
Piripiri	*A negative, signifying a covetous man, who gives nothing.*
Po	*Day, (light.)*

 Póe

Póe — *Pearl, ornament for the ears.*
Poi — *For, to.*
Poiri — *Obscure*
Poria — *Fat, lusty, of a good stature.*
Porotata — *Dog-kennel.*
Pouaa — *Hog, boar.*
Pouerata — *Flowers.*
Poupoui — *Under sail.*
Pouta — *Wound.*
Poto — *Little, minute.*

I know no word that begins with Q.

R.

Rai — *Great, big, considerable.*
Ratira — *Old, aged.*
Roa. — *Big, very fat.*
Roea — *Thread.*

No word is come to my knowledge beginning with S.

T.

Taitai — *Salted.*
Taio — *Friend.*
Tamai — *Enemy, at war.*
Tane — *Man, husband.*
Tao-titi — *Name of the high priestess, who is obliged to perpetual virginity. She has the highest consideration in the country.*
Taoa tane — *Married woman.*
Taporai — *To beat, abuse.*
Taoua-mai — *Physician.*
Taoumi — *Gorget of ceremony.*
Taoumta — *Covering of the head.*
Taoura — *Cord.*
Tata — *Man.*
Tatoue — *The act of generation.*
Tearea — *Yellow.*
Teouteou — *Servant, slave.*
Tero — *Black.*
Tetouarn — *Femme barée?*
Tiarai — *White flowers, which they wear in their ears instead of ornaments.*
Titi — *A peg, or pin.*
Tinatore — *A serpent.*
Twa — *Strong, malignant, powerful.*
Tomaiti — *Child.*
Toni — *Exclamation to call the girls. They add* Peio *lengthened, or* Pijo *softly pronounced, like the Spanish* j. *If the girl slaps her hand on the outside of the knee, it is a refusal, but if she says* enemoi, *she thereby expresses her consent.*
Toto — *Blood.*
Toua-pouou — *Hump-backed.*
Touaine — *Brother or sister, by adding the word which distinguishes the sex.*
Toubabaou — *To weep.*
Touie — *Lean.*
Toumany — *Action of fencing; this they do with a piece of wood, armed with a point, made of harder materials than wood. They put themselves in the same posture as we do for fencing.*
Toura — *Without.*

Toutai

Toutai	*To make the natural evacuations.*
Toutn	*Excrements.*
Toupanoa	*To open a window or door.*
Touroutoto	*A decrepit old man.*
Toutoi-papa	*Light or fire of the great people;* niao-papa, *light of the common people.*

V.

Vereva	*Flag which is carried before the king and the principal people.*

I know no words beginning with the letters U, X, Y, Z.

Names of different parts of the body.

Auopo	*The crown of the head.*
Boho	*The skull.*
Eouttou	*The face.*
Mata	*The eyes.*
Taria	*The ears.*
Etaa	*The jaw.*
Eiou	*The nose.*
Lamolou	*The lips.*
Ourou	*The hairs.*
Allelo	*The tongue.*
Eniou	*Teeth.*
Eniaou.	*Tooth-picks; they make them of wood.*
Oumi	*The beard.*
Papaourou	*The cheeks.*
Arapoa	*The throat.*
Taah	*Chin.*
Eou	*Teats, nipples.*
Aſao	*The heart.*
Erima	*The hand.*
Apourima	*The inside of the hand.*
Eaiou	*The nails.*
Etoua	*The back.*
Etapono	*The ſhoulders.*
Obou	*The bowels.*
Tinai	*The belly.*
Pito	*The navel.*
Toutaba	*The glands of the groin.*
Etoe	*Buttocks.*
Aoua	*Thighs.*
Eanai	*Legs.*
Etapoué	*The foot.*
Eoua	*Teſticles.*
Eoure	*The male parts.*
Erao	*The female parts.*
Eomo	*The clitoris.*

Numerals.

Atai	*One.*
Aroua.	*Two.*
Atorou	*Three.*
Aheho	*Four.*
Erima	*Five.*
Aouno	*Six.*
Ahitou	*Seven.*
Awarou	*Eight.*
Ahiva	*Nine.*
Aourou	*Ten.*

They have no words to expreſs eleven, twelve, &c. They repeat Atai, Aroua, *&c. till to twenty, which they call* ataitao.

Ataitáo-mala-atai	*Twenty, more one, or twenty-one, &c.*
Ataitao-mala-aourou	*Thirty, i. e.* 20 *more* 10.
Aroua-tao	*Forty.*
Aroua-tao mala atorou	*Forty-three, &c.*
Aroua-tao mala aourou	*Fifty, or* 40 *more* 10.

I could not make Aotourou count beyond his laſt number.

Names

Names of Plants.

Amiami	*Cotyledon.*
Amoa	*Fern.*
Aoute	*Rose.*
Eaaeo	*Sugar-cane.*
Eaere	*Weeping-willow, or Babylonian willow.*
Eaia	*Pears.*
Eape	*Virginian arum.*
Eatou	*Lys de S. Jaques, a species of lily.*
Eoe	*Bamboo.*
Eóai	*Indigo.*
Eora	*Indian saffron.*
Eotonoutou	*Figs.*
Eoui	*Yams.*
Epoua	*Rhubarb.*
Eraca	*Chesnuts.*
Erea	*Ginger.*
Etaro	*Purple arum.*
Eti	*Dragon's blood.*
Etiare	*Grenadille, or passion-flower.*
Etoutou	*Rivina.*
Maireraro	*Three leaved sumach.*
Mati	*Raisins.*
Oporo-maa	*Pepper.*
Pouraou	*Cayenne-rose.*
Toroire	*Heliotropium, or tournesol.*

They have a kind of article answering to our articles *of* and *to (de & à)*. This is the word *te*. Thus they say *parouai te Aotourou*; the clothes of or (belonging) to Aotourou; *maa-te-eri*, the eating of kings.

THE END.

ERRATA.

Page 2. line last, *for* main yards, *read* lower-yards—p. 9. l. 3. *for* one quarter, *read* by—ib. l. ib. *for* one quarter, *read* by—ib. l. 6. *for* one quarter, *read* by—p. 17. l. 16. *for* the river of Plate, *read* Rio de la Plata—p. 33. l. 15. *for* top-masts handed, and main-yards lowered, *read* yards and top-masts struck—ib. l. 19. *for* main-sails *read* courses and top-sails—p. 34. l. 12. *for* one quarter, *read* by—p. 245. note, l. 1. *for* cooes nutifera, *read* cocos nucifera—ib. l. 2. *for* parasidiaca, *read* paradisiaca.

Printed in the United States
By Bookmasters